Edited by Editorial Board of UTokyo Engineering Course

Linear Algebra I

Basic Concepts

UTokyo Engineering Course/Basic Mathematics

Linear Algebra II: Advanced Topics for Applications
by Kazuo Murota and Masaaki Sugihara
ISBN: 978-981-125-705-6
ISBN: 978-981-125-798-8 (pbk)

UTokyo Engineering Course/ Basic Mathematics

Edited by Editorial Board of UTokyo Engineering Course

Linear Algebra I

Basic Concepts

Kazuo Murota

The Institute of Statistical Mathematics, Japan;

Professor Emeritus at

The University of Tokyo, Japan & Kyoto University, Japan

& Tokyo Metropolitan University, Japan

Masaaki Sugihara

Professor Emeritus at The University of Tokyo, Japan

& Nagoya University, Japan

MARUZEN PUBLISHING

World Scientific

Published by

World Scientific Publishing Co. Pte. Ltd.
5 Toh Tuck Link, Singapore 596224
USA office: 27 Warren Street, Suite 401-402, Hackensack, NJ 07601
UK office: 57 Shelton Street, Covent Garden, London WC2H 9HE

and

Maruzen Publishing Co., Ltd.
Kanda Jimbo-cho Bldg. 6F, Kanda Jimbo-cho 2-17
Chiyoda-ku, Tokyo 101-0051, Japan ('MP')

Library of Congress Control Number: 2022030855

British Library Cataloguing-in-Publication Data
A catalogue record for this book is available from the British Library.

LINEAR ALGEBRA I
Basic Concept

ISBN 978-981-125-702-5 (hardcover)
ISBN 978-981-125-797-1 (paperback)
ISBN 978-981-125-703-2 (ebook for institutions)
ISBN 978-981-125-704-9 (ebook for individuals)

For any available supplementary material, please visit
https://www.worldscientific.com/worldscibooks/10.1142/12866#t=suppl

Desk Editor: Tan Rok Ting

Printed in Singapore

UTokyo Engineering Course

About This Compilation

What is the purpose of engineering education at the University of Tokyo's Undergraduate and Graduate School of Engineering? This School was established 125 years ago, therefore we feel it is an appropriate time to ask this question again. More than a century has passed since Japan embarked on a path to introduce and negotiate Western knowledge and practices. Japan and the world are very different places now, and today our university stands as a world leading institute in engineering research and education. As such, it is our duty and mission to build a firm foundation of education that will support the creation and dissemination of engineering knowledge, practices and resources. Our School of Engineering must not only teach outstanding students from Japan but also those from throughout the world. Put another way, the engineering that we teach students is not only a responsibility of this School, but an imperative placed on us by society and the age in which we live. It is in this changed context, where we have gone from follower to leader, that we present this curriculum, The University of Tokyo (UTokyo) Engineering Course. The course is a reflection of the School's desire to engage with those outside the walls of the Ivory Tower, and to spread the best of engineering knowledge to the world outside our institution. At the same time, the course is also designed for the undergraduate and graduate students of the School. As such, the course contains the knowledge that should be learnt by our students, taught by our instructors and critically explored by all.

February 2012

Takehiko Kitamori
Dean, Undergraduate and Graduate Schools of Engineering
The University of Tokyo
(April 2010–March 2012)

UTokyo Engineering Course

The Purpose of This Publication

Modern engineering is composed of the academic discipline of fundamental engineering and the academic discipline of integrated engineering that deals with specific systems and subjects. Interdisciplinary disciplines and multidisciplinary disciplines are amalgamations of multiple academic disciplines that result in new academic disciplines when the academic pursuit in question does not fit within one traditional fundamental discipline. Such interdisciplinary disciplines and multidisciplinary disciplines, once established, often develop into integrated engineering. Moreover, the movement toward interdisciplinarity and multidisciplinarity is well underway within both fundamental engineering and advanced research.

These circumstances are producing a variety of challenges in engineering. That is, the scope of research of integrated engineering is gradually growing larger, with economics, medicine and society converging into an enormously complex social system, which is resulting in the trend of connotative academic disciplines growing larger and becoming self-contained research fields, which, in turn, is resulting in a trend of neglect toward fundamental engineering. The challenge of fundamental engineering is how to connect engineering education that is built upon traditional disciplines with that of advanced engineering research in which interdisciplinarity and multidisciplinarity is continuing at a rapid pace. Truly this is an educational challenge shared by all the top engineering schools in the world. Without having a solid understanding of engineering, however, education related to learning state-of-the-art research methodologies will not hold up. This is the dichotomy of higher education in engineering; that is, higher education in engineering simply will not work out if either side of the equation is missing.

In the meantime, the internationalization of universities is going forward in routine fashion. In fact, here at the University of Tokyo (UTokyo), one quarter of the graduate students enrolled in engineering fields are of foreign nationality and the percentage of foreign undergraduate students is expected to increase more and more. On top of that, Japan is experiencing a reduction in the population of its youth. Therefore, the time is ripe to ramp up efforts to look outside of Japan in order to secure the human resources to sustain the future of advanced science and technology here in Japan. It is clear that the internationalization of engineering education is rapidly underway. As such, the need for a curriculum that is firmly rooted in engineering knowledge needs to be oriented toward both local and foreign students.

Due to these circumstances surrounding modern engineering, we at UTokyo's School of Engineering have systematically organized an engineering curriculum of fundamental engineering knowledge that will not be unduly influenced by the times, with the goal of firmly establishing a benchmark suitable for that of the top schools of engineering of science and technology for students to learn and teachers to teach. This engineering curriculum clarifies the disciplines and instruction policy of UTokyo's School of Engineering and is composed of three layers: Fundamental (sophomores (second semester) and juniors), Intermediate (seniors and graduates) and Advanced (graduates). Therefore, this engineering course is a policy for the thorough education of the engineering knowledge necessary for forming the foundation of our doctorate program as well. The following is an outline of the expected effect of this engineering course:

- Surveying the total outline of this engineering course will assist students in understanding which studies they should undertake for each field they are pursuing, and provide an overall image by which the students will know what fundamentals they should be studying in relation to their field.
- This course will build the foundation of education at UTokyo's School of Engineering and clarify the standard for what instructors should be teaching and what students need to know.
- As students progress in their major it may be necessary for them to go back and study a new fundamental course. Therefore the textbooks are designed with such considerations as well.

- By incorporating explanations from the viewpoint of engineering departments, the courses will make it possible for students to learn the fundamentals with a constant awareness of their application to engineering.

Takao Someya, Board Chair
Yukitoshi Motome, Shinobu Yoshimura, Executive Secretary
Editorial Board of UTokyo Engineering Course

UTokyo Engineering Course

Editorial Board

Preface

In the University of Tokyo (UTokyo) Engineering Course there are two volumes on linear algebra, *Linear Algebra I: Basic Concepts* and *Linear Algebra II: Advanced Topics for Applications*. The objective of this first volume is to show the standard mathematical results in linear algebra from the engineering viewpoint. The objective of the second volume is to branch out from the first volume to illustrate useful specific topics pertaining to applications. These two volumes were originally written in Japanese and published from Maruzen Publishing in 2013 and 2015. The present English version is their almost faithful translations by the authors.

Linear algebra is primarily concerned with systems of equations and eigenvalue problems for matrices and vectors with real or complex entries, which are treated in this volume, *Linear Algebra I*. The contents of the chapters of this volume are described below.

Chapter 1, "Matrices," introduces the concept of matrices together with basic operations, inverse matrices, and norms. It also mentions nonzero-patterns of matrices and diagonal dominance, which are convenient and useful in applications.

Chapter 2, "Determinants," explains the definition and role of determinants with emphasis on multilinearity and various identities like the Laplace expansion. Adjoint matrices are also introduced in this chapter.

Chapter 3, "Elementary Transformations and Elimination," describes the elimination procedure using elementary transformations as a fundamental tool for computing and transforming matrices. The rank normal form and the reduced echelon form can be computed by means of elementary transformations. Elementary transformations are also used to compute inverse matrices.

Chapter 4, "Rank," introduces three different definitions of the rank of a matrix, which turn out to be equivalent. Fundamental properties of the rank are presented. Examples from different disciples of engineering demonstrate the role of this concept to describe degrees of freedom of engineering systems.

Chapter 5, "Systems of Linear Equations," discusses the existence and uniqueness of solutions to a system of linear equations with the aid of the rank normal form and the reduced echelon form. A computational procedure based on elimination operations and a solution formula called Cramer's rule are also presented. The solvability of a system of equations arising from discretization of a differential equation (the Laplace equation) is investigated. This chapter also covers the Sylvester equations and the Lyapunov equations, which are important in engineering applications.

Chapter 6, "Eigenvalues," presents fundamental facts about eigenvalues and eigenvectors, as well as their significances in engineering. Diagonalization by orthogonal (resp., unitary) transformations, minimax theorems, and perturbation theorems are given for symmetric (resp., Hermitian) matrices. For general (non-symmetric) matrices the Schur decomposition and the Jordan normal form are presented.

Chapter 7, "Quadratic Forms," deals with fundamental facts about quadratic forms defined by symmetric (or Hermitian) matrices, including positive definiteness, completion of squares, and Sylvester's law of inertia. Examples of positive definite matrices in several engineering disciples are presented. Physically, a quadratic form defined by a positive definite matrix corresponds to energy.

Chapter 8, "Singular Values and the Method of Least Squares," introduces singular values in comparison with eigenvalues. The theoretical basis of the method of least squares is presented as an application of this concept.

Chapter 9, "Vector Spaces," presents fundamental concepts and standard results related to vector spaces such as subspace, linear independence, basis, linear mapping, inner product, and dual space. This last chapter, dealing with the general mathematical theory including infinite dimensional spaces, is (partly) more sophisticated than other chapters. Although it is put at the end of this volume, some of the facts presented there are used in engineering applications. In this chapter, theorems are most often stated without proofs.

This book, assuming that the reader is already familiar with the very basics of linear algebra, aims at presenting a further step of linear algebra toward engineering, not toward mathematics. For successful engineering use of linear algebra it will be crucial to recognize the following four aspects from applications, in addition to mathematical theorems and formulas.

(1) How matrices arise.
For example: discretization of differential equations, description of system structures, description of transition probability.
(2) What kinds of matrices arise.
For example: sparse matrices, positive definite matrices, diagonally dominant matrices, nonnegative matrices, integer matrices, polynomial matrices.
(3) What characteristics we are interested in.
For example: rank, eigenvalues, singular values, positive definiteness, and their engineering significances.
(4) How we can compute.
For example: expansion formulas of determinants, elementary transformations, estimates of eigenvalues, various numerical methods.

Serious effort has been made to relate mathematical concepts and theorems to engineering applications, although no attempt has been made to provide actual case studies that demonstrate the use of linear algebra. For numerical methods, the reader is referred to textbooks such as [36–44] listed at the end of this volume.

In this book, it is also intended to describe the mathematical reasoning precisely. Sometimes alternative proofs are given to a single theorem to enhance multifaceted understanding of the result. Proofs in mathematics serve not only as a formal verification process of the correctness but also as an effective means to capture the essence. Learning different proofs leads to multifaceted and systematic understanding. Moreover, they often provide clues for solutions of other problems. The theory of linear algebra is linked to engineering applications and numerical computations through representations by matrices. Accordingly this book takes the approach of developing theories through concrete matrix representations rather than through abstract concepts.

In writing this book, we received lot of help from many people. In particular, we would like to express our appreciation to Yoshihiro Kanno for providing information about structural mechanics, Koji Tsumura for information about control engineering, Ken'ichiro Tanaka for checking the numerical examples, and Naonori Kakimura and Mizuyo Takamatsu for reading the whole manuscript. We are also thankful to Yusuke Hasegawa, Ken Hayami, Rintaro Ikeshita, Ai Ishikawa, Shinji Ito, Satoru Iwata, Fumiyasu Komaki, Yusuke Kuroki, Takayasu Matsuo, Ryuhei Mizutani, Satoko Moriguchi, Taihei Oki, Nobutaka Shimizu, Akihisa Tamura, Itta Toda, Takaharu Yaguchi, and Tomohiko Yokoyama for a variety of feedback.

Kazuo Murota
Masaaki Sugihara

Contents

List of Figures

Chapter 1

Matrices

In this chapter we introduce the concept of matrices together with basic operations, inverse matrices, and norms. We also mention nonzero-patterns of matrices and diagonal dominance, which are convenient and useful in applications.

1.1 Matrices

1.1.1 *Definition*

We denote the set of all real numbers by $\mathbb{R}$ and the set of all complex numbers by $\mathbb{C}$.

Definition 1.1. Let K denote $\mathbb{R}$ or $\mathbb{C}$, and let m and n be natural numbers. An array of mn numbers $a_{ij} \in K$ $(i = 1, 2, \ldots, m; j = 1, 2, \ldots, n)$ like

$$A = \begin{bmatrix} a_{11} & a_{12} & \cdots & a_{1n} \\ a_{21} & a_{22} & \cdots & a_{2n} \\ \vdots & \vdots & & \vdots \\ a_{m1} & a_{m2} & \cdots & a_{mn} \end{bmatrix} \tag{1.1}$$

is called an $m \times n$ *matrix* or a *matrix of type* (m, n).[1] We speak of a *matrix over* K when we want to emphasize that a_{ij}'s are members of K. We also speak of a *real matrix* if $K = \mathbb{R}$, and a *complex matrix* if $K = \mathbb{C}$. An $m \times 1$ matrix (a vertical arrangement of m numbers) is called an m*-dimensional column vector* or a *column* m*-vector*. Similarly, a $1 \times n$ matrix (a horizontal arrangement of n numbers) is called an n*-dimensional row vector* or a *row* n*-vector*. ∎

[1] Here we assume that a_{ij}'s are real or complex numbers, but they can also be integers or polynomials. Mathematically, a matrix is an array of elements belonging to some algebraic system.

Remark 1.1. There are a variety of notations for matrices. In this book we use brackets [ ] to embrace numbers, but it is also customary to use parentheses as

$$A = \begin{pmatrix} a_{11} & a_{12} & \cdots & a_{1n} \\ a_{21} & a_{22} & \cdots & a_{2n} \\ \vdots & \vdots & & \vdots \\ a_{m1} & a_{m2} & \cdots & a_{mn} \end{pmatrix}.$$

To save space we sometimes write

$$A = (a_{ij} \mid i = 1, 2, \dots, m;\ j = 1, 2, \dots, n)$$

or $A = (a_{ij})$ for short. ■

Definition 1.2. Each a_{ij} is called an *entry* (*element, component*) of matrix A. An entry at the ith row and jth column is called the (i, j) entry of A. ■

Definition 1.3. A horizontal array

$$\begin{bmatrix} a_{i1} & a_{i2} & \cdots & a_{in} \end{bmatrix} \tag{1.2}$$

in the expression (1.1) is called a *row* of matrix A. We speak of the ith row to specify which row. A vertical array

$$\begin{bmatrix} a_{1j} \\ a_{2j} \\ \vdots \\ a_{mj} \end{bmatrix} \tag{1.3}$$

in (1.1) is called a *column* of matrix A. We speak of the jth column to specify which column. ■

Remark 1.2. Denoting the column vector in (1.3) by $\boldsymbol{a}_j$, we can think of matrix A as a collection of column vectors as

$$A = [\boldsymbol{a}_1, \boldsymbol{a}_2, \dots, \boldsymbol{a}_n].$$

Similarly, denoting the row vector in (1.2) by $\hat{\boldsymbol{a}}_i^\top$, we can think of matrix A as a collection of row vectors as[2]

$$A = \begin{bmatrix} \hat{\boldsymbol{a}}_1^\top \\ \hat{\boldsymbol{a}}_2^\top \\ \vdots \\ \hat{\boldsymbol{a}}_m^\top \end{bmatrix}.$$

■

[2]Here $\hat{\boldsymbol{a}}_i$ is a column vector and $\hat{\boldsymbol{a}}_i^\top$ denotes its transpose, which is a row vector. The notation $^\top$ for transposition will be introduced in Sec. 1.4.

Definition 1.4. A matrix is called a *zero matrix*, denoted by O, if all of its entries are equal to 0. That is,

$$O = \begin{bmatrix} 0 & 0 & \cdots & 0 \\ 0 & 0 & \cdots & 0 \\ \vdots & \vdots & & \vdots \\ 0 & 0 & \cdots & 0 \end{bmatrix}.$$

The $m \times n$ zero matrix is denoted by $O_{m,n}$, and $O_{n,n}$ is also denoted by O_n. ■

Definition 1.5. A matrix is called a *square matrix* if it is an $n \times n$ matrix for some n. An $n \times n$ square matrix is called a (square) matrix of *order* n. ■

Definition 1.6. For a square matrix $A = (a_{ij})$, an entry a_{ij} is called a *diagonal element* if $i = j$, and an *off-diagonal element* if $i \neq j$.[3] ■

Definition 1.7. A square matrix is called a *diagonal matrix* if all of its off-diagonal elements are equal to 0.[4] A diagonal matrix of order n with diagonal elements $d_1, d_2, \ldots, d_n$ is denoted by $\mathrm{diag}\,(d_1, d_2, \ldots, d_n)$.[5] That is,

$$\mathrm{diag}\,(d_1, d_2, \ldots, d_n) = \begin{bmatrix} d_1 & 0 & \cdots & 0 \\ 0 & d_2 & \cdots & 0 \\ \vdots & \vdots & \ddots & \vdots \\ 0 & 0 & \cdots & d_n \end{bmatrix}.$$

■

Definition 1.8. A matrix is called a *unit matrix* (or *identity matrix*), denoted by I, if it is a diagonal matrix whose diagonal elements are all equal to 1. That is,

$$I = \begin{bmatrix} 1 & 0 & \cdots & 0 \\ 0 & 1 & \cdots & 0 \\ \vdots & \vdots & \ddots & \vdots \\ 0 & 0 & \cdots & 1 \end{bmatrix}.$$

---

[3] Sometimes this definition is extended to a non-square matrix $A = (a_{ij})$. That is, an entry a_{ii} is referred to as a diagonal element and a_{ij} with $i \neq j$ as an off-diagonal element.

[4] Sometimes a non-square matrix $A = (a_{ij})$ is called a diagonal matrix if $a_{ij} = 0$ for all (i, j) with $i \neq j$.

[5] It should be obvious that "diag" stands for "diagonal".

The unit matrix of order n is denoted by I_n. The (i,j) entry of a unit matrix is given by the Kronecker delta δ_{ij} (cf., Remark 1.3). Each column vector $(0,\ldots,0,1,0,\ldots,0)^\top$ of a unit matrix is called a *(standard) unit vector.* ■

Remark 1.3. The symbol δ_{ij}, called the *Kronecker delta*, is defined by

$$\delta_{ij} = \begin{cases} 1 \ (i = j), \\ 0 \ (i \neq j). \end{cases} \tag{1.4}$$

■

Definition 1.9. Let $A = (a_{ij} \mid i = 1,2,\ldots,m;\ j = 1,2,\ldots,n)$ be an $m \times n$ matrix. For r row indices $i_1 < i_2 < \cdots < i_r$ and s column indices $j_1 < j_2 < \cdots < j_s$, where $1 \leqq r \leqq m$ and $1 \leqq s \leqq n$, we can consider an $r \times s$ matrix $B = (b_{pq} \mid p = 1,2,\ldots,r;\ q = 1,2,\ldots,s)$ with $b_{pq} = a_{i_p j_q}$ as its (p,q) entry. Such a matrix is called a *submatrix* of A. ■

Definition 1.10. Let $A = (a_{ij} \mid i,j = 1,2,\ldots,n)$ be a square matrix of order n. A submatrix is called a *principal submatrix* of order k if it is determined by the same indices $i_1 < i_2 < \cdots < i_k$ for rows and columns. That is, a principal submatrix of order k is given by $B = (b_{pq} \mid p,q = 1,2,\ldots,k)$ with $b_{pq} = a_{i_p i_q}$ for some $i_1 < i_2 < \cdots < i_k$. If $i_1 = 1, i_2 = 2, \ldots, i_k = k$, it is called the *leading principal submatrix* of order k. That is, the submatrix $(a_{ij} \mid i,j = 1,2,\ldots,k)$ is the leading principal submatrix of A of order k. ■

1.1.2 *Block Matrices*

Consider, for example, a 5×7 matrix $A = (a_{ij})$ with rows and columns partitioned as

$$A = \left[\begin{array}{ccc|cc|cc} a_{11} & a_{12} & a_{13} & a_{14} & a_{15} & a_{16} & a_{17} \\ a_{21} & a_{22} & a_{23} & a_{24} & a_{25} & a_{26} & a_{27} \\ \hline a_{31} & a_{32} & a_{33} & a_{34} & a_{35} & a_{36} & a_{37} \\ a_{41} & a_{42} & a_{43} & a_{44} & a_{45} & a_{46} & a_{47} \\ \hline a_{51} & a_{52} & a_{53} & a_{54} & a_{55} & a_{56} & a_{57} \end{array}\right].$$

By defining matrices

$$A_{11} = \begin{bmatrix} a_{11} & a_{12} & a_{13} \\ a_{21} & a_{22} & a_{23} \end{bmatrix}, \quad A_{12} = \begin{bmatrix} a_{14} & a_{15} \\ a_{24} & a_{25} \end{bmatrix}, \quad A_{22} = \begin{bmatrix} a_{34} & a_{35} \\ a_{44} & a_{45} \end{bmatrix},$$

etc., we may regard A as a 3×3 matrix

$$A = \begin{bmatrix} A_{11} & A_{12} & A_{13} \\ A_{21} & A_{22} & A_{23} \\ A_{31} & A_{32} & A_{33} \end{bmatrix}$$

whose entries are matrices A_{pq} $(1 \leqq p, q \leqq 3)$. Such a matrix is called a block matrix.

Definition 1.11. Let k and l be natural numbers. A matrix consisting of kl matrices A_{pq} $(p = 1, 2, \ldots, k; q = 1, 2, \ldots, l)$ as

$$A = \begin{bmatrix} A_{11} & A_{12} & \cdots & A_{1l} \\ A_{21} & A_{22} & \cdots & A_{2l} \\ \vdots & \vdots & & \vdots \\ A_{k1} & A_{k2} & \cdots & A_{kl} \end{bmatrix} = (A_{pq} \mid p = 1, 2, \ldots, k; q = 1, 2, \ldots, l)$$

is called a *block matrix*. Here the matrices A_{pq} must be consistent in size. Specifically, for some $m_1, m_2, \ldots, m_k$ $(\geqq 1)$ and $n_1, n_2, \ldots, n_l$ $(\geqq 1)$, each matrix A_{pq} is an $m_p \times n_q$ matrix for $p = 1, 2, \ldots, k$ and $q = 1, 2, \ldots, l$. The tuple $(m_1, m_2, \ldots, m_k; n_1, n_2, \ldots, n_l)$ is called the *type* of a block matrix A. ∎

Definition 1.12. A matrix is called a *bordered matrix* if it is a block matrix with $k = l = 2$ and $m_2 = n_2 = 1$ in Definition 1.11. A bordered matrix is a matrix of the form:

$$\left[\begin{array}{c|c} A & \boldsymbol{b} \\ \hline \boldsymbol{c}^\top & d \end{array}\right]. \tag{1.5}$$

∎

Definition 1.13. In a block matrix $(A_{pq} \mid p, q = 1, 2, \ldots, k)$ with the same number of blocks in rows and columns, we call each A_{pp} a *diagonal block*, where $1 \leqq p \leqq k$, and A_{pq} with $p \neq q$ an *off-diagonal block*.[6] ∎

Definition 1.14. A block matrix $A = (A_{pq} \mid p, q = 1, 2, \ldots, k)$ is called a *block-diagonal matrix* if all the off-diagonal blocks are zero matrices. That is, a block-diagonal matrix takes the following form:[7]

$$A = \begin{bmatrix} B_1 & O & \cdots & O \\ O & B_2 & \cdots & O \\ \vdots & \vdots & \ddots & \vdots \\ O & O & \cdots & B_k \end{bmatrix},$$

which is also expressed as $A = \operatorname{diag}(B_1, B_2, \ldots, B_k)$. ∎

---

[6] Usually the diagonal blocks $A_{11}, A_{22}, \ldots, A_{kk}$ are square matrices, but not always.

[7] The diagonal blocks $B_1, B_2, \ldots, B_k$ are usually a square matrix, but not always.

1.1.3 *Non-zero Patterns*

In engineering applications, matrices of special patterns often play important roles. By a matrix of a special pattern we mean a matrix that has a pre-specified set of positions for non-zero entries. Figure 1.1 shows some of such patterns used frequently.[8]

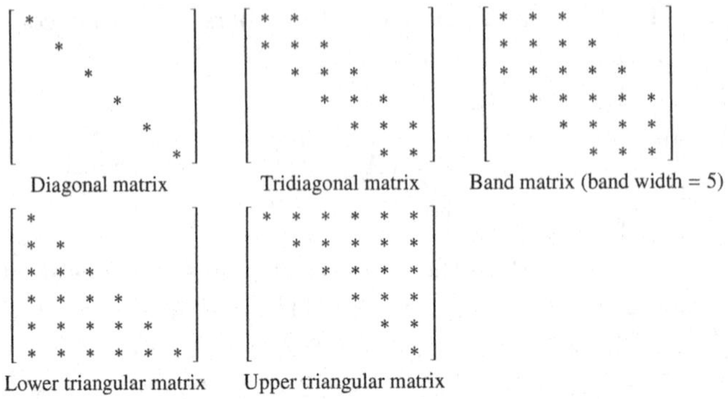

Fig. 1.1 Patterns of matrices.

Definition 1.15. A matrix $A = (a_{ij})$ is called a *tridiagonal matrix* if $a_{ij} = 0$ for all (i, j) with $|i - j| > 1$. ∎

Definition 1.16. A matrix $A = (a_{ij})$ is called a *band matrix* if there is some k such that $a_{ij} = 0$ for all (i, j) with $|i - j| > k$. Here k is an integer ($k \geqq 1$) and the number $2k + 1$ is called the *band width.* A band matrix of band width 3 is nothing but a tridiagonal matrix. ∎

Definition 1.17. A matrix $A = (a_{ij})$ is called a *lower triangular matrix* if $a_{ij} = 0$ for all (i, j) with $i < j$, and a *strictly lower triangular matrix* if $a_{ij} = 0$ for all (i, j) with $i \leqq j$. A strictly lower triangular matrix is a lower triangular matrix whose diagonal entries are all equal to zero. A *lower block-triangular matrix* is defined in an obvious manner. ∎

Definition 1.18. A matrix $A = (a_{ij})$ is called an *upper triangular matrix* if $a_{ij} = 0$ for all (i, j) with $i > j$, and a *strictly upper triangular matrix* if $a_{ij} = 0$ for all (i, j) with $i \geqq j$. A strictly upper triangular matrix is an

[8]We deal with square matrices here, but the definitions of these special forms can naturally be extended to (general) $m \times n$ matrices.

upper triangular matrix whose diagonal entries are all equal to zero. An *upper block-triangular matrix* is defined in an obvious manner. ■

Definition 1.19. A matrix is called a *triangular matrix* if it is a lower triangular matrix or an upper triangular matrix. A *block-triangular matrix* is defined as a lower or upper block-triangular matrix. ■

In general, matrices appearing in engineering problems have the following two characteristics:

(1) The size of a matrix (the number of rows and columns) is large, and
(2) Not many entries are non-zero.

The latter property is called *sparsity.*

Definition 1.20. A matrix is called a *large-scale matrix* if it has many rows and columns. ■

Definition 1.21. A matrix is called a *sparse matrix* if it has a small number of non-zero entries. A matrix is called a *dense matrix* if the majority of the entries are non-zero. ■

The concepts of "large-scale matrix," "sparse matrix," and "dense matrix" are not mathematically rigorous terms but rather are practically useful concepts. It is emphasized that they are crucial properties to be considered in dealing with matrices in engineering applications, especially in numerical computations.

1.2 Operations on Matrices

1.2.1 *Sum and Scalar Multiple*

Definition 1.22. For matrices $A = (a_{ij})$ and $B = (b_{ij})$ of the same type (size), their *sum*, denoted by $A + B$, is defined as the matrix whose (i, j) entry is equal to $a_{ij} + b_{ij}$. That is, $A + B = (a_{ij} + b_{ij})$. The *difference* of A and B, denoted by $A - B$, is defined as the matrix whose (i, j) entry is equal to $a_{ij} - b_{ij}$. That is, $A - B = (a_{ij} - b_{ij})$. If A and B are $m \times n$ matrices, both $A + B$ and $A - B$ are $m \times n$ matrices. ■

Example 1.1. The sum of two 3×2 matrices is given as

$$\begin{bmatrix} a_{11} & a_{12} \\ a_{21} & a_{22} \\ a_{31} & a_{32} \end{bmatrix} + \begin{bmatrix} b_{11} & b_{12} \\ b_{21} & b_{22} \\ b_{31} & b_{32} \end{bmatrix} = \begin{bmatrix} a_{11} + b_{11} & a_{12} + b_{12} \\ a_{21} + b_{21} & a_{22} + b_{22} \\ a_{31} + b_{31} & a_{32} + b_{32} \end{bmatrix}.$$

■

Proposition 1.1. *The following identities hold concerning the sum of matrices.*

(1) $(A+B)+C=A+(B+C)$.
(2) $A+B=B+A$.
(3) $A+O=A$.
(4) $A-A=O$.

The identity in Proposition 1.1(1) is called the *associative law* and the identity in (2) is the *commutative law.*

Definition 1.23. For a matrix $A=(a_{ij})$ and a scalar c ($\in K$), the matrix whose (i,j) entry is equal to $c\,a_{ij}$ is called the *scalar multiple* of A by c, and is denoted by $c\,A$. That is, $c\,A=(c\,a_{ij})$. If A is an $m\times n$ matrix, then $c\,A$ is also an $m\times n$ matrix. ■

Example 1.2. The scalar multiple of a 2×3 matrix is given as

$$c\begin{bmatrix} a_{11} & a_{12} & a_{13} \\ a_{21} & a_{22} & a_{23} \end{bmatrix} = \begin{bmatrix} c\,a_{11} & c\,a_{12} & c\,a_{13} \\ c\,a_{21} & c\,a_{22} & c\,a_{23} \end{bmatrix}.$$ ■

Proposition 1.2. *The following identities hold concerning the scalar multiple of matrices.*[9]

(1) $c(A+B)=cA+cB$.
(2) $(c+d)A=cA+dA$.
(3) $(cd)A=c(dA)$.
(4) $1\,A=A$.
(5) $0\,A=O$.
(6) $cO=O$.

Remark 1.4. The set of all $m\times n$ matrices forms a vector space with respect to the sum and scalar multiple operations (cf., Example 9.4 in Sec. 9.1.2). Propositions 1.1 and 1.2 show that all the conditions for a vector space (Definition 9.1 in Sec. 9.1.1) are satisfied indeed. ■

Remark 1.5. The sum of block matrices can be calculated by the formal operation that treats each block as a matrix element. For example,

$$\begin{bmatrix} A_{11} & A_{12} \\ A_{21} & A_{22} \\ A_{31} & A_{32} \end{bmatrix} + \begin{bmatrix} B_{11} & B_{12} \\ B_{21} & B_{22} \\ B_{31} & B_{32} \end{bmatrix} = \begin{bmatrix} A_{11}+B_{11} & A_{12}+B_{12} \\ A_{21}+B_{21} & A_{22}+B_{22} \\ A_{31}+B_{31} & A_{32}+B_{32} \end{bmatrix}.$$

[9]In Proposition 1.2, A and B denote matrices while c and d are scalars. Such notational conventions will be adopted implicitly in the following propositions.

Note, however, that the two block matrices must have a common type, that is, must be compatible with respect to the partition into blocks. The scalar multiple of a block matrix can be computed similarly. For example,

$$c \begin{bmatrix} A_{11} & A_{12} & A_{13} \\ A_{21} & A_{22} & A_{23} \end{bmatrix} = \begin{bmatrix} cA_{11} & cA_{12} & cA_{13} \\ cA_{21} & cA_{22} & cA_{23} \end{bmatrix}.$$

■

1.2.2 *Product*

Definition 1.24. For an $l \times m$ matrix $A = (a_{ij})$ and an $m \times n$ matrix $B = (b_{jk})$, their *product* $C = AB$ is defined as an $l \times n$ matrix $C = (c_{ik})$ with

$$c_{ik} = \sum_{j=1}^{m} a_{ij} b_{jk} \qquad (i = 1, 2, \ldots, l;\ k = 1, 2, \ldots, n).$$

Note that AB is defined (only) when the number of columns of A is equal to the number of rows of B. ■

Example 1.3. The product of a 2×2 matrix and a 2×3 matrix is given as

$$\begin{bmatrix} a_{11} & a_{12} \\ a_{21} & a_{22} \end{bmatrix} \begin{bmatrix} b_{11} & b_{12} & b_{13} \\ b_{21} & b_{22} & b_{23} \end{bmatrix}$$
$$= \begin{bmatrix} a_{11}b_{11} + a_{12}b_{21} & a_{11}b_{12} + a_{12}b_{22} & a_{11}b_{13} + a_{12}b_{23} \\ a_{21}b_{11} + a_{22}b_{21} & a_{21}b_{12} + a_{22}b_{22} & a_{21}b_{13} + a_{22}b_{23} \end{bmatrix}.$$

■

Remark 1.6. Let A be an $m_1 \times n_1$ matrix and B an $m_2 \times n_2$ matrix. The product AB is defined if (and only if) $n_1 = m_2$, whereas the product BA is defined if (and only if) $n_2 = m_1$. Therefore, it can happen that AB is defined while BA is not. It can also happen that BA is defined while AB is not. Moreover, even if both AB and BA are defined, they may have different types (sizes). For example, for a 2×3 matrix A and a 3×2 matrix B, both AB and BA are defined, where AB is a 2×2 matrix and BA is a 3×3 matrix. ■

Proposition 1.3. *The following identities hold concerning the product of matrices.*

(1) $(AB)C = A(BC)$.
(2) $A(B + C) = AB + AC$.

(3) $(A+B)C = AC + BC$.
(4) $AI = IA = A$.
(5) $AO = OA = O$.
(6) $c(AB) = (cA)B = A(cB)$.

The identity in Proposition 1.3(1) is called the *associative law* and the identities (2) and (3) are the *distributive law.*

Remark 1.7. For square matrices A and B of the same order, both AB and BA are defined. However, they are not always equal. In other words, the commutative law does not hold for product operations. For example, for

$$A = \begin{bmatrix} 1 & 0 \\ 1 & 1 \end{bmatrix}, \qquad B = \begin{bmatrix} 1 & 1 \\ 0 & 1 \end{bmatrix}$$

we have

$$AB = \begin{bmatrix} 1 & 1 \\ 1 & 2 \end{bmatrix}, \qquad BA = \begin{bmatrix} 2 & 1 \\ 1 & 1 \end{bmatrix}.$$

■

Definition 1.25. Let A and B be square matrices of the same order. We say that A and B are *commutative* if

$$AB = BA.$$

■

Remark 1.8. For real or complex numbers a and b, if $a \neq 0$ and $b \neq 0$, then $ab \neq 0$. For matrices, however, we may have $AB = O$ when $A \neq O$ and $B \neq O$. For example,

$$A = \begin{bmatrix} 1 & 0 \\ 0 & 0 \end{bmatrix}, \qquad B = \begin{bmatrix} 0 & 0 \\ 1 & 0 \end{bmatrix}, \qquad AB = \begin{bmatrix} 0 & 0 \\ 0 & 0 \end{bmatrix}.$$

■

Definition 1.26. A square matrix A is said to be *nilpotent* if

$$A^k = O$$

for some k $(\geqq 1)$. ■

Remark 1.9. The set of all square matrices of order n forms a ring with respect to the addition (sum) in Definition 1.22 and the multiplication (product) in Definition 1.24; see Remark 9.2 in Sec. 9.1.1 for the definition of a ring. Propositions 1.1 and 1.3 show that all the conditions defining a ring are indeed satisfied. In particular, the zero matrix serves as the zero element of the ring and the unit matrix as the unit element of the ring. When $n \geqq 2$, this ring is not commutative (Remark 1.7) and has zero divisors (Remark 1.8). ■

Remark 1.10. The product of block matrices can be calculated by the formal operation that treats each block as a matrix element. For example,

$$\begin{bmatrix} A_{11} & A_{12} \\ A_{21} & A_{22} \end{bmatrix} \begin{bmatrix} B_{11} & B_{12} & B_{13} \\ B_{21} & B_{22} & B_{23} \end{bmatrix}$$
$$= \begin{bmatrix} A_{11}B_{11} + A_{12}B_{21} & A_{11}B_{12} + A_{12}B_{22} & A_{11}B_{13} + A_{12}B_{23} \\ A_{21}B_{11} + A_{22}B_{21} & A_{21}B_{12} + A_{22}B_{22} & A_{21}B_{13} + A_{22}B_{23} \end{bmatrix}.$$

Note, however, that the two block matrices must be partitioned consistently so that the product $A_{pq}B_{qr}$ is defined for all (p, q, r). ■

1.2.3 *Direct Sum*

Definition 1.27. For matrices A and B, their *direct sum*, denoted as $A \oplus B$, is defined by

$$A \oplus B = \begin{bmatrix} A & O \\ O & B \end{bmatrix}.$$

If A is an $m_1 \times n_1$ matrix and B is an $m_2 \times n_2$ matrix, $A \oplus B$ is an $(m_1 + m_2) \times (n_1 + n_2)$ matrix. ■

1.2.4 *Kronecker Product*

Definition 1.28. For an $m_1 \times n_1$ matrix $A = (a_{ij})$ and an $m_2 \times n_2$ matrix B, their *Kronecker product*, denoted by $A \otimes B$, is defined as

$$A \otimes B = \begin{bmatrix} a_{11}B & a_{12}B & \cdots & a_{1n_1}B \\ a_{21}B & a_{22}B & \cdots & a_{2n_1}B \\ \vdots & \vdots & & \vdots \\ a_{m_1 1}B & a_{m_1 2}B & \cdots & a_{m_1 n_1}B \end{bmatrix},$$

which is an $(m_1 m_2) \times (n_1 n_2)$ matrix. The Kronecker product is sometimes called the *direct product.* ■

Proposition 1.4. *The following identities hold concerning the Kronecker product of matrices.*

(1) $(A \otimes B) \otimes C = A \otimes (B \otimes C)$.
(2) $A \otimes (B + C) = A \otimes B + A \otimes C$.
(3) $(A + B) \otimes C = A \otimes C + B \otimes C$.
(4) $(AB) \otimes (CD) = (A \otimes C)(B \otimes D)$.

Proof. Parts (1), (2), and (3) follow easily from the definition of the Kronecker product.

(4) Let $A = (a_{ij})$ be an $l \times m$ matrix and $B = (b_{jk})$ an $m \times n$ matrix, and define $S = AB = (s_{ik} \mid i = 1, 2, \ldots, l;\ k = 1, 2, \ldots, n)$. Since

$$A \otimes C = \begin{bmatrix} a_{11}C & \cdots & a_{1m}C \\ \vdots & & \vdots \\ a_{l1}C & \cdots & a_{lm}C \end{bmatrix}, \quad B \otimes D = \begin{bmatrix} b_{11}D & \cdots & b_{1n}D \\ \vdots & & \vdots \\ b_{m1}D & \cdots & b_{mn}D \end{bmatrix},$$

the (i,k) block of $(A \otimes C)(B \otimes D)$ is given by

$$\sum_j a_{ij}C \cdot b_{jk}D = \left(\sum_j a_{ij}b_{jk} \right) CD = s_{ik}CD,$$

which is equal to the (i,k) block of $S \otimes (CD)$. □

Remark 1.11. In general, $A \otimes B$ and $B \otimes A$ are different matrices. However, they are related by permutations of rows and columns. For example, for a 3×3 matrix A and a 2×2 matrix B, we have

$$A \otimes B = \left[\begin{array}{cc|cc|cc} a_{11}b_{11} & a_{11}b_{12} & a_{12}b_{11} & a_{12}b_{12} & a_{13}b_{11} & a_{13}b_{12} \\ a_{11}b_{21} & a_{11}b_{22} & a_{12}b_{21} & a_{12}b_{22} & a_{13}b_{21} & a_{13}b_{22} \\ \hline a_{21}b_{11} & a_{21}b_{12} & a_{22}b_{11} & a_{22}b_{12} & a_{23}b_{11} & a_{23}b_{12} \\ a_{21}b_{21} & a_{21}b_{22} & a_{22}b_{21} & a_{22}b_{22} & a_{23}b_{21} & a_{23}b_{22} \\ \hline a_{31}b_{11} & a_{31}b_{12} & a_{32}b_{11} & a_{32}b_{12} & a_{33}b_{11} & a_{33}b_{12} \\ a_{31}b_{21} & a_{31}b_{22} & a_{32}b_{21} & a_{32}b_{22} & a_{33}b_{21} & a_{33}b_{22} \end{array} \right],$$

$$B \otimes A = \left[\begin{array}{ccc|ccc} a_{11}b_{11} & a_{12}b_{11} & a_{13}b_{11} & a_{11}b_{12} & a_{12}b_{12} & a_{13}b_{12} \\ a_{21}b_{11} & a_{22}b_{11} & a_{23}b_{11} & a_{21}b_{12} & a_{22}b_{12} & a_{23}b_{12} \\ a_{31}b_{11} & a_{32}b_{11} & a_{33}b_{11} & a_{31}b_{12} & a_{32}b_{12} & a_{33}b_{12} \\ \hline a_{11}b_{21} & a_{12}b_{21} & a_{13}b_{21} & a_{11}b_{22} & a_{12}b_{22} & a_{13}b_{22} \\ a_{21}b_{21} & a_{22}b_{21} & a_{23}b_{21} & a_{21}b_{22} & a_{22}b_{22} & a_{23}b_{22} \\ a_{31}b_{21} & a_{32}b_{21} & a_{33}b_{21} & a_{31}b_{22} & a_{32}b_{22} & a_{33}b_{22} \end{array} \right].$$

Through the permutation $(1, 2, 3, 4, 5, 6) \mapsto (1, 3, 5, 2, 4, 6)$ of rows and columns, the matrix $A \otimes B$ is brought to the matrix $B \otimes A$. ■

1.2.5 *Hadamard Product*

Definition 1.29. For matrices $A = (a_{ij})$ and $B = (b_{ij})$ of the same type (size), their *Hadamard product*, denoted by $A \odot B$, is defined as the matrix whose (i,j) entry is equal to $a_{ij}b_{ij}$. That is, $A \odot B = (a_{ij}b_{ij})$. If A and B are $m \times n$ matrices, then $A \odot B$ is also an $m \times n$ matrix. The Hadamard product is also called the *Schur product*. ■

Proposition 1.5. *The following identities hold concerning the Hadamard product of matrices.*

(1) $A \odot B = B \odot A$.
(2) $(A \odot B) \odot C = A \odot (B \odot C)$.
(3) $A \odot (B + C) = A \odot B + A \odot C$.

Remark 1.12. The Hadamard product is closely related to integral equations and the Fourier series [4, 5], and also used in machine learning. ■

1.3 Inverse Matrix

1.3.1 *Definition*

Definition 1.30. A square matrix A is called a *nonsingular matrix* (or *regular matrix*, *invertible matrix*) if there exists a square matrix X of the same order satisfying

$$XA = AX = I,$$

where I denotes the unit matrix. Such X is called the *inverse* of A and denoted by A^{-1}. A matrix that is not nonsingular is called a *singular matrix*. ■

Theorem 1.1.

(1) *The inverse of a nonsingular matrix A is uniquely determined.*
(2) *If A is nonsingular, its inverse A^{-1} is also nonsingular, and $(A^{-1})^{-1} = A$.*
(3) *If A and B are nonsingular, their product AB is also nonsingular, and then $(AB)^{-1} = B^{-1}A^{-1}$.*

Proof. (1) Suppose that $XA = AX = I$ and $YA = AY = I$. By multiplying $I = AY$ with X from the left we obtain

$$X = X(AY) = (XA)Y = IY = Y.$$

(2) Letting $B = A^{-1}$ and $X = A$, we obtain $XB = BX = I$. Therefore, $X = B^{-1}$, that is, $A = (A^{-1})^{-1}$.

(3) Letting $C = AB$ and $X = B^{-1}A^{-1}$, we obtain

$$XC = (B^{-1}A^{-1})(AB) = B^{-1}(A^{-1}A)B = B^{-1}B = I.$$

Similarly we can show $CX = I$. Therefore, $X = C^{-1}$. □

Theorem 1.2. *Let A be a square matrix.*

(1) *If there exists a square matrix X satisfying $XA = I$, then A is nonsingular and $X = A^{-1}$.*
(2) *If there exists a square matrix X satisfying $AX = I$, then A is nonsingular and $X = A^{-1}$.*

Proof. The proof is given in Remark 2.11 in Sec. 2.6. □

Remark 1.13. The converse of Theorem 1.1(3) is also true. That is, if the product AB of square matrices A and B is nonsingular, then both A and B are nonsingular. Indeed, by nonsingularity of AB, there exists a matrix X for which $(AB)X = X(AB) = I$. This implies $A(BX) = I$ and $(XA)B = I$, from which A and B are nonsingular by Theorem 1.2. ■

The following theorem offers a preview of mutually equivalent conditions for a matrix to be nonsingular. They are discussed in detail later.

Theorem 1.3. *The following seven conditions* (a)–(g) *are equivalent for a square matrix A of order n.*[10]

(a) *A is nonsingular, that is, it has the inverse A^{-1}.*
(b) $\det A \neq 0$.
(c) $\operatorname{rank} A = n$.
(d) $\operatorname{Im} A = K^n$.
(e) $\operatorname{Ker} A = \{\mathbf{0}\}$.
(f) *The column vectors of A are linearly independent.*
(g) *The row vectors of A are linearly independent.*

Proof. The equivalence of (a) and (b) is given in Theorem 2.16 in Sec. 2.6. The equivalence among (b), (c), (f), and (g) is shown in Remark 4.1 in Sec. 4.1.4. The conditions (d) and (e) are considered in Remark 5.2 in Sec. 5.1. □

[10] See Sec. 2.2 for "det" in (b), Sec. 4.1 for "rank" in (c), and Sec. 5.1 for "Im" and "Ker" in (d) and (e). Linear independence in (f) and (g) is defined in Sec. 4.1.2.

1.3.2 *Inverse of Block Matrices*

In this section we consider the inverse of block matrices. Theorem 1.4 deals with simple basic cases, whereas the general case is treated in Theorem 1.5.

Theorem 1.4. *Let A and B be square matrices of orders n and m, respectively.*[11]

(1) *For a block-diagonal matrix we have:*

$$\left[\begin{array}{c|c} A & O \\ \hline O & B \end{array}\right] \text{ is nonsingular} \iff A \text{ and } B \text{ are nonsingular.} \tag{1.6}$$

If this condition is satisfied, we have

$$\left[\begin{array}{c|c} A & O \\ \hline O & B \end{array}\right]^{-1} = \left[\begin{array}{c|c} A^{-1} & O \\ \hline O & B^{-1} \end{array}\right]. \tag{1.7}$$

(2) *For a block-triangular matrix we have:*

$$\left[\begin{array}{c|c} A & C \\ \hline O & B \end{array}\right] \text{ is nonsingular} \iff A \text{ and } B \text{ are nonsingular.} \tag{1.8}$$

If this condition is satisfied, we have

$$\left[\begin{array}{c|c} A & C \\ \hline O & B \end{array}\right]^{-1} = \left[\begin{array}{c|c} A^{-1} & -A^{-1}CB^{-1} \\ \hline O & B^{-1} \end{array}\right]. \tag{1.9}$$

Proof. Part (1) is a special case of (2) with $C = O$. We will prove (2). Suppose that A and B are nonsingular and note the identity[12]

$$\left[\begin{array}{c|c} A^{-1} & -A^{-1}CB^{-1} \\ \hline O & B^{-1} \end{array}\right] \left[\begin{array}{c|c} A & C \\ \hline O & B \end{array}\right] = \left[\begin{array}{c|c} I & O \\ \hline O & I \end{array}\right].$$

By Theorem 1.2(1), this equation implies that $\left[\begin{array}{c|c} A & C \\ \hline O & B \end{array}\right]$ is nonsingular and its inverse is given by (1.9). Conversely, suppose that $\left[\begin{array}{c|c} A & C \\ \hline O & B \end{array}\right]$ is nonsingular. Then there exist matrices X, Y, Z, and W such that

$$\left[\begin{array}{c|c} X & Y \\ \hline Z & W \end{array}\right] \left[\begin{array}{c|c} A & C \\ \hline O & B \end{array}\right] = \left[\begin{array}{c|c} I & O \\ \hline O & I \end{array}\right], \quad \left[\begin{array}{c|c} A & C \\ \hline O & B \end{array}\right] \left[\begin{array}{c|c} X & Y \\ \hline Z & W \end{array}\right] = \left[\begin{array}{c|c} I & O \\ \hline O & I \end{array}\right].$$

The $(1, 1)$ block of the first equation shows $XA = I$, whereas the $(2, 2)$ block of the second equation shows $BW = I$. By Theorem 1.2, both A and B are nonsingular. □

[11] The matrix C in (2) must be an $n \times m$ matrix for the block matrix to be defined. Such consistency in size will be a tacit assumption throughout.

[12] A derivation of this identity is explained in Remark 3.5 in Sec. 3.4.

For the inverse of bordered matrices and block matrices, we have the following formulas. The matrix $S = D - CA^{-1}B$ in (1.12) below is called the *Schur complement* with respect to pivot A.

Theorem 1.5.

(1) *Let A be a nonsingular matrix, $\mathbf{b}$ and $\mathbf{c}$ vectors, and d a scalar.*

$$\left[\begin{array}{c|c} A & \mathbf{b} \\ \hline \mathbf{c}^\top & d \end{array}\right] \text{ is nonsingular} \iff s = d - \mathbf{c}^\top A^{-1}\mathbf{b} \text{ is non-zero.} \quad (1.10)$$

If these conditions are satisfied, we have

$$\left[\begin{array}{c|c} A & \mathbf{b} \\ \hline \mathbf{c}^\top & d \end{array}\right]^{-1} = \left[\begin{array}{c|c} A^{-1} + (A^{-1}\mathbf{b})(\mathbf{c}^\top A^{-1})/s & -(A^{-1}\mathbf{b})/s \\ \hline -(\mathbf{c}^\top A^{-1})/s & 1/s \end{array}\right]. \quad (1.11)$$

(2) *Let A be a nonsingular matrix of order n and D a square matrix of order m.*

$$\left[\begin{array}{c|c} A & B \\ \hline C & D \end{array}\right] \text{ is nonsingular}$$
$$\iff S = D - CA^{-1}B \text{ is a nonsingular matrix.} \quad (1.12)$$

If these conditions are satisfied, we have

$$\left[\begin{array}{c|c} A & B \\ \hline C & D \end{array}\right]^{-1} = \left[\begin{array}{c|c} A^{-1} + A^{-1}BS^{-1}CA^{-1} & -A^{-1}BS^{-1} \\ \hline -S^{-1}CA^{-1} & S^{-1} \end{array}\right]. \quad (1.13)$$

Proof. Part (1) is a special case of (2). We will prove (2). As is easily verified, we have

$$\left[\begin{array}{c|c} A & B \\ \hline C & D \end{array}\right] = \left[\begin{array}{c|c} A & O_{n,m} \\ \hline C & I_m \end{array}\right] \left[\begin{array}{c|c} I_n & A^{-1}B \\ \hline O_{m,n} & D - CA^{-1}B \end{array}\right]. \quad (1.14)$$

Then the equivalence in (1.12) follows from (1.8) in Theorem 1.4(2), Theorem 1.1(3), and Remark 1.13. It follows from the decomposition (1.14) and Theorem 1.1(3) that

$$\left[\begin{array}{c|c} A & B \\ \hline C & D \end{array}\right]^{-1} = \left[\begin{array}{c|c} I_n & A^{-1}B \\ \hline O_{m,n} & D - CA^{-1}B \end{array}\right]^{-1} \left[\begin{array}{c|c} A & O_{n,m} \\ \hline C & I_m \end{array}\right]^{-1},$$

where the formula (1.9) gives

$$\left[\begin{array}{c|c} I_n & A^{-1}B \\ \hline O_{m,n} & D - CA^{-1}B \end{array}\right]^{-1} = \left[\begin{array}{c|c} I_n & A^{-1}B \\ \hline O_{m,n} & S \end{array}\right]^{-1} = \left[\begin{array}{c|c} I_n & -A^{-1}BS^{-1} \\ \hline O_{m,n} & S^{-1} \end{array}\right],$$

$$\begin{aligned} \left[\begin{array}{c|c} A & O_{n,m} \\ \hline C & I_m \end{array}\right]^{-1} &= \left[\begin{array}{c|c} O_{n,m} & I_n \\ \hline I_m & O_{m,n} \end{array}\right] \left[\begin{array}{c|c} I_m & C \\ \hline O_{n,m} & A \end{array}\right]^{-1} \left[\begin{array}{c|c} O_{m,n} & I_m \\ \hline I_n & O_{n,m} \end{array}\right] \\ &= \left[\begin{array}{c|c} O_{n,m} & I_n \\ \hline I_m & O_{m,n} \end{array}\right] \left[\begin{array}{c|c} I_m & -CA^{-1} \\ \hline O_{n,m} & A^{-1} \end{array}\right] \left[\begin{array}{c|c} O_{m,n} & I_m \\ \hline I_n & O_{n,m} \end{array}\right] \\ &= \left[\begin{array}{c|c} A^{-1} & O_{n,m} \\ \hline -CA^{-1} & I_m \end{array}\right]. \end{aligned}$$

The product of these matrices is equal to the right-hand side of (1.13). Thus the proof is completed.[13] □

Remark 1.14. Suppose that $\left[\begin{array}{c|c} A & B \\ \hline C & D \end{array}\right]$ is nonsingular and let $\left[\begin{array}{c|c} A & B \\ \hline C & D \end{array}\right]^{-1} = \left[\begin{array}{c|c} X & Y \\ \hline Z & W \end{array}\right]$. Then A is nonsingular if and only if W is nonsingular. We can prove this by applying Theorem 1.5(2) to $\left[\begin{array}{c|c} A & B \\ \hline C & D \end{array}\right]$ and $\left[\begin{array}{c|c} X & Y \\ \hline Z & W \end{array}\right]$. ■

The following formulas show how to modify the inverse when a matrix is modified.[14]

Theorem 1.6.

(1) Sherman–Morrison formula: *Let A be a nonsingular matrix, and $\boldsymbol{b}$ and $\boldsymbol{c}$ be vectors. If $\boldsymbol{c}^\top A^{-1}\boldsymbol{b} \neq -1$, then*

$$(A + \boldsymbol{b}\boldsymbol{c}^\top)^{-1} = A^{-1} - (A^{-1}\boldsymbol{b})(\boldsymbol{c}^\top A^{-1})/(1 + \boldsymbol{c}^\top A^{-1}\boldsymbol{b}).$$

(2) Sherman–Morrison–Woodbury formula: *Let A and D be nonsingular matrices of orders n and m, respectively, and B an $n \times m$ matrix and*

[13]The present proof of the formula (1.13) is not truly constructive as the decomposition (1.14) is given out of the blue. A more natural derivation of the formula (1.13) is shown in Remark 3.5 in Sec. 3.4.

[14]If $\boldsymbol{b} \neq \boldsymbol{0}$ and $\boldsymbol{c} \neq \boldsymbol{0}$, for example, the matrix $A + \boldsymbol{b}\boldsymbol{c}^\top$ represents a modification of A with a rank-one matrix $\boldsymbol{b}\boldsymbol{c}^\top$; see Sec. 4.1 for "rank." The formulas in Theorem 1.6 are used effectively in optimization (the quasi-Newton method) and statistics (computation of covariance matrices and solution of the normal equation).

C an $m \times n$ matrix. If $D^{-1} + CA^{-1}B$ is nonsingular, then[15]

$$(A + BDC)^{-1} = A^{-1} - A^{-1}B(D^{-1} + CA^{-1}B)^{-1}CA^{-1} \quad (1.15)$$
$$= A^{-1} - A^{-1}BD(D + DCA^{-1}BD)^{-1}DCA^{-1}. \quad (1.16)$$

Proof. Part (1) is a special case of (2). We will prove (2), in which the formula (1.16) is an easy consequence of (1.15) because

$$\begin{aligned}(D^{-1} + CA^{-1}B)^{-1} &= DD^{-1} \cdot (D^{-1} + CA^{-1}B)^{-1} \cdot D^{-1}D \\ &= D(D + DCA^{-1}BD)^{-1}D.\end{aligned}$$

In the following we show two alternative proofs of (1.15).

[Proof 1: By Theorem 1.5] We apply Theorem 1.5 to two matrices

$$\left[\begin{array}{c|c} A & -B \\ \hline C & D^{-1} \end{array}\right], \qquad \left[\begin{array}{c|c} D^{-1} & C \\ \hline -B & A \end{array}\right].$$

Define X and Y by

$$\left[\begin{array}{c|c} A & -B \\ \hline C & D^{-1} \end{array}\right]^{-1} = \left[\begin{array}{c|c} X & * \\ \hline * & * \end{array}\right], \qquad \left[\begin{array}{c|c} D^{-1} & C \\ \hline -B & A \end{array}\right]^{-1} = \left[\begin{array}{c|c} * & * \\ \hline * & Y \end{array}\right],$$

where $*$ denotes an appropriate matrix. By Theorem 1.5(2) we obtain

$$X = A^{-1} - A^{-1}B(D^{-1} + CA^{-1}B)^{-1}CA^{-1}, \qquad Y = (A + BDC)^{-1}.$$

On the other hand, it follows from

$$\left[\begin{array}{c|c} O & I \\ \hline I & O \end{array}\right] \left[\begin{array}{c|c} A & -B \\ \hline C & D^{-1} \end{array}\right] \left[\begin{array}{c|c} O & I \\ \hline I & O \end{array}\right] = \left[\begin{array}{c|c} D^{-1} & C \\ \hline -B & A \end{array}\right]$$

that

$$\left[\begin{array}{c|c} O & I \\ \hline I & O \end{array}\right] \left[\begin{array}{c|c} A & -B \\ \hline C & D^{-1} \end{array}\right]^{-1} \left[\begin{array}{c|c} O & I \\ \hline I & O \end{array}\right] = \left[\begin{array}{c|c} D^{-1} & C \\ \hline -B & A \end{array}\right]^{-1}.$$

Therefore we have $X = Y$.

[Proof 2: By direct verification] Letting

$$Z = (D^{-1} + CA^{-1}B)^{-1}CA^{-1},$$

we can express the right-hand side of (1.15) as $A^{-1} - A^{-1}BZ$. A direct calculation shows

$$\begin{aligned}&(A + BDC)(A^{-1} - A^{-1}BZ) = (I - BZ) + (BDCA^{-1} - BDCA^{-1}BZ) \\ &= I - BZ + BD(D^{-1} + CA^{-1}B)Z - BDCA^{-1}BZ = I.\end{aligned}$$

This implies (1.15) by Theorem 1.2(2). □

---

[15]Since D and $D^{-1} + CA^{-1}B$ are nonsingular, the matrix $D + DCA^{-1}BD = D(D^{-1} + CA^{-1}B)D$ is also nonsingular. Note also that the formula (1.16) does not involve the inverse of D.

In this section we have presented the definition and formulas concerning the inverse matrices. The inverse of a matrix can be represented in terms of the determinants of submatrices (cofactors), to be discussed in Sec. 2.6. The inverse can also be computed by repeated application of elimination operations, to be discussed in Sec. 3.4.

1.4 Transpose and Conjugate

1.4.1 *Definition*

Let $A = (a_{ij})$ be an $m \times n$ matrix. When A is a complex matrix, the complex conjugate of a_{ij} is denoted by $\overline{a_{ij}}$.

Definition 1.31. The *transpose* of A (or the *transposed matrix* for A), denoted[16] by $A^\top$, is the $n \times m$ matrix whose (i, j) entry is equal to a_{ji} for all (i, j). ∎

Definition 1.32. The *complex conjugate* of A (or the *complex conjugate matrix* for A), denoted by $\overline{A}$, is the $m \times n$ matrix whose (i, j) entry is equal to $\overline{a_{ij}}$ for all (i, j). ∎

Definition 1.33. The *adjoint* of A (or the *adjoint matrix* for A), denoted by A^*, is the $n \times m$ matrix whose (i, j) entry is equal to $\overline{a_{ji}}$ for all (i, j). Sometimes it is also called the *conjugate transpose* of A because of the relation $A^* = (\overline{A})^\top = \overline{A^\top}$. ∎

Proposition 1.6. *The following identities hold, where A and B are matrices (of appropriate sizes) and c is a scalar.*

(1)
$$(A^\top)^\top = A, \qquad (A + B)^\top = A^\top + B^\top, \qquad (cA)^\top = cA^\top,$$
$$(AB)^\top = B^\top A^\top, \qquad (A^{-1})^\top = (A^\top)^{-1}, \qquad (A \otimes B)^\top = A^\top \otimes B^\top,$$
$$\begin{bmatrix} A_{11} & A_{12} & \cdots & A_{1l} \\ A_{21} & A_{22} & \cdots & A_{2l} \\ \vdots & \vdots & & \vdots \\ A_{k1} & A_{k2} & \cdots & A_{kl} \end{bmatrix}^\top = \begin{bmatrix} {A_{11}}^\top & {A_{21}}^\top & \cdots & {A_{k1}}^\top \\ {A_{12}}^\top & {A_{22}}^\top & \cdots & {A_{k2}}^\top \\ \vdots & \vdots & & \vdots \\ {A_{1l}}^\top & {A_{2l}}^\top & \cdots & {A_{kl}}^\top \end{bmatrix}.$$

[16] Several different symbols are commonly used to denote the transposition, such as A^{T}, A^{t}, and ${}^{\mathrm{t}}A$ (or A^T, A^t, and ${}^tA$).

(2)
$$\overline{\overline{A}} = A, \qquad \overline{A+B} = \overline{A} + \overline{B}, \qquad \overline{(cA)} = \overline{c}\,\overline{A},$$
$$\overline{(AB)} = \overline{A}\,\overline{B}, \qquad \overline{(A^{-1})} = \left(\overline{A}\right)^{-1}, \qquad \overline{(A \otimes B)} = \overline{A} \otimes \overline{B},$$
$$\overline{\begin{bmatrix} A_{11} & A_{12} & \cdots & A_{1l} \\ A_{21} & A_{22} & \cdots & A_{2l} \\ \vdots & \vdots & & \vdots \\ A_{k1} & A_{k2} & \cdots & A_{kl} \end{bmatrix}} = \begin{bmatrix} \overline{A_{11}} & \overline{A_{12}} & \cdots & \overline{A_{1l}} \\ \overline{A_{21}} & \overline{A_{22}} & \cdots & \overline{A_{2l}} \\ \vdots & \vdots & & \vdots \\ \overline{A_{k1}} & \overline{A_{k2}} & \cdots & \overline{A_{kl}} \end{bmatrix}.$$

(3)
$$(A^*)^* = A, \qquad (A+B)^* = A^* + B^*, \qquad (cA)^* = \overline{c}A^*,$$
$$(AB)^* = B^*A^*, \qquad (A^{-1})^* = (A^*)^{-1}, \qquad (A \otimes B)^* = A^* \otimes B^*,$$
$$\begin{bmatrix} A_{11} & A_{12} & \cdots & A_{1l} \\ A_{21} & A_{22} & \cdots & A_{2l} \\ \vdots & \vdots & & \vdots \\ A_{k1} & A_{k2} & \cdots & A_{kl} \end{bmatrix}^* = \begin{bmatrix} {A_{11}}^* & {A_{21}}^* & \cdots & {A_{k1}}^* \\ {A_{12}}^* & {A_{22}}^* & \cdots & {A_{k2}}^* \\ \vdots & \vdots & & \vdots \\ {A_{1l}}^* & {A_{2l}}^* & \cdots & {A_{kl}}^* \end{bmatrix}.$$

1.4.2 *Symmetric Matrices and Hermitian Matrices*

Definition 1.34. A real square matrix $A = (a_{ij})$ is called a *symmetric matrix* if

$$A^\top = A \qquad (i.e.,\ a_{ji} = a_{ij} \text{ for all } (i,j)). \tag{1.17}$$

■

Definition 1.35. A real square matrix $A = (a_{ij})$ is called an *anti-symmetric matrix* (or *skew-symmetric matrix, alternating matrix*) if

$$A^\top = -A \qquad (i.e.,\ a_{ji} = -a_{ij} \text{ for all } (i,j)). \tag{1.18}$$

■

Definition 1.36. A complex square matrix $A = (a_{ij})$ is called a *Hermitian matrix* if

$$A^* = A \qquad (i.e.,\ \overline{a_{ji}} = a_{ij} \text{ for all } (i,j)). \tag{1.19}$$

A symmetric matrix is a Hermitian matrix, and a real matrix that is Hermitian is a symmetric matrix.[17] ■

[17] A Hermitian matrix should not be confused with a symmetric complex matrix, which, by definition, is a matrix $A = (a_{ij})$ satisfying $a_{ji} = a_{ij} \in \mathbb{C}$ for all (i,j). In this book, we do not treat a symmetric matrix of complex numbers, and a symmetric matrix always means a real symmetric matrix.

Definition 1.37. A complex square matrix $A = (a_{ij})$ is called a *skew-Hermitian matrix* if

$$A^* = -A \qquad (i.e., \overline{a_{ji}} = -a_{ij} \text{ for all } (i,j)). \tag{1.20}$$

An anti-symmetric matrix is a skew-Hermitian matrix, and a real matrix that is skew-Hermitian is an anti-symmetric matrix. ■

1.4.3 *Orthogonal Matrices and Unitary Matrices*

Definition 1.38. A real square matrix A is called an *orthogonal matrix* if

$$A^\top A = AA^\top = I, \tag{1.21}$$

where I denotes the unit matrix. ■

Definition 1.39.

- Real vectors $\boldsymbol{x}$ and $\boldsymbol{y}$ are said to be *orthogonal* to each other if $\boldsymbol{x}^\top \boldsymbol{y} = 0$.
- The *length* of a real vector $\boldsymbol{x}$ is defined to be $\sqrt{\boldsymbol{x}^\top \boldsymbol{x}}$.
- A set of real vectors $\boldsymbol{x}_1, \boldsymbol{x}_2, \ldots, \boldsymbol{x}_k$ is called an *orthonormal system* if each vector has a unit length and every pair of vectors are orthogonal. We can express the condition of orthonormality as $\boldsymbol{x}_i^\top \boldsymbol{x}_j = \delta_{ij}$ $(i, j = 1, 2, \ldots, k)$.[18] ■

Theorem 1.7. *Let* $A = [\boldsymbol{a}_1, \boldsymbol{a}_2, \ldots, \boldsymbol{a}_n]$ *be a real square matrix of order* n. *The column vectors* $\boldsymbol{a}_1, \boldsymbol{a}_2, \ldots, \boldsymbol{a}_n$ *form an orthonormal system if and only if* A *is an orthogonal matrix.*

Definition 1.40. A complex square matrix A is called a *unitary matrix* if

$$A^* A = AA^* = I, \tag{1.22}$$

where I denotes the unit matrix. An orthogonal matrix is a unitary matrix, and a real matrix that is unitary is an orthogonal matrix. ■

Definition 1.41.

- Complex vectors $\boldsymbol{x}$ and $\boldsymbol{y}$ are said to be *orthogonal* to each other if $\boldsymbol{x}^\top \overline{\boldsymbol{y}} = 0$. (The condition $\boldsymbol{x}^\top \overline{\boldsymbol{y}} = 0$ is equivalent to $\boldsymbol{x}^* \boldsymbol{y} = 0$.)
- The *length* of a complex vector $\boldsymbol{x}$ is defined to be $\sqrt{\boldsymbol{x}^\top \overline{\boldsymbol{x}}}$ $(= \sqrt{\boldsymbol{x}^* \boldsymbol{x}}$).

[18]The symbol δ_{ij} is the Kronecker delta introduced in Remark 1.3 in Sec. 1.1.1.

- A set of complex vectors $\boldsymbol{x}_1, \boldsymbol{x}_2, \ldots, \boldsymbol{x}_k$ is called an *orthonormal system* if each vector has a unit length and every pair of vectors are orthogonal. We can express the condition of orthonormality as $\boldsymbol{x}_i^\top \overline{\boldsymbol{x}_j} = \delta_{ij}$ $(i, j = 1, 2, \ldots, k)$, or equivalently as $\boldsymbol{x}_i^* \boldsymbol{x}_j = \delta_{ij}$ $(i, j = 1, 2, \ldots, k)$. ■

Theorem 1.8. *Let $A = [\boldsymbol{a}_1, \boldsymbol{a}_2, \ldots, \boldsymbol{a}_n]$ be a complex square matrix of order n. The column vectors $\boldsymbol{a}_1, \boldsymbol{a}_2, \ldots, \boldsymbol{a}_n$ form an orthonormal system if and only if A is a unitary matrix.*

1.5 Trace

Definition 1.42. For a square matrix A, the sum of its diagonal entries is called the *trace* of A and denoted by $\mathrm{Tr}\, A$. That is,

$$\mathrm{Tr}\, A = \sum_{i=1}^{n} a_{ii}$$

if $A = (a_{ij})$ is a square matrix of order n. ■

Theorem 1.9. *The following identities hold concerning the trace of matrices.*

(1) $\mathrm{Tr}\,(A + B) = \mathrm{Tr}\, A + \mathrm{Tr}\, B$.
(2) $\mathrm{Tr}\,(cA) = c \cdot \mathrm{Tr}\, A$.
(3) $\mathrm{Tr}\,(AB) = \mathrm{Tr}\,(BA)$.
(4) $\mathrm{Tr}\,(ABC) = \mathrm{Tr}\,(BCA) = \mathrm{Tr}\,(CAB)$.
(5) $\mathrm{Tr}\,(P^{-1}AP) = \mathrm{Tr}\, A$, where P is a nonsingular matrix.

Proof. Parts (1) and (2) follow easily from the definition. (3) is proved as

$$\mathrm{Tr}\,(AB) = \sum_i \Big(\sum_k a_{ik} b_{ki} \Big) = \sum_k \Big(\sum_i b_{ki} a_{ik} \Big) = \mathrm{Tr}\,(BA).$$

The first identity in (4) follows from (3) as

$$\mathrm{Tr}\,(ABC) = \mathrm{Tr}\,(A(BC)) = \mathrm{Tr}\,((BC)A) = \mathrm{Tr}\,(BCA),$$

and similarly for the second identity in (4). By (4) we have $\mathrm{Tr}\,(P^{-1}AP) = \mathrm{Tr}\,(APP^{-1}) = \mathrm{Tr}\, A$, which proves (5). □

Remark 1.15. The identities in (3) and (4) of Theorem 1.9 are true for non-square matrices A, B, and C, as long as their products are defined. ■

Remark 1.16. In general,

$$\operatorname{Tr}(ABC) \neq \operatorname{Tr}(BAC).$$

For instance, let

$$A = \begin{bmatrix} 1 & 0 \\ 0 & 0 \end{bmatrix}, \qquad B = \begin{bmatrix} 0 & 0 \\ 1 & 0 \end{bmatrix}, \qquad C = \begin{bmatrix} 0 & 1 \\ 1 & 0 \end{bmatrix},$$

for which

$$\operatorname{Tr}(ABC) = \operatorname{Tr}\left(\begin{bmatrix} 0 & 0 \\ 0 & 0 \end{bmatrix}\right) = 0, \quad \operatorname{Tr}(BAC) = \operatorname{Tr}\left(\begin{bmatrix} 0 & 0 \\ 0 & 1 \end{bmatrix}\right) = 1.$$

■

Remark 1.17. More generally, the trace of a matrix product involving more than three matrices is preserved under cyclic permutations of the matrices. That is, for any integer $k \geqq 2$, we have

$$\operatorname{Tr}(A_1 A_2 \cdots A_{k-1} A_k) = \operatorname{Tr}(A_2 \cdots A_{k-1} A_k A_1),$$
$$\operatorname{Tr}(A_1 A_2 \cdots A_{k-1} A_k) = \operatorname{Tr}(A_k A_1 A_2 \cdots A_{k-1}).$$

■

1.6 Norms

Definition 1.43. Let $p \geqq 1$ be a real number or $p = \infty$. The *p-norm* of a vector $\boldsymbol{x} = (x_1, x_2, \ldots, x_n)^\top$, denoted by $\|\boldsymbol{x}\|_p$, is defined by

$$\|\boldsymbol{x}\|_p = \begin{cases} \left(\sum_{j=1}^{n} |x_j|^p\right)^{1/p} & (1 \leqq p < \infty), \\ \max_{1 \leqq j \leqq n} |x_j| & (p = \infty). \end{cases} \tag{1.23}$$

The norm $\|\cdot\|_2$ (with $p = 2$) is called the *Euclidean norm*. ■

Theorem 1.10.[19]

(1) $\|\boldsymbol{x}\|_p \geqq 0.$ *If* $\|\boldsymbol{x}\|_p = 0$, *then* $\boldsymbol{x} = \boldsymbol{0}$.
(2) $\|\boldsymbol{x} + \boldsymbol{y}\|_p \leqq \|\boldsymbol{x}\|_p + \|\boldsymbol{y}\|_p$ (triangle inequality).[20]
(3) $\|c\boldsymbol{x}\|_p = |c| \cdot \|\boldsymbol{x}\|_p$.

[19] In Theorems 1.10, 1.11, and 1.14, c denotes an arbitrary scalar.
[20] This inequality is called *Minkowski's inequality*.

The p-norm of a matrix is defined using the p-norm of a vector.

Definition 1.44. Let $p \geqq 1$ be a real number or $p = \infty$. The *p-norm* of a matrix A, denoted by $\|A\|_p$, is defined by[21]

$$\|A\|_p = \sup_{\boldsymbol{x} \neq \boldsymbol{0}} \frac{\|A\boldsymbol{x}\|_p}{\|\boldsymbol{x}\|_p} = \sup_{\|\boldsymbol{x}\|_p = 1} \|A\boldsymbol{x}\|_p, \tag{1.24}$$

where A is not necessarily a square matrix. ■

Theorem 1.11.

(1) $\|A\boldsymbol{x}\|_p \leqq \|A\|_p \cdot \|\boldsymbol{x}\|_p$.
(2) $\|A\|_p \geqq 0$. *If* $\|A\|_p = 0$, *then* $A = O$.
(3) $\|A + B\|_p \leqq \|A\|_p + \|B\|_p$.
(4) $\|cA\|_p = |c| \cdot \|A\|_p$.
(5) $\|AB\|_p \leqq \|A\|_p \cdot \|B\|_p$.

Proof. Parts (1), (2) and (4) follow easily from the definition.

(3) Using the triangle inequality of the p-norm for vectors (Theorem 1.10 (2)) we can show

$$\begin{aligned}\|A + B\|_p &= \sup_{\boldsymbol{x}} \|(A + B)\boldsymbol{x}\|_p = \sup_{\boldsymbol{x}} \|A\boldsymbol{x} + B\boldsymbol{x}\|_p \\ &\leqq \sup_{\boldsymbol{x}} \left(\|A\boldsymbol{x}\|_p + \|B\boldsymbol{x}\|_p\right) \\ &\leqq \left(\sup_{\boldsymbol{x}} \|A\boldsymbol{x}\|_p\right) + \left(\sup_{\boldsymbol{x}} \|B\boldsymbol{x}\|_p\right) = \|A\|_p + \|B\|_p,\end{aligned}$$

where $\sup_{\boldsymbol{x}}$ means the supremum taken over all $\boldsymbol{x}$ with $\|\boldsymbol{x}\|_p = 1$.

(5) By (1) above, we have

$$\|(AB)\boldsymbol{x}\|_p = \|A(B\boldsymbol{x})\|_p \leqq \|A\|_p \cdot \|B\boldsymbol{x}\|_p \leqq \|A\|_p \cdot \|B\|_p \cdot \|\boldsymbol{x}\|_p$$

for any $\boldsymbol{x}$. By taking the supremum over all $\boldsymbol{x}$ with $\|\boldsymbol{x}\|_p = 1$, we obtain the desired inequality. □

Theorem 1.12. *Let* $A = (a_{ij})$ *be an* $m \times n$ *matrix.*

(1) $\|A\|_1 = \max_{1 \leqq j \leqq n} \sum_{i=1}^{m} |a_{ij}|$,
where the right-hand side means the maximum column-sum of the absolute element values.

[21] The 2-norm $\|\cdot\|_2$ of a matrix is sometimes called the *spectral norm*.

(2) $\|A\|_\infty = \max_{1 \leqq i \leqq m} \sum_{j=1}^{n} |a_{ij}|$,

where the right-hand side means the maximum row-sum of the absolute element values.

Proof. (1) Let $\alpha = \max_{1 \leqq j \leqq n} \sum_{i=1}^{m} |a_{ij}| = \sum_{i=1}^{m} |a_{ik}|$, where k is the column-index j for which the maximum is attained. For any $\boldsymbol{x}$ satisfying $\|\boldsymbol{x}\|_1 = \sum_{j=1}^{n} |x_j| = 1$, we have

$$\begin{aligned}\|A\boldsymbol{x}\|_1 &= \sum_{i=1}^{m} \Big| \sum_{j=1}^{n} a_{ij} x_j \Big| \\ &\leqq \sum_{i=1}^{m} \sum_{j=1}^{n} |a_{ij}||x_j| = \sum_{j=1}^{n} \Big(|x_j| \sum_{i=1}^{m} |a_{ij}| \Big) \leqq \alpha \sum_{j=1}^{n} |x_j| = \alpha,\end{aligned}$$

where the two inequalities simultaneously turn into inequalities for $\boldsymbol{x} = (0, \ldots, 0, \overset{\overset{k}{\vee}}{1}, 0, \ldots, 0)^\top$ (the kth unit vector). Therefore, we have $\sup_{\boldsymbol{x}} \|A\boldsymbol{x}\|_1 = \alpha$.

(2) Let $\beta = \max_{1 \leqq i \leqq m} \sum_{j=1}^{n} |a_{ij}| = \sum_{j=1}^{n} |a_{kj}|$, where k is the row-index i for which the maximum is attained. For any $\boldsymbol{x}$ satisfying $\|\boldsymbol{x}\|_\infty = \max_{1 \leqq j \leqq n} |x_j| = 1$, we have

$$\|A\boldsymbol{x}\|_\infty = \max_{1 \leqq i \leqq m} \Big| \sum_{j=1}^{n} a_{ij} x_j \Big| \leqq \max_{1 \leqq i \leqq m} \sum_{j=1}^{n} |a_{ij} x_j| \leqq \max_{1 \leqq i \leqq m} \sum_{j=1}^{n} |a_{ij}| = \beta.$$

For the vector $\boldsymbol{x}$ defined by[22] $x_j = \exp(-\mathrm{i} \cdot \arg(a_{kj}))$ for $j = 1, 2, \ldots, n$, the two inequalities in the above turn into inequalities because $a_{kj} x_j = |a_{kj}|$ for all j. Therefore, we have $\sup_{\boldsymbol{x}} \|A\boldsymbol{x}\|_\infty = \beta$. □

Definition 1.45. The *Frobenius norm* of an $m \times n$ matrix $A = (a_{ij})$, denoted by $\|A\|_{\mathrm{F}}$, is defined by

$$\|A\|_{\mathrm{F}} = \Big(\sum_{i=1}^{m} \sum_{j=1}^{n} |a_{ij}|^2 \Big)^{1/2}. \tag{1.25}$$

■

[22]Here i stands for the imaginary unit and $\arg(z)$ denotes the argument of a complex number z, where we define the argument of $z = 0$ to be 0.

Theorem 1.13.

(1) $\|A\|_{\mathrm{F}} = \left(\sum_{j=1}^{n} \|\boldsymbol{a}_j\|_2{}^2 \right)^{1/2}$, *where* $A = [\boldsymbol{a}_1, \boldsymbol{a}_2, \ldots, \boldsymbol{a}_n]$.

(2) $\|A\|_{\mathrm{F}} = \sqrt{\mathrm{Tr}\,(A^*A)} = \sqrt{\mathrm{Tr}\,(AA^*)}$.

Theorem 1.14.

(1) $\|A\boldsymbol{x}\|_2 \leqq \|A\|_{\mathrm{F}} \cdot \|\boldsymbol{x}\|_2$.
(2) $\|A\|_{\mathrm{F}} \geqq 0$. *If* $\|A\|_{\mathrm{F}} = 0$, *then* $A = O$.
(3) $\|A + B\|_{\mathrm{F}} \leqq \|A\|_{\mathrm{F}} + \|B\|_{\mathrm{F}}$.
(4) $\|cA\|_{\mathrm{F}} = |c| \cdot \|A\|_{\mathrm{F}}$.
(5) $\|AB\|_{\mathrm{F}} \leqq \|A\|_{\mathrm{F}} \cdot \|B\|_{\mathrm{F}}$.

Proof. Parts (2) and (4) follow easily from the definition.

(1) Let $\hat{\boldsymbol{a}}_i^\top$ denote the ith row vector of A. By using the Cauchy–Schwarz inequality (cf., Remark 1.18 below), we obtain

$$\begin{aligned} \|A\boldsymbol{x}\|_2{}^2 &= \sum_i |\hat{\boldsymbol{a}}_i^\top \boldsymbol{x}|^2 \\ &\leqq \sum_i \left(\|\hat{\boldsymbol{a}}_i\|_2{}^2 \|\boldsymbol{x}\|_2{}^2\right) = \left(\sum_i \|\hat{\boldsymbol{a}}_i\|_2{}^2\right) \|\boldsymbol{x}\|_2{}^2 = \|A\|_{\mathrm{F}}{}^2 \|\boldsymbol{x}\|_2{}^2. \end{aligned}$$

(3) Using the entries of A and B, define mn-dimensional vectors

$$\begin{aligned} \boldsymbol{a} &= (a_{ij} \mid i = 1, 2, \ldots, m;\ j = 1, 2, \ldots, n), \\ \boldsymbol{b} &= (b_{ij} \mid i = 1, 2, \ldots, m;\ j = 1, 2, \ldots, n). \end{aligned}$$

For their Euclidean norms, we have the triangle inequality $\|\boldsymbol{a} + \boldsymbol{b}\|_2 \leqq \|\boldsymbol{a}\|_2 + \|\boldsymbol{b}\|_2$, which is equivalent to the desired inequality $\|A + B\|_{\mathrm{F}} \leqq \|A\|_{\mathrm{F}} + \|B\|_{\mathrm{F}}$.

(5) Let $\boldsymbol{b}_j$ denote the jth column vector of B for $j = 1, 2, \ldots, n$. By (1) we obtain

$$\begin{aligned} \|AB\|_{\mathrm{F}}{}^2 &= \sum_j \|A\boldsymbol{b}_j\|_2{}^2 \\ &\leqq \sum_j \|A\|_{\mathrm{F}}{}^2 \|\boldsymbol{b}_j\|_2{}^2 = \|A\|_{\mathrm{F}}{}^2 \sum_j \|\boldsymbol{b}_j\|_2{}^2 = \|A\|_{\mathrm{F}}{}^2 \cdot \|B\|_{\mathrm{F}}{}^2. \end{aligned}$$

□

Remark 1.18. For any vectors $\boldsymbol{x} = (x_1, x_2, \ldots, x_n)^\top$ and $\boldsymbol{y} = (y_1, y_2, \ldots, y_n)^\top$, the following inequality holds:

$$|\boldsymbol{x}^\top \boldsymbol{y}| \leqq \|\boldsymbol{x}\|_2 \cdot \|\boldsymbol{y}\|_2,$$

which is expressed in terms of the components as

$$\left|\sum_{i=1}^{n} x_i y_i\right| \leqq \left(\sum_{i=1}^{n} |x_i|^2\right)^{1/2} \left(\sum_{i=1}^{n} |y_i|^2\right)^{1/2}.$$

This is called the *Cauchy–Schwarz inequality.* For complex vectors, we usually adopt another (equivalent) form

$$|\overline{\boldsymbol{x}}^{\top} \boldsymbol{y}| \leqq \|\boldsymbol{x}\|_2 \cdot \|\boldsymbol{y}\|_2,$$

which is expressed in terms of the components as

$$\left|\sum_{i=1}^{n} \overline{x_i} y_i\right| \leqq \left(\sum_{i=1}^{n} |x_i|^2\right)^{1/2} \left(\sum_{i=1}^{n} |y_i|^2\right)^{1/2}.$$

■

Remark 1.19. We have

$$\|A\|_2 \leqq \|A\|_{\mathrm{F}} \tag{1.26}$$

by Theorem 1.14(1) and the definition (1.24). ■

Remark 1.20. The 2-norm and the Frobenius norm of a matrix remain invariant under unitary transformations. That is, for any unitary matrices U and V, we have

$$\|A\|_2 = \|U^* A V\|_2, \qquad \|A\|_{\mathrm{F}} = \|U^* A V\|_{\mathrm{F}}. \tag{1.27}$$

The 2-norm $\|A\|_2$ of matrix A is equal to the maximum singular value of A, and the Frobenius norm $\|A\|_{\mathrm{F}}$ is equal to the square root of the sum of the squared singular values of A; see Theorem 8.7 in Sec. 8.2. ■

Remark 1.21. Let A be an $m \times n$ matrix. For any norm[23] $\|\boldsymbol{x}\|_X$ for n-dimensional vectors $\boldsymbol{x}$ and any norm $\|\boldsymbol{y}\|_Y$ for m-dimensional vectors $\boldsymbol{y}$ we can define a matrix norm $\|A\|$, similarly to (1.24), as

$$\|A\| = \sup_{\boldsymbol{x} \neq \boldsymbol{0}} \frac{\|A\boldsymbol{x}\|_Y}{\|\boldsymbol{x}\|_X} = \sup_{\|\boldsymbol{x}\|_X = 1} \|A\boldsymbol{x}\|_Y. \tag{1.28}$$

In parallel with Theorem 1.11, we have the following.

(1) $\|A\boldsymbol{x}\|_Y \leqq \|A\| \cdot \|\boldsymbol{x}\|_X$.
(2) $\|A\| \geqq 0$. If $\|A\| = 0$, then $A = O$.
(3) $\|A + B\| \leqq \|A\| + \|B\|$.
(4) $\|cA\| = |c| \cdot \|A\|$.
(5) $\|AB\| \leqq \|A\| \cdot \|B\|$. ■

[23] The definition of a norm in general can be found in Definition 9.36 in Sec. 9.6.7.

1.7 Matrices of Special Characteristics

1.7.1 *Diagonally Dominant Matrices*

Definition 1.46. A square matrix is called a *diagonally dominant matrix* if, in each row, the absolute value of the diagonal entry is greater than or equal to the sum of the absolute values of the off-diagonal entries in that row. That is, a square matrix $A = (a_{ij})$ of order n is said to be diagonally dominant if

$$|a_{ii}| \geqq \sum_{j \neq i} |a_{ij}| \qquad (i = 1, 2, \ldots, n). \tag{1.29}$$

If the condition (1.29) is satisfied in strict inequalities, that is, if

$$|a_{ii}| > \sum_{j \neq i} |a_{ij}| \qquad (i = 1, 2, \ldots, n), \tag{1.30}$$

we speak of a *strictly diagonally dominant matrix.* ■

Theorem 1.15. *A strictly diagonally dominant matrix is nonsingular.*

Proof. Suppose that A is singular (*i.e.*, not nonsingular). Then $A\boldsymbol{x} = \mathbf{0}$ for some non-zero vector $\boldsymbol{x} = (x_1, x_2, \ldots, x_n)^\top \neq \mathbf{0}$ (cf., condition (e) of Theorem 1.3 in Sec. 1.3.1). Let $|x_k|$ be the maximum of $|x_1|, |x_2|, \ldots, |x_n|$. Then $|x_k| \neq 0$ and the equation

$$a_{kk}x_k = -\sum_{j \neq k} a_{kj}x_j$$

implies

$$|a_{kk}| \leqq \sum_{j \neq k} |a_{kj}| \frac{|x_j|}{|x_k|} \leqq \sum_{j \neq k} |a_{kj}|,$$

which contradicts (1.30). Therefore, A is nonsingular. □

Definition 1.47. A square matrix is called a *generalized diagonally dominant matrix* if there exist positive numbers $d_1, d_2, \ldots, d_n$ such that

$$|a_{ii}|\, d_i \geqq \sum_{j \neq i} |a_{ij}|\, d_j \qquad (i = 1, 2, \ldots, n). \tag{1.31}$$

We speak of a *generalized strictly diagonally dominant matrix* if

$$|a_{ii}|\, d_i > \sum_{j \neq i} |a_{ij}|\, d_j \qquad (i = 1, 2, \ldots, n). \tag{1.32}$$

In other words, A is called a generalized (strictly) diagonally dominant matrix if an appropriate choice of a diagonal matrix $D = \operatorname{diag}(d_1, d_2, \ldots, d_n)$ with positive diagonal entries makes the matrix AD (strictly) diagonally dominant. ■

Theorem 1.16. *A generalized strictly diagonally dominant matrix is nonsingular.*

Proof. Let D be a diagonal matrix with positive diagonal entries such that AD is strictly diagonally dominant. The matrix AD is nonsingular by Theorem 1.15. Therefore, A is nonsingular. □

Remark 1.22. The diagonal dominance, as defined in Definitions 1.46 and 1.47, refers to the off-diagonal entries in the row determined by each diagonal entry. It is possible to define a variant of this concept by referring to the off-diagonal entries in the column (instead of row) determined by each diagonal entry. This variant amounts to applying Definitions 1.46 and 1.47 to the transposed matrix $A^\top$. ■

Diagonally dominant matrices naturally arise in discretization of differential equations (Sec. 5.5) and description of electric circuits (Sec. 7.4.4).

1.7.2 *Circulant Matrices*

Definition 1.48. A square matrix A of order n is called a *circulant matrix* if the value of the (i, j) entry a_{ij} is determined by the value of $j - i \pmod n$. ■

Example 1.4. A circulant matrix of order 5 looks like

$$\begin{bmatrix} a_0 & a_1 & a_2 & a_3 & a_4 \\ a_4 & a_0 & a_1 & a_2 & a_3 \\ a_3 & a_4 & a_0 & a_1 & a_2 \\ a_2 & a_3 & a_4 & a_0 & a_1 \\ a_1 & a_2 & a_3 & a_4 & a_0 \end{bmatrix}.$$

■

Definition 1.49. A square matrix A of order n is called a *Toeplitz matrix* if the value of the (i, j) entry a_{ij} is determined by the value of $j - i$. ■

Example 1.5. A Toeplitz matrix of order 5 looks like

$$\begin{bmatrix} a_0 & b_1 & b_2 & b_3 & b_4 \\ c_1 & a_0 & b_1 & b_2 & b_3 \\ c_2 & c_1 & a_0 & b_1 & b_2 \\ c_3 & c_2 & c_1 & a_0 & b_1 \\ c_4 & c_3 & c_2 & c_1 & a_0 \end{bmatrix}.$$

This matrix is a circulant matrix if $c_1 = b_4$, $c_2 = b_3$, $c_3 = b_2$, and $c_4 = b_1$. In general, a circulant matrix is a special case of a Toeplitz matrix. ■

Definition 1.50. A square matrix A of order n is called a *Hankel matrix* if the value of the (i, j) entry a_{ij} is determined by the value of $i + j$. ■

Example 1.6. A Hankel matrix of order 5 looks like

$$\begin{bmatrix} a_1 & a_2 & a_3 & a_4 & a_5 \\ a_2 & a_3 & a_4 & a_5 & a_6 \\ a_3 & a_4 & a_5 & a_6 & a_7 \\ a_4 & a_5 & a_6 & a_7 & a_8 \\ a_5 & a_6 & a_7 & a_8 & a_9 \end{bmatrix}.$$

■

Circulant, Toeplitz, and Hankel matrices appear in signal processing, time-series analysis, control theory, statistics, *etc.* ([1], [3], Example 7.2 in Sec. 7.4.1).

Chapter 2

Determinants

In this chapter we explain the definition and role of determinants with emphasis on multilinearity and various identities. Adjoint matrices are also introduced.

2.1 Permutations

As a preparation for defining determinants, we first introduce permutations.

Definition 2.1. A rearranging of n objects is called a *permutation* of order n. That is, a permutation of order n is a bijective mapping σ from the set $\{1, 2, \ldots, n\}$ to itself. A bijective mapping σ represents a one-to-one correspondence of the elements of $\{1, 2, \ldots, n\}$. We use notation

$$\sigma = \begin{pmatrix} 1 & 2 & \cdots & n \\ i_1 & i_2 & \cdots & i_n \end{pmatrix}$$

to mean the permutation σ defined by

$$\sigma(1) = i_1, \quad \sigma(2) = i_2, \quad \ldots, \quad \sigma(n) = i_n.$$

In this notation, the numbers in the upper row are not required to be in the ascending order. For example, we may write

$$\sigma = \begin{pmatrix} 2 & 1 & \cdots & n \\ i_2 & i_1 & \cdots & i_n \end{pmatrix}$$

to denote the same permutation. ■

Definition 2.2. A *transposition* means a permutation that exchanges exactly two elements without moving other elements. That is, a transposition is a permutation σ such that

$$\sigma(i) = j, \quad \sigma(j) = i, \quad \sigma(k) = k \quad (k \neq i, j)$$

for distinct i and j. ■

Example 2.1. There are six permutations of order 3. Three of them

$$\begin{pmatrix} 1 & 2 & 3 \\ 1 & 3 & 2 \end{pmatrix}, \quad \begin{pmatrix} 1 & 2 & 3 \\ 3 & 2 & 1 \end{pmatrix}, \quad \begin{pmatrix} 1 & 2 & 3 \\ 2 & 1 & 3 \end{pmatrix}$$

are transpositions and the others

$$\begin{pmatrix} 1 & 2 & 3 \\ 1 & 2 & 3 \end{pmatrix}, \quad \begin{pmatrix} 1 & 2 & 3 \\ 2 & 3 & 1 \end{pmatrix}, \quad \begin{pmatrix} 1 & 2 & 3 \\ 3 & 1 & 2 \end{pmatrix}$$

are not. In general, the number of permutations of order n is equal to $n! = n \times (n-1) \times \cdots \times 2 \times 1$, and $n(n-1)/2$ of them are transpositions. ■

Definition 2.3. The *product* of permutations τ and σ, denoted by $\tau\sigma$, is defined as the composition of the mappings associated with τ and σ. That is, $(\tau\sigma)(k) = \tau(\sigma(k))$ for $k = 1, 2, \ldots, n$. If

$$\sigma = \begin{pmatrix} 1 & 2 & \cdots & n \\ i_1 & i_2 & \cdots & i_n \end{pmatrix}, \qquad \tau = \begin{pmatrix} i_1 & i_2 & \cdots & i_n \\ j_1 & j_2 & \cdots & j_n \end{pmatrix},$$

the product $\tau\sigma$ is given by

$$\tau\sigma = \begin{pmatrix} 1 & 2 & \cdots & n \\ j_1 & j_2 & \cdots & j_n \end{pmatrix}.$$

■

Definition 2.4. The *inverse* of a permutation σ is defined as the permutation τ that satisfies $\tau(\sigma(k)) = \sigma(\tau(k)) = k$ for all $k = 1, 2, \ldots, n$. If σ is given as

$$\sigma = \begin{pmatrix} 1 & 2 & \cdots & n \\ i_1 & i_2 & \cdots & i_n \end{pmatrix},$$

the inverse of σ is given by

$$\tau = \begin{pmatrix} i_1 & i_2 & \cdots & i_n \\ 1 & 2 & \cdots & n \end{pmatrix}.$$

■

Theorem 2.1. *Every permutation can be represented as a product of transpositions. Although this representation is not unique, the parity (i.e., being even or odd) of the number of transpositions used in this representation is determined by the given permutation.*

Definition 2.5. A permutation is called an *even permutation* if it can be represented as a product of even number of transpositions. A permutation is called an *odd permutation* if it can be represented as a product of odd number of transpositions. The *signature* of a permutation σ, denoted as $\operatorname{sgn}\sigma$, is defined by

$$\operatorname{sgn}\sigma = \begin{cases} +1 \text{ (if } \sigma \text{ is an even permutation)}, \\ -1 \text{ (if } \sigma \text{ is an odd permutation)}. \end{cases}$$

■

Example 2.2. For permutations

$$\sigma = \begin{pmatrix} 1 & 2 & 3 \\ 2 & 3 & 1 \end{pmatrix}, \qquad \tau = \begin{pmatrix} 1 & 2 & 3 \\ 2 & 1 & 3 \end{pmatrix},$$

their products $\tau\sigma$ and $\sigma\tau$ are given, respectively, by

$$\tau\sigma = \begin{pmatrix} 1 & 2 & 3 \\ 1 & 3 & 2 \end{pmatrix}, \qquad \sigma\tau = \begin{pmatrix} 1 & 2 & 3 \\ 3 & 2 & 1 \end{pmatrix}.$$

Note that $\tau\sigma \neq \sigma\tau$ in this case. The permutation σ can be represented as a product of two transpositions in two different ways as

$$\sigma = \begin{pmatrix} 2 & 1 & 3 \\ 2 & 3 & 1 \end{pmatrix}\begin{pmatrix} 1 & 2 & 3 \\ 2 & 1 & 3 \end{pmatrix}, \qquad \sigma = \begin{pmatrix} 1 & 3 & 2 \\ 2 & 3 & 1 \end{pmatrix}\begin{pmatrix} 1 & 2 & 3 \\ 1 & 3 & 2 \end{pmatrix}.$$

Therefore, σ is an even permutation. The permutation τ, which is itself a transposition, is an odd permutation. We can also represent τ as a product of three transpositions as

$$\tau = \begin{pmatrix} 2 & 3 & 1 \\ 2 & 1 & 3 \end{pmatrix}\begin{pmatrix} 1 & 3 & 2 \\ 2 & 3 & 1 \end{pmatrix}\begin{pmatrix} 1 & 2 & 3 \\ 1 & 3 & 2 \end{pmatrix}.$$

■

2.2 Definition of Determinants

Definition 2.6. For a square matrix $A = (a_{ij} \mid i, j = 1, 2, \ldots, n)$ of order n,

$$\det A = \sum_{\sigma} \operatorname{sgn} \sigma \cdot a_{1\sigma(1)} a_{2\sigma(2)} \cdots a_{n\sigma(n)} \tag{2.1}$$

is called the *determinant* of A, where the summation $\sum_{\sigma}$ is taken over all permutations σ of order n. The symbol "det" stands for "determinant." Sometimes $\det A$ is written as $|A|$. ■

Theorem 2.2. *The following hold for a square matrix $A = (a_{ij})$ of order n.*

(1)

$$\det A = \sum_{\tau} \operatorname{sgn} \tau \cdot a_{\tau(1)1} a_{\tau(2)2} \cdots a_{\tau(n)n}, \tag{2.2}$$

where the summation $\sum_{\tau}$ is taken over all permutations τ of order n.

(2)

$$\det A^\top = \det A.$$

Proof. (1) For each σ in (2.1), let τ denote the inverse permutation of σ. Since

$$\operatorname{sgn}\sigma = \operatorname{sgn}\tau, \qquad a_{1\sigma(1)}a_{2\sigma(2)}\cdots a_{n\sigma(n)} = a_{\tau(1)1}a_{\tau(2)2}\cdots a_{\tau(n)n},$$

the term in (2.1) corresponding to σ is equal to the term in (2.2) corresponding to τ. Moreover, when σ runs over all permutations of order n, the inverse permutation τ also runs over all permutations of order n. Therefore, the right-hand side of (2.1) is equal to that of (2.2).

(2) By (2.1) applied to $A^\top$, the right-hand side of (2.2) is equal to $\det A^\top$. □

Example 2.3. The determinant of a matrix of order 2 is given by

$$\det\begin{bmatrix} a_{11} & a_{12} \\ a_{21} & a_{22} \end{bmatrix} = \begin{vmatrix} a_{11} & a_{12} \\ a_{21} & a_{22} \end{vmatrix} = a_{11}a_{22} - a_{12}a_{21}.$$

■

Example 2.4. The determinant of a matrix of order 3 is given by

$$\det\begin{bmatrix} a_{11} & a_{12} & a_{13} \\ a_{21} & a_{22} & a_{23} \\ a_{31} & a_{32} & a_{33} \end{bmatrix} = \begin{vmatrix} a_{11} & a_{12} & a_{13} \\ a_{21} & a_{22} & a_{23} \\ a_{31} & a_{32} & a_{33} \end{vmatrix}$$

$$= a_{11}a_{22}a_{33} + a_{12}a_{23}a_{31} + a_{13}a_{21}a_{32} - a_{11}a_{23}a_{32} - a_{12}a_{21}a_{33} - a_{13}a_{22}a_{31}.$$

These six terms correspond to the permutations shown in Example 2.1; the three terms preceded by "−" (negative signature) correspond to the transpositions. ■

Example 2.5. A square matrix is called a *permutation matrix* if the entries are 0 or 1 and there is exactly one entry of 1 in each row and each column. For a permutation σ of order n, the square matrix $A = (a_{ij})$ of order n defined by

$$a_{ij} = \begin{cases} 1 \text{ (if } j = \sigma(i)), \\ 0 \text{ (otherwise)} \end{cases}$$

is a permutation matrix. Conversely, every permutation matrix A uniquely corresponds to a permutation σ in this way. We have $\det A = \operatorname{sgn}\sigma$ if A and σ correspond to each other. ■

Example 2.6. The determinant of a triangular matrix is equal to the product of the diagonal entries, that is,

$$\det \begin{bmatrix} a_{11} & a_{12} & \cdots & a_{1n} \\ 0 & a_{22} & \cdots & a_{2n} \\ \vdots & \vdots & \ddots & \vdots \\ 0 & 0 & \cdots & a_{nn} \end{bmatrix} = a_{11}a_{22}\cdots a_{nn},$$

$$\det \begin{bmatrix} a_{11} & 0 & \cdots & 0 \\ a_{21} & a_{22} & \cdots & 0 \\ \vdots & \vdots & \ddots & \vdots \\ a_{n1} & a_{n2} & \cdots & a_{nn} \end{bmatrix} = a_{11}a_{22}\cdots a_{nn}.$$

The determinant of a diagonal matrix is equal to the product of the diagonal entries, that is,

$$\det \begin{bmatrix} d_1 & 0 & \cdots & 0 \\ 0 & d_2 & \cdots & 0 \\ \vdots & \vdots & \ddots & \vdots \\ 0 & 0 & \cdots & d_n \end{bmatrix} = d_1 d_2 \cdots d_n.$$

In particular, the determinant of a unit matrix is equal to 1. ■

Definition 2.7. Let $A = (a_{ij} \mid i = 1, 2, \ldots, m;\ j = 1, 2, \ldots, n)$ be an $m \times n$ matrix. The determinant of a square submatrix of A is called a *minor* (or *subdeterminant*) of A. A minor of *order* k means a minor corresponding to a submatrix of order k. If the submatrix has row indices $i_1 < i_2 < \cdots < i_k$ and column indices $j_1 < j_2 < \cdots < j_k$, the minor is denoted by $|A|^{i_1, i_2, \ldots, i_k}_{j_1, j_2, \ldots, j_k}$, that is,

$$|A|^{i_1, i_2, \ldots, i_k}_{j_1, j_2, \ldots, j_k} = \det\ (a_{i_p j_q} \mid p, q = 1, 2, \ldots, k).$$

■

Definition 2.8. A minor is called a *principal minor* if it corresponds to the same set of indices for rows and columns. A principal minor of order k is expressed as $|A|^{i_1, i_2, \ldots, i_k}_{i_1, i_2, \ldots, i_k}$ for some $i_1 < i_2 < \cdots < i_k$. If $i_1 = 1, i_2 = 2, \ldots, i_k = k$, it is called the *leading principal minor.* ■

Remark 2.1. The determinant has a geometric meaning of area or volume.

(1) In the 2-dimensional space $\mathbb{R}^2$, two vectors, say, $\boldsymbol{a} = (a_1, a_2)^\top$ and $\boldsymbol{b} = (b_1, b_2)^\top$ determine a *parallelogram*, which has vertices at $\boldsymbol{0}$, $\boldsymbol{a}$, $\boldsymbol{b}$, and $\boldsymbol{a} + \boldsymbol{b}$. The *area* of this parallelogram is equal to $|\det[\boldsymbol{a}, \boldsymbol{b}]|$.

(2) In the 3-dimensional space $\mathbb{R}^3$, three vectors, say, $\boldsymbol{a} = (a_1, a_2, a_3)^\top$, $\boldsymbol{b} = (b_1, b_2, b_3)^\top$, and $\boldsymbol{c} = (c_1, c_2, c_3)^\top$ determine a *parallelepiped* (also spelt *parallelopiped*). The *volume* of this parallelepiped is equal to $|\det[\boldsymbol{a}, \boldsymbol{b}, \boldsymbol{c}]|$.

(3) In the n-dimensional space $\mathbb{R}^n$, n vectors $\boldsymbol{a}_1, \boldsymbol{a}_2, \ldots, \boldsymbol{a}_n$ determine a domain

$$\{\boldsymbol{x} = c_1\boldsymbol{a}_1 + c_2\boldsymbol{a}_2 + \cdots + c_n\boldsymbol{a}_n \mid 0 \leqq c_i \leqq 1 \ (i = 1, 2, \ldots, n)\}.$$

The volume of this domain is equal to $|\det[\boldsymbol{a}_1, \boldsymbol{a}_2, \ldots, \boldsymbol{a}_n]|$. We have the *Hadamard inequality*[1]

$$|\det[\boldsymbol{a}_1, \boldsymbol{a}_2, \ldots, \boldsymbol{a}_n]| \leqq \|\boldsymbol{a}_1\|_2 \cdot \|\boldsymbol{a}_2\|_2 \cdot \cdots \cdot \|\boldsymbol{a}_n\|_2.$$

■

2.3 Multilinearity

2.3.1 *Multilinearity and Alternating Properties of Determinants*

In this section we discuss multilinearity and alternating properties of determinants, which are characteristic properties of determinants. The following theorem shows *multilinearity* in the columns.

Theorem 2.3 (Multilinearity).

(1) $\det[\boldsymbol{a}_1, \ldots, \boldsymbol{a}_j + \boldsymbol{b}, \ldots, \boldsymbol{a}_n]$

$$= \det[\boldsymbol{a}_1, \ldots, \boldsymbol{a}_j, \ldots, \boldsymbol{a}_n] + \det[\boldsymbol{a}_1, \ldots, \overset{\overset{j}{\vee}}{\boldsymbol{b}}, \ldots, \boldsymbol{a}_n].$$

(2) $\det[\boldsymbol{a}_1, \ldots, c\,\boldsymbol{a}_j, \ldots, \boldsymbol{a}_n] = c \cdot \det[\boldsymbol{a}_1, \ldots, \boldsymbol{a}_j, \ldots, \boldsymbol{a}_n].$

(3) $\det[\boldsymbol{a}_1, \ldots, \boldsymbol{a}_j + c\,\boldsymbol{b}, \ldots, \boldsymbol{a}_n]$

$$= \det[\boldsymbol{a}_1, \ldots, \boldsymbol{a}_j, \ldots, \boldsymbol{a}_n] + c \cdot \det[\boldsymbol{a}_1, \ldots, \overset{\overset{j}{\vee}}{\boldsymbol{b}}, \ldots, \boldsymbol{a}_n].$$

Remark 2.2. In Theorem 2.3, Part (3) is a combination of (1) and (2) in one equation. ■

Example 2.7. When $n = 3$ and $j = 1$, Theorem 2.3(3) reads:

$$\det \begin{bmatrix} a_{11} + c\,b_1 & a_{12} & a_{13} \\ a_{21} + c\,b_2 & a_{22} & a_{23} \\ a_{31} + c\,b_3 & a_{32} & a_{33} \end{bmatrix} = \det \begin{bmatrix} a_{11} & a_{12} & a_{13} \\ a_{21} & a_{22} & a_{23} \\ a_{31} & a_{32} & a_{33} \end{bmatrix} + c \cdot \det \begin{bmatrix} b_1 & a_{12} & a_{13} \\ b_2 & a_{22} & a_{23} \\ b_3 & a_{32} & a_{33} \end{bmatrix}.$$

■

[1] $\|\boldsymbol{x}\|_2 = \left(|x_1|^2 + |x_2|^2 + \cdots + |x_n|^2\right)^{1/2}$; see Definition 1.43 in Sec. 1.6.

The following theorem shows the *alternating property* in the columns.

Theorem 2.4 (Alternating property).

(1) *For $j < k$ we have*

$$\det[\boldsymbol{a}_1, \ldots, \boldsymbol{a}_j, \ldots, \boldsymbol{a}_k, \ldots, \boldsymbol{a}_n] = -\det[\boldsymbol{a}_1, \ldots, \overset{\overset{j}{\vee}}{\boldsymbol{a}_k}, \ldots, \overset{\overset{k}{\vee}}{\boldsymbol{a}_j}, \ldots, \boldsymbol{a}_n].$$

(2) *For any permutation σ of order n we have*

$$\det[\boldsymbol{a}_{\sigma(1)}, \boldsymbol{a}_{\sigma(2)}, \ldots, \boldsymbol{a}_{\sigma(n)}] = \operatorname{sgn}\sigma \cdot \det[\boldsymbol{a}_1, \boldsymbol{a}_2, \ldots, \boldsymbol{a}_n].$$

Remark 2.3. In Theorem 2.4, Part (1) is a special case of (2). Conversely, Part (2) follows from (1), since every permutation can be represented as a product of transpositions (Theorem 2.1 in Sec. 2.1). ■

Example 2.8. When $n = 3$, $j = 1$, and $k = 2$, Theorem 2.4(1) reads:

$$\det \begin{bmatrix} a_{11} & a_{12} & a_{13} \\ a_{21} & a_{22} & a_{23} \\ a_{31} & a_{32} & a_{33} \end{bmatrix} = -\det \begin{bmatrix} a_{12} & a_{11} & a_{13} \\ a_{22} & a_{21} & a_{23} \\ a_{32} & a_{31} & a_{33} \end{bmatrix}.$$

■

Multilinearity and alternating properties immediately imply the following.

Proposition 2.1.

(1) *The determinant of a matrix containing the zero vector* $\boldsymbol{0}$ *in a column is equal to* 0. *That is, if $\boldsymbol{a}_j = \boldsymbol{0}$ for some j, then* $\det[\boldsymbol{a}_1, \boldsymbol{a}_2, \ldots, \boldsymbol{a}_n] = 0$.
(2) *The determinant of a matrix containing two identical column vectors is equal to* 0. *That is, if $\boldsymbol{a}_j = \boldsymbol{a}_k$ for some $j \neq k$, then* $\det[\boldsymbol{a}_1, \boldsymbol{a}_2, \ldots, \boldsymbol{a}_n] = 0$.
(3) *The determinant of a matrix remains unchanged when a scalar multiple of one column is added to another column. That is, for $j < k$ we have*
$\det[\boldsymbol{a}_1, \ldots, \boldsymbol{a}_j, \ldots, \boldsymbol{a}_k, \ldots, \boldsymbol{a}_n] = \det[\boldsymbol{a}_1, \ldots, \boldsymbol{a}_j, \ldots, \boldsymbol{a}_k + c\,\boldsymbol{a}_j, \ldots, \boldsymbol{a}_n]$,
$\det[\boldsymbol{a}_1, \ldots, \boldsymbol{a}_j, \ldots, \boldsymbol{a}_k, \ldots, \boldsymbol{a}_n] = \det[\boldsymbol{a}_1, \ldots, \boldsymbol{a}_j + c\,\boldsymbol{a}_k, \ldots, \boldsymbol{a}_k, \ldots, \boldsymbol{a}_n]$.

Example 2.9. When $n = 3$, $j = 1$, and $k = 2$, the first equation of Proposition 2.1(3) reads:

$$\det \begin{bmatrix} a_{11} & a_{12} & a_{13} \\ a_{21} & a_{22} & a_{23} \\ a_{31} & a_{32} & a_{33} \end{bmatrix} = \det \begin{bmatrix} a_{11} & a_{12} + c\,a_{11} & a_{13} \\ a_{21} & a_{22} + c\,a_{21} & a_{23} \\ a_{31} & a_{32} + c\,a_{31} & a_{33} \end{bmatrix}.$$

■

Remark 2.4. The multilinearity and alternating properties given in Theorems 2.3 and 2.4 and Proposition 2.1 can also be formulated for row vectors. For example,

$$\det \begin{bmatrix} a_{11}+cb_1 & a_{12}+cb_2 & a_{13}+cb_3 \\ a_{21} & a_{22} & a_{23} \\ a_{31} & a_{32} & a_{33} \end{bmatrix}$$

$$= \det \begin{bmatrix} a_{11} & a_{12} & a_{13} \\ a_{21} & a_{22} & a_{23} \\ a_{31} & a_{32} & a_{33} \end{bmatrix} + c \cdot \det \begin{bmatrix} b_1 & b_2 & b_3 \\ a_{21} & a_{22} & a_{23} \\ a_{31} & a_{32} & a_{33} \end{bmatrix},$$

$$\det \begin{bmatrix} a_{11} & a_{12} & a_{13} \\ a_{21} & a_{22} & a_{23} \\ a_{31} & a_{32} & a_{33} \end{bmatrix} = -\det \begin{bmatrix} a_{21} & a_{22} & a_{23} \\ a_{11} & a_{12} & a_{13} \\ a_{31} & a_{32} & a_{33} \end{bmatrix},$$

$$\det \begin{bmatrix} a_{11} & a_{12} & a_{13} \\ a_{21} & a_{22} & a_{23} \\ a_{31} & a_{32} & a_{33} \end{bmatrix} = \det \begin{bmatrix} a_{11} & a_{12} & a_{13} \\ a_{21}+ca_{11} & a_{22}+ca_{12} & a_{23}+ca_{13} \\ a_{31} & a_{32} & a_{33} \end{bmatrix}.$$

■

Example 2.10. A matrix of the form

$$V_n(x_1, x_2, \dots, x_n) = \begin{bmatrix} 1 & 1 & \cdots & 1 \\ x_1 & x_2 & \cdots & x_n \\ {x_1}^2 & {x_2}^2 & \cdots & {x_n}^2 \\ \vdots & \vdots & & \vdots \\ {x_1}^{n-1} & {x_2}^{n-1} & \cdots & {x_n}^{n-1} \end{bmatrix} \tag{2.3}$$

is called the *Vandermonde matrix* of order n, where $x_1, x_2, \dots, x_n$ denote independent variables.[2] It is known that the determinant of this matrix is given as

$$\det V_n(x_1, x_2, \dots, x_n) = \prod_{j>i} (x_j - x_i). \tag{2.4}$$

This is proved as follows. The determinant of V_n is a polynomial of degree $n(n-1)/2$ in variables $x_1, x_2, \dots, x_n$. If $x_i = x_j$, the ith and jth columns are identical, and hence $\det V_n = 0$ by Proposition 2.1(2). Therefore, $\det V_n$ is divisible by $(x_j - x_i)$ for all $j > i$. The expression (2.4) follows from this fact, with additional observations on the degree and the leading coefficient of the polynomial $\det V_n$. ■

[2]Sometimes the transpose of V_n is called a Vandermonde matrix.

2.3.2 Characterization of Determinants

Multilinearity and alternating properties are essential properties of determinants, and determinants are, in turn, characterized by these two properties.

Theorem 2.5. *Let $F(\boldsymbol{a}_1, \boldsymbol{a}_2, \ldots, \boldsymbol{a}_n)$ be a scalar-valued (K-valued) function in n-tuple $(\boldsymbol{a}_1, \boldsymbol{a}_2, \ldots, \boldsymbol{a}_n)$ of n-dimensional vectors. If F has the following two properties:*

- Multilinearity: *For any j and any $c \in K$,*

$$F(\boldsymbol{a}_1, \ldots, \boldsymbol{a}_j + c\,\boldsymbol{b}, \ldots, \boldsymbol{a}_n) = F(\boldsymbol{a}_1, \ldots, \boldsymbol{a}_j, \ldots, \boldsymbol{a}_n) + c \cdot F(\boldsymbol{a}_1, \ldots, \overset{\overset{j}{\vee}}{\boldsymbol{b}}, \ldots, \boldsymbol{a}_n), \tag{2.5}$$

- Alternating property: *For any $j < k$,*

$$F(\boldsymbol{a}_1, \ldots, \boldsymbol{a}_j, \ldots, \boldsymbol{a}_k, \ldots, \boldsymbol{a}_n) = -F(\boldsymbol{a}_1, \ldots, \overset{\overset{j}{\vee}}{\boldsymbol{a}_k}, \ldots, \overset{\overset{k}{\vee}}{\boldsymbol{a}_j}, \ldots, \boldsymbol{a}_n), \tag{2.6}$$

then F is equal to the determinant of the matrix $[\boldsymbol{a}_1, \boldsymbol{a}_2, \ldots, \boldsymbol{a}_n]$ up to a multiplicative constant, that is,

$$F(\boldsymbol{a}_1, \boldsymbol{a}_2, \ldots, \boldsymbol{a}_n) = \tilde{c} \cdot \det[\boldsymbol{a}_1, \boldsymbol{a}_2, \ldots, \boldsymbol{a}_n] \tag{2.7}$$

for some constant $\tilde{c}$. The constant $\tilde{c}$ is given by

$$\tilde{c} = F(\boldsymbol{e}_1, \boldsymbol{e}_2, \ldots, \boldsymbol{e}_n),$$

where $\boldsymbol{e}_j$ denotes the jth unit vector, i.e., $\boldsymbol{e}_j = (0, \ldots, 0, \overset{\overset{j}{\vee}}{1}, 0, \ldots, 0)^\top$.

Proof. For $j = 1, 2, \ldots, n$, let $\boldsymbol{a}_j = (a_{1j}, a_{2j}, \ldots, a_{nj})^\top$. Then $\boldsymbol{a}_j = \sum_{i=1}^{n} a_{ij}\boldsymbol{e}_i$. By the assumed multilinearity of F we have

$$\begin{aligned} F(\boldsymbol{a}_1, \boldsymbol{a}_2, \ldots, \boldsymbol{a}_n) &= F\left(\sum_{i_1=1}^{n} a_{i_1 1}\boldsymbol{e}_{i_1}, \sum_{i_2=1}^{n} a_{i_2 2}\boldsymbol{e}_{i_2}, \ldots, \sum_{i_n=1}^{n} a_{i_n n}\boldsymbol{e}_{i_n}\right) \\ &= \sum_{i_1=1}^{n}\sum_{i_2=1}^{n}\cdots\sum_{i_n=1}^{n} a_{i_1 1}a_{i_2 2}\cdots a_{i_n n} \cdot F(\boldsymbol{e}_{i_1}, \boldsymbol{e}_{i_2}, \ldots, \boldsymbol{e}_{i_n}). \end{aligned}$$

If the indices $i_1, i_2, \ldots, i_n$ are not all distinct, the assumed alternating property of F implies $F(\boldsymbol{e}_{i_1}, \boldsymbol{e}_{i_2}, \ldots, \boldsymbol{e}_{i_n}) = 0$. Therefore we assume that $(i_1, i_2, \ldots, i_n)$ is a permutation of $(1, 2, \ldots, n)$, and let $(i_1, i_2, \ldots, i_n) = (\tau(1), \tau(2), \ldots, \tau(n))$. Then we have

$$F(\boldsymbol{e}_{i_1}, \boldsymbol{e}_{i_2}, \ldots, \boldsymbol{e}_{i_n}) = F(\boldsymbol{e}_{\tau(1)}, \boldsymbol{e}_{\tau(2)}, \cdots, \boldsymbol{e}_{\tau(n)}) = \operatorname{sgn}\tau \cdot F(\boldsymbol{e}_1, \boldsymbol{e}_2, \ldots, \boldsymbol{e}_n)$$

from the alternating property of F (cf., Remark 2.5 below). Substituting this into the above expression we obtain

$$F(\boldsymbol{a}_1, \boldsymbol{a}_2, \ldots, \boldsymbol{a}_n) = F(\boldsymbol{e}_1, \boldsymbol{e}_2, \ldots, \boldsymbol{e}_n) \cdot \sum_{\tau} \operatorname{sgn} \tau \cdot a_{\tau(1)1} a_{\tau(2)2} \cdots a_{\tau(n)n}$$
$$= F(\boldsymbol{e}_1, \boldsymbol{e}_2, \ldots, \boldsymbol{e}_n) \cdot \det[\boldsymbol{a}_1, \boldsymbol{a}_2, \ldots, \boldsymbol{a}_n],$$

which is (2.7) with $\tilde{c} = F(\boldsymbol{e}_1, \boldsymbol{e}_2, \ldots, \boldsymbol{e}_n)$. □

Remark 2.5. For any permutation τ we obtain

$$F(\boldsymbol{a}_{\tau(1)}, \boldsymbol{a}_{\tau(2)}, \ldots, \boldsymbol{a}_{\tau(n)}) = \operatorname{sgn} \tau \cdot F(\boldsymbol{a}_1, \boldsymbol{a}_2, \ldots, \boldsymbol{a}_n) \tag{2.8}$$

from the alternating property (2.6), since τ can be represented as a product of transpositions (Theorem 2.1 in Sec. 2.1). ■

Theorem 2.5 can be generalized as follows.[3]

Theorem 2.6. *Let $F(\boldsymbol{a}_1, \boldsymbol{a}_2, \ldots, \boldsymbol{a}_n)$ be a scalar-valued (K-valued) function in n-tuple $(\boldsymbol{a}_1, \boldsymbol{a}_2, \ldots, \boldsymbol{a}_n)$ of m-dimensional vectors, where $m \geqq n$. If F is equipped with multilinearity* (2.5) *and the alternating property* (2.6), *then F is a linear combination of minors of order n of the matrix $A = [\boldsymbol{a}_1, \boldsymbol{a}_2, \ldots, \boldsymbol{a}_n]$, that is,*[4]

$$F(\boldsymbol{a}_1, \boldsymbol{a}_2, \ldots, \boldsymbol{a}_n) = \sum_{k_1 < k_2 < \cdots < k_n} c_{k_1 k_2 \cdots k_n} \cdot |A|_{1,2,\ldots,n}^{k_1, k_2, \ldots, k_n}, \tag{2.9}$$

where the right-hand side is a summation over all $(k_1, k_2, \ldots, k_n)$ satisfying $1 \leqq k_1 < k_2 < \cdots < k_n \leqq m$. The coefficients $c_{k_1 k_2 \cdots k_n}$ are given by

$$c_{k_1 k_2 \cdots k_n} = F(\boldsymbol{e}_{k_1}, \boldsymbol{e}_{k_2}, \ldots, \boldsymbol{e}_{k_n}),$$

where $\boldsymbol{e}_j$ denotes the jth unit vector.

Proof. For $j = 1, 2, \ldots, n$, let $\boldsymbol{a}_j = (a_{1j}, a_{2j}, \ldots, a_{mj})^\top$. Then $\boldsymbol{a}_j = \sum_{i=1}^{m} a_{ij} \boldsymbol{e}_i$. By the assumed multilinearity of F we have

$$F(\boldsymbol{a}_1, \boldsymbol{a}_2, \ldots, \boldsymbol{a}_n) = F\left(\sum_{i_1=1}^{m} a_{i_1 1} \boldsymbol{e}_{i_1}, \sum_{i_2=1}^{m} a_{i_2 2} \boldsymbol{e}_{i_2}, \ldots, \sum_{i_n=1}^{m} a_{i_n n} \boldsymbol{e}_{i_n} \right)$$
$$= \sum_{i_1=1}^{m} \sum_{i_2=1}^{m} \cdots \sum_{i_n=1}^{m} a_{i_1 1} a_{i_2 2} \cdots a_{i_n n} \cdot F(\boldsymbol{e}_{i_1}, \boldsymbol{e}_{i_2}, \ldots, \boldsymbol{e}_{i_n}).$$

---

[3] Theorem 2.6 will be used for the proof of the generalized Laplace expansion (Theorem 2.12) in Sec. 2.5.2.

[4] $|A|_{1,2,\ldots,n}^{k_1,k_2,\ldots,k_n}$ denotes the minor (subdeterminant) of A with row indices $k_1 < k_2 < \cdots < k_n$ and column indices $1, 2, \ldots, n$. See Definition 2.7 in Sec. 2.2 for this notation.

If the indices $i_1, i_2, \ldots, i_n$ are not all distinct, the assumed alternating property of F implies $F(\boldsymbol{e}_{i_1}, \boldsymbol{e}_{i_2}, \ldots, \boldsymbol{e}_{i_n}) = 0$. Therefore we assume that $(i_1, i_2, \ldots, i_n) = (\tau(k_1), \tau(k_2), \ldots, \tau(k_n))$ for a set of n distinct indices $k_1 < k_2 < \cdots < k_n$ and a permutation τ of the set $\{k_1, k_2, \ldots, k_n\}$. Then we have

$$\begin{aligned} F(\boldsymbol{e}_{i_1}, \boldsymbol{e}_{i_2}, \ldots, \boldsymbol{e}_{i_n}) &= F(\boldsymbol{e}_{\tau(k_1)}, \boldsymbol{e}_{\tau(k_2)}, \cdots, \boldsymbol{e}_{\tau(k_n)}) \\ &= \operatorname{sgn} \tau \cdot F(\boldsymbol{e}_{k_1}, \boldsymbol{e}_{k_2}, \ldots, \boldsymbol{e}_{k_n}) \\ &= \operatorname{sgn} \tau \cdot c_{k_1 k_2 \cdots k_n} \end{aligned}$$

from the alternating property of F (cf., Remark 2.5). Substituting this into the above expression we obtain

$$\begin{aligned} &F(\boldsymbol{a}_1, \boldsymbol{a}_2, \ldots, \boldsymbol{a}_n) \\ &= \sum_{k_1 < k_2 < \cdots < k_n} c_{k_1 k_2 \cdots k_n} \cdot \sum_{\tau} \operatorname{sgn} \tau \cdot a_{\tau(k_1)1} a_{\tau(k_2)2} \cdots a_{\tau(k_n)n} \\ &= \sum_{k_1 < k_2 < \cdots < k_n} c_{k_1 k_2 \cdots k_n} \cdot |A|^{k_1, k_2, \ldots, k_n}_{1,2,\ldots,n}, \end{aligned}$$

which is (2.9). □

2.4 Basic Formulas

Theorem 2.7. *The following identities hold for square matrices A and B.*

(1) $\det(cA) = c^n \cdot (\det A)$, *where A is a matrix of order n.*
(2) $\det(AB) = (\det A) \cdot (\det B)$.
(3) $\det(A^{-1}) = 1/(\det A)$, *where A is a nonsingular matrix.*
(4) $\det(A \otimes B) = (\det A)^m \cdot (\det B)^n$,
where A and B are matrices of orders n and m, respectively.
(5) $\det(A^\top) = \det A$.
(6) $\det(\overline{A}) = \overline{\det A}$.
(7) $\det(A^*) = \overline{\det A}$.

Proof. Parts (1), (5), (6), and (7) follow easily from the definition and Theorem 2.2 in Sec. 2.2.

(2) Let $A = [\boldsymbol{a}_1, \boldsymbol{a}_2, \ldots, \boldsymbol{a}_n]$ and $B = (b_{ij} \mid i, j = 1, 2, \ldots, n)$. By

multilinearity we have

$$\det(AB) = \det\left[\sum_{i_1=1}^{n} b_{i_1 1}\boldsymbol{a}_{i_1}, \sum_{i_2=1}^{n} b_{i_2 2}\boldsymbol{a}_{i_2}, \ldots, \sum_{i_n=1}^{n} b_{i_n n}\boldsymbol{a}_{i_n}\right]$$

$$= \sum_{i_1=1}^{n}\sum_{i_2=1}^{n}\cdots\sum_{i_n=1}^{n} b_{i_1 1}b_{i_2 2}\cdots b_{i_n n}\det[\boldsymbol{a}_{i_1}, \boldsymbol{a}_{i_2}, \ldots, \boldsymbol{a}_{i_n}].$$

If the indices $i_1, i_2, \ldots, i_n$ are not all distinct, it follows from the alternating property that $\det[\boldsymbol{a}_{i_1}, \boldsymbol{a}_{i_2}, \ldots, \boldsymbol{a}_{i_n}] = 0$. Hence we may restrict ourselves to the terms with $(i_1, i_2, \ldots, i_n) = (\tau(1), \tau(2), \ldots, \tau(n))$ for a permutation τ, for which we have

$$\det[\boldsymbol{a}_{i_1}, \boldsymbol{a}_{i_2}, \ldots, \boldsymbol{a}_{i_n}] = \det[\boldsymbol{a}_{\tau(1)}, \boldsymbol{a}_{\tau(2)}, \cdots, \boldsymbol{a}_{\tau(n)}]$$
$$= \operatorname{sgn}\tau \cdot \det[\boldsymbol{a}_1, \boldsymbol{a}_2, \ldots, \boldsymbol{a}_n]$$

by the alternating property. Substituting this into the above expression we obtain

$$\det(AB) = \det[\boldsymbol{a}_1, \boldsymbol{a}_2, \ldots, \boldsymbol{a}_n] \cdot \sum_{\tau} \operatorname{sgn}\tau \cdot b_{\tau(1)1}b_{\tau(2)2}\cdots b_{\tau(n)n}$$
$$= (\det A) \cdot (\det B).$$

(3) Let $B = A^{-1}$. The application of (2) to $AB = I$ shows $(\det A) \cdot (\det B) = \det I = 1$. Therefore, $\det(A^{-1}) = \det B = 1/(\det A)$.

(4) It follows from $A \otimes B = (I_n \otimes B) \cdot (A \otimes I_m)$ that $\det(A \otimes B) = \det(I_n \otimes B) \cdot \det(A \otimes I_m)$. Here we have

$$\det(I_n \otimes B) = \det\begin{bmatrix} B & 0 & \cdots & 0 \\ 0 & B & \cdots & 0 \\ \vdots & \vdots & \ddots & \vdots \\ 0 & 0 & \cdots & B \end{bmatrix} = (\det B)^n$$

(cf., Theorem 2.8(1)). Similarly, we have $\det(A \otimes I_m) = \det(I_m \otimes A) = (\det A)^m$, where the first equality holds since $A \otimes I_m$ is obtained from $I_m \otimes A$ by a simultaneous permutation of rows and columns, as in Remark 1.11 in Sec. 1.2.4. □

We next consider the determinants of block-matrices. Theorem 2.8 deals with simpler cases, where the general cases are treated in Theorem 2.9.

Theorem 2.8. *Let A and B are square matrices of orders n and m, respectively.*

(1) $\det \left[\begin{array}{c|c} A & O \\ \hline O & B \end{array}\right] = (\det A) \cdot (\det B)$.

(2) $\det \left[\begin{array}{c|c} A & C \\ \hline O & B \end{array}\right] = (\det A) \cdot (\det B)$.

Proof. Part (1) is a special case of (2). We will prove (2). Let $F = \left[\begin{array}{c|c} A & C \\ \hline O & B \end{array}\right] = (f_{ij})$, which is a square matrix of order $(n+m)$. In the defining expansion of the determinant

$$\det F = \sum_{\sigma} \operatorname{sgn} \sigma \cdot f_{1\sigma(1)} f_{2\sigma(2)} \cdots f_{n+m,\sigma(n+m)}, \tag{2.10}$$

we may only consider permutations σ of order $(n+m)$ that map $\{n+1, n+2, \ldots, n+m\}$ to $\{n+1, n+2, \ldots, n+m\}$ (and hence map $\{1, 2, \ldots, n\}$ to $\{1, 2, \ldots, n\}$). For such σ define permutations τ and ρ of orders n and m, respectively, by

$$\tau(i) = \sigma(i) \quad (i = 1, 2, \ldots, n), \qquad \rho(i) = \sigma(n+i) - n \quad (i = 1, 2, \ldots, m).$$

Then we have

$$f_{1\sigma(1)} f_{2\sigma(2)} \cdots f_{n+m,\sigma(n+m)} = a_{1\tau(1)} a_{2\tau(2)} \cdots a_{n\tau(n)} \times b_{1\rho(1)} b_{2\rho(2)} \cdots b_{m\rho(m)}$$

as well as $\operatorname{sgn} \sigma = \operatorname{sgn} \tau \cdot \operatorname{sgn} \rho$. Therefore, the right-hand side of (2.10) is equal to

$$\sum_{\tau} \operatorname{sgn} \tau \cdot a_{1\tau(1)} a_{2\tau(2)} \cdots a_{n\tau(n)} \sum_{\rho} \operatorname{sgn} \rho \cdot b_{1\rho(1)} b_{2\rho(2)} \cdots b_{m\rho(m)}$$
$$= (\det A) \cdot (\det B),$$

and hence $\det F = (\det A) \cdot (\det B)$, as desired. □

Theorem 2.9. *Let A and D are square matrices of orders n and m, respectively.*

(1) *If A is nonsingular, then*

$$\det \left[\begin{array}{c|c} A & B \\ \hline C & D \end{array}\right] = (\det A) \cdot \det(D - CA^{-1}B).$$

(2) *If D is nonsingular, then*

$$\det \left[\begin{array}{c|c} A & B \\ \hline C & D \end{array}\right] = (\det D) \cdot \det(A - BD^{-1}C).$$

Proof. (1) We can easily verify the factorization:

$$\left[\begin{array}{c|c} A & B \\ \hline C & D \end{array}\right] = \left[\begin{array}{c|c} A & O \\ \hline C & I \end{array}\right] \left[\begin{array}{c|c} I & A^{-1}B \\ \hline O & D - CA^{-1}B \end{array}\right]$$

and apply Theorem 2.8(2) to derive the desired identity of the determinants.

(2) We can prove this similarly as (1). Alternatively, we permute rows and columns to obtain

$$\det \left[\begin{array}{c|c} A & B \\ \hline C & D \end{array}\right] = \det \left[\begin{array}{c|c} D & C \\ \hline B & A \end{array}\right]$$

and apply the formula of (1) to the right-hand side. □

The following formulas show how to modify the determinant when a matrix is modified.[5]

Theorem 2.10. *Let A be a nonsingular matrix of order n.*

(1) *For vectors $\boldsymbol{b}$ and $\mathbf{c}$,*

$$\det(A + \boldsymbol{b}\mathbf{c}^\top) = (\det A) \cdot (1 + \mathbf{c}^\top A^{-1}\boldsymbol{b}).$$

(2) *For an $n \times m$ matrix B and an $m \times n$ matrix C,*

$$\det(A + BC) = (\det A) \cdot \det(I + CA^{-1}B).$$

Proof. Part (1) is a special case of (2). We will prove (2). In Theorem 2.9 we choose $D = I$ and change B to $-B$, to obtain

$$\det(A + BC) = \det \left[\begin{array}{c|c} A & -B \\ \hline C & I \end{array}\right] = (\det A) \cdot \det(I + CA^{-1}B),$$

where the first equality is by Theorem 2.9(2) and the second is by Theorem 2.9(1). □

2.5 Expansion Formulas

2.5.1 *Laplace Expansion*

Let A be a square matrix of order n. We denote by Δ_{ij} the determinant of the submatrix of A obtained by deleting row i and column j. The following formulas are known as the *Laplace expansions.*

[5] If $\boldsymbol{b} \neq \mathbf{0}$ and $\mathbf{c} \neq \mathbf{0}$, for example, the matrix $A + \boldsymbol{b}\mathbf{c}^\top$ represents a modification of A with a rank-one matrix $\boldsymbol{b}\mathbf{c}^\top$; see Sec. 4.1 for "rank."

Theorem 2.11.

(1) *The Laplace expansion by minors along row i:*

$$\det A = \sum_{j=1}^{n} (-1)^{i+j} a_{ij} \Delta_{ij}. \tag{2.11}$$

(2) *The Laplace expansion by minors along column j:*

$$\det A = \sum_{i=1}^{n} (-1)^{i+j} a_{ij} \Delta_{ij}. \tag{2.12}$$

Proof. It is easy to see (cf., Example 2.11 below) that, each term $a_{1\sigma(1)}a_{2\sigma(2)}\cdots a_{n\sigma(n)}$ in the defining expansion (2.1) of determinant appears exactly once on the right-hand side of (2.11) with the same sign as $\operatorname{sgn}\sigma$. Moreover, the right-hand side of (2.11) contains no other terms. Therefore, the identity (2.11) of Part (1) holds. Part (2) can be proved by a similar argument. An alternative proof based on multilinearity and alternating properties can be found in Remark 2.8 in Sec. 2.5.2, which is actually a proof of a more general theorem (Theorem 2.12). □

Example 2.11. When $n = 3$, the Laplace expansion for row $i = 1$ reads:

$$\begin{vmatrix} a_{11} & a_{12} & a_{13} \\ a_{21} & a_{22} & a_{23} \\ a_{31} & a_{32} & a_{33} \end{vmatrix} = a_{11} \begin{vmatrix} a_{22} & a_{23} \\ a_{32} & a_{33} \end{vmatrix} - a_{12} \begin{vmatrix} a_{21} & a_{23} \\ a_{31} & a_{33} \end{vmatrix} + a_{13} \begin{vmatrix} a_{21} & a_{22} \\ a_{31} & a_{32} \end{vmatrix}.$$

As we have seen in Example 2.4 in Sec. 2.2, the left-hand side is, by definition, equal to

$$a_{11}a_{22}a_{33} + a_{12}a_{23}a_{31} + a_{13}a_{21}a_{32} - a_{11}a_{23}a_{32} - a_{12}a_{21}a_{33} - a_{13}a_{22}a_{31}. \tag{2.13}$$

This expression can be represented as a linear combination of the variables a_{1j} $(j = 1, 2, 3)$ in the first row as

$$a_{11}(a_{22}a_{33} - a_{23}a_{32}) - a_{12}(a_{21}a_{33} - a_{23}a_{31}) + a_{13}(a_{21}a_{32} - a_{22}a_{31}),$$

which coincides with the right-hand side of the Laplace expansion above. ■

Example 2.12. When $n = 3$, the Laplace expansion for column $j = 2$ reads:

$$\begin{vmatrix} a_{11} & a_{12} & a_{13} \\ a_{21} & a_{22} & a_{23} \\ a_{31} & a_{32} & a_{33} \end{vmatrix} = -a_{12} \begin{vmatrix} a_{21} & a_{23} \\ a_{31} & a_{33} \end{vmatrix} + a_{22} \begin{vmatrix} a_{11} & a_{13} \\ a_{31} & a_{33} \end{vmatrix} - a_{32} \begin{vmatrix} a_{11} & a_{13} \\ a_{21} & a_{23} \end{vmatrix}.$$

This expansion is obtained from (2.13) by representing it as a linear combination of the variables a_{i2} $(i = 1, 2, 3)$ in the second column as

$$-a_{12}(a_{21}a_{33} - a_{23}a_{31}) + a_{22}(a_{11}a_{33} - a_{13}a_{31}) - a_{32}(a_{11}a_{23} - a_{13}a_{21}).$$

Different expansions are obtained depending on the chosen variables. ■

Example 2.13. For a monic polynomial[6] of degree n represented as

$$P(x) = x^n + c_1x^{n-1} + \cdots + c_{n-1}x + c_n,$$

the matrix

$$C = \begin{bmatrix} 0 & \cdots & \cdots & 0 & -c_n \\ 1 & 0 & \cdots & 0 & -c_{n-1} \\ 0 & \ddots & \ddots & \vdots & \vdots \\ \vdots & \ddots & 1 & 0 & -c_2 \\ 0 & \cdots & 0 & 1 & -c_1 \end{bmatrix} \tag{2.14}$$

is called the *companion matrix* associated with $P(x)$, where C is a square matrix of order n. For any monic polynomial $P(x)$, the determinant[7] of the matrix $xI - C$ coincides with $P(x)$ itself. That is,

$$\begin{vmatrix} x & 0 & \cdots & \cdots & 0 & c_n \\ -1 & x & \cdots & \cdots & 0 & c_{n-1} \\ 0 & -1 & \ddots & & \vdots & c_{n-2} \\ \vdots & \ddots & \ddots & \ddots & \vdots & \vdots \\ \vdots & & \ddots & -1 & x & c_2 \\ 0 & \cdots & \cdots & 0 & -1 & x + c_1 \end{vmatrix} = x^n + c_1x^{n-1} + \cdots + c_{n-1}x + c_n. \tag{2.15}$$

We can prove this by the Laplace expansion. Let $d_n = \det(xI - C)$, which is the determinant on the left-hand side. The Laplace expansion along the first row gives the recurrence relation

$$d_n = x\begin{vmatrix} x & \cdots & \cdots & 0 & c_{n-1} \\ -1 & x & \cdots & 0 & c_{n-2} \\ 0 & \ddots & \ddots & \vdots & \vdots \\ & \ddots & -1 & x & c_2 \\ 0 & \cdots & 0 & -1 & x + c_1 \end{vmatrix} + (-1)^{n+1}c_n \begin{vmatrix} -1 & x & \cdots & \cdots & 0 \\ 0 & -1 & \ddots & & \vdots \\ \vdots & \ddots & \ddots & \ddots & \vdots \\ \vdots & & \ddots & -1 & x \\ 0 & \cdots & \cdots & 0 & -1 \end{vmatrix}$$

$$= xd_{n-1} + c_n.$$

[6]A polynomial in one variable is called *monic* if the coefficient of the highest-degree term is equal to 1.

[7]$\det(xI - C)$ is called the *characteristic polynomial* of the companion matrix C; see Sec. 6.1.2.

Together with the initial condition $d_1 = x + c_1$, this recurrence relation shows that d_n is given by $d_n = x^n + c_1 x^{n-1} + \cdots + c_{n-1}x + c_n$. ■

2.5.2 *Generalized Laplace Expansion*

The Laplace expansion (Theorem 2.11) expresses the determinant of A as a sum of the products of an entry of A (minor of order 1) and a minor of order $n-1$ at the complementary position. We can generalize this by expressing $\det A$ as a sum of the products of a minor of A of order k and a minor of order $n-k$ at the complementary position, where k is an integer satisfying $1 \leqq k \leqq n-1$. This is called the *generalized Laplace expansion.* For row indices $i_1 < i_2 < \cdots < i_k$ and column indices $j_1 < j_2 < \cdots < j_k$, the corresponding minor is denoted as

$$|A|^{i_1,i_2,\ldots,i_k}_{j_1,j_2,\ldots,j_k}$$

(cf., Definition 2.7 in Sec. 2.2). Let

$$\Delta^{i_1,i_2,\ldots,i_k}_{j_1,j_2,\ldots,j_k}$$

denote the determinant of the submatrix of order $n-k$ obtained from A by deleting rows $i_1 < i_2 < \cdots < i_k$ and columns $j_1 < j_2 < \cdots < j_k$.

Theorem 2.12 (Generalized Laplace expansion). *Let A be a square matrix of order n and k be an integer satisfying $1 \leqq k \leqq n-1$.*

(1) *Expansion along rows $i_1 < i_2 < \cdots < i_k$:*

$$\det A = \sum_{j_1<j_2<\cdots<j_k} (-1)^{(i_1+\cdots+i_k)+(j_1+\cdots+j_k)} |A|^{i_1,i_2,\ldots,i_k}_{j_1,j_2,\ldots,j_k} \cdot \Delta^{i_1,i_2,\ldots,i_k}_{j_1,j_2,\ldots,j_k}, \tag{2.16}$$

where the summation is taken over all $(j_1, j_2, \ldots, j_k)$ with $1 \leqq j_1 < j_2 < \cdots < j_k \leqq n$.

(2) *Expansion along columns $j_1 < j_2 < \cdots < j_k$:*

$$\det A = \sum_{i_1<i_2<\cdots<i_k} (-1)^{(i_1+\cdots+i_k)+(j_1+\cdots+j_k)} |A|^{i_1,i_2,\ldots,i_k}_{j_1,j_2,\ldots,j_k} \cdot \Delta^{i_1,i_2,\ldots,i_k}_{j_1,j_2,\ldots,j_k}, \tag{2.17}$$

where the summation is taken over all $(i_1, i_2, \ldots, i_k)$ with $1 \leqq i_1 < i_2 < \cdots < i_k \leqq n$.

Proof. Two different proofs are given in Remarks 2.7 and 2.8. □

Example 2.14. When $n = 4$, the generalized Laplace expansion along columns for $k = 2$ and $(j_1, j_2) = (1, 2)$ is given by

$$\begin{aligned}
&\begin{vmatrix} a_{11} & a_{12} & a_{13} & a_{14} \\ a_{21} & a_{22} & a_{23} & a_{24} \\ a_{31} & a_{32} & a_{33} & a_{34} \\ a_{41} & a_{42} & a_{43} & a_{44} \end{vmatrix} \\
&= \begin{vmatrix} a_{11} & a_{12} \\ a_{21} & a_{22} \end{vmatrix} \begin{vmatrix} a_{33} & a_{34} \\ a_{43} & a_{44} \end{vmatrix} - \begin{vmatrix} a_{11} & a_{12} \\ a_{31} & a_{32} \end{vmatrix} \begin{vmatrix} a_{23} & a_{24} \\ a_{43} & a_{44} \end{vmatrix} \\
&\quad + \begin{vmatrix} a_{11} & a_{12} \\ a_{41} & a_{42} \end{vmatrix} \begin{vmatrix} a_{23} & a_{24} \\ a_{33} & a_{34} \end{vmatrix} + \begin{vmatrix} a_{21} & a_{22} \\ a_{31} & a_{32} \end{vmatrix} \begin{vmatrix} a_{13} & a_{14} \\ a_{43} & a_{44} \end{vmatrix} \\
&\quad - \begin{vmatrix} a_{21} & a_{22} \\ a_{41} & a_{42} \end{vmatrix} \begin{vmatrix} a_{13} & a_{14} \\ a_{33} & a_{34} \end{vmatrix} + \begin{vmatrix} a_{31} & a_{32} \\ a_{41} & a_{42} \end{vmatrix} \begin{vmatrix} a_{13} & a_{14} \\ a_{23} & a_{24} \end{vmatrix} .
\end{aligned}$$

The terms on the right-hand side correspond, respectively, to the terms in (2.17) for $(i_1, i_2) = (1, 2), (1, 3), (1, 4), (2, 3), (2, 4), (3, 4)$. ■

Remark 2.6. Here we consider the number of the terms in the generalized Laplace expansion (2.16) or (2.17). The sum on the right-hand side consists of $\binom{n}{k}$ terms, and each term, being a product of minors of orders k and $n-k$, gives rise to $k! \times (n-k)!$ terms in its expansion. Therefore, the total number of terms amounts to

$$\binom{n}{k} \times k! \times (n-k)! = \frac{n!}{k!(n-k)!} \times k! \times (n-k)! = n! \,,$$

which coincides with the number of terms contained in the defining expansion of $\det A$. ■

Remark 2.7. Here is the first proof of Theorem 2.12 (generalized Laplace expansion), which is a straightforward comparison of terms. We consider the expansion (2.17) along columns. It suffices to show that each term $a_{\tau(1)1} a_{\tau(2)2} \cdots a_{\tau(n)n}$ in the expansion (2.2) of $\det A$ appears exactly once on the right-hand side of (2.17) with the same sign as $\operatorname{sgn} \tau$ and that the right-hand side of (2.17) contains no other terms.

In Example 2.14 with $n = 4$, for instance, the term $a_{11}a_{22}a_{33}a_{44}$ is contained in the expansion on the right-hand side in the form of $(a_{11}a_{22}) \cdot (a_{33}a_{44})$ arising from the first term. The other term $-a_{41}a_{12}a_{23}a_{34}$ is contained in the form of $(-a_{41}a_{12}) \cdot (a_{23}a_{34})$ arising from the third term.

First we assume $(j_1, j_2, \ldots, j_k) = (1, 2, \ldots, k)$ in (2.17). Let $a_{\tau(1)1} a_{\tau(2)2} \cdots a_{\tau(n)n}$ be a term in (2.2) and rearrange the row indices $\tau(1), \tau(2), \ldots, \tau(k)$ to the ascending order $i_1 < i_2 < \cdots < i_k$. In the summation on the right-hand side of (2.17), this term $a_{\tau(1)1}a_{\tau(2)2}\cdots a_{\tau(n)n}$ is contained in the term for $(i_1, i_2, \ldots, i_k)$. We can verify the coincidence of the signs as follows. Let $\bar{i}_1 < \bar{i}_2 < \cdots < \bar{i}_{n-k}$ be the ascending rearrangement of $\{1, 2, \ldots, n\} \setminus \{i_1, i_2, \ldots, i_k\}$. Then

$$\operatorname{sgn}\begin{pmatrix} 1 & \cdots & k & k+1 & \cdots & n \\ i_1 & \cdots & i_k & \bar{i}_1 & \cdots & \bar{i}_{n-k} \end{pmatrix} = (-1)^{(i_1-1)+\cdots+(i_k-k)}. \tag{2.18}$$

Since the permutation τ can be represented as

$$\begin{aligned} \tau &= \begin{pmatrix} 1 & \cdots & k & k+1 & \cdots & n \\ \tau(1) & \cdots & \tau(k) & \tau(k+1) & \cdots & \tau(n) \end{pmatrix} \\ &= \begin{pmatrix} i_1 & \cdots & i_k \\ \tau(1) & \cdots & \tau(k) \end{pmatrix} \begin{pmatrix} \bar{i}_1 & \cdots & \bar{i}_{n-k} \\ \tau(k+1) & \cdots & \tau(n) \end{pmatrix} \\ &\quad \times \begin{pmatrix} 1 & \cdots & k & k+1 & \cdots & n \\ i_1 & \cdots & i_k & \bar{i}_1 & \cdots & \bar{i}_{n-k} \end{pmatrix}, \end{aligned}$$

we have

$$\begin{aligned} \operatorname{sgn}\tau =& (-1)^{(i_1-1)+\cdots+(i_k-k)} \times \operatorname{sgn}\begin{pmatrix} i_1 & \cdots & i_k \\ \tau(1) & \cdots & \tau(k) \end{pmatrix} \\ & \times \operatorname{sgn}\begin{pmatrix} \bar{i}_1 & \cdots & \bar{i}_{n-k} \\ \tau(k+1) & \cdots & \tau(n) \end{pmatrix}, \end{aligned}$$

which shows that the term $a_{\tau(1)1}a_{\tau(2)2}\cdots a_{\tau(n)n}$ has the same sign on both sides of (2.17). In addition, the comparison of the numbers of terms (cf., Remark 2.6) shows that no extra terms are contained on the right-hand side of (2.17).

The proof for general $(j_1, j_2, \ldots, j_k)$ can be reduced to the above special case of $(j_1, j_2, \ldots, j_k) = (1, 2, \ldots, k)$ through a rearrangement of columns. Let $\bar{j}_1 < \bar{j}_2 < \cdots < \bar{j}_{n-k}$ be the ascending reordering of $\{1, 2, \ldots, n\} \setminus \{j_1, j_2, \ldots, j_k\}$. Then

$$\operatorname{sgn}\begin{pmatrix} 1 & \cdots & k & k+1 & \cdots & n \\ j_1 & \cdots & j_k & \bar{j}_1 & \cdots & \bar{j}_{n-k} \end{pmatrix} = (-1)^{(j_1-1)+\cdots+(j_k-k)}, \tag{2.19}$$

from which it follows that the term $a_{\tau(1)1}a_{\tau(2)2}\cdots a_{\tau(n)n}$ is accompanied by the same sign on both sides of (2.17). ∎

Remark 2.8. Here is the second proof of Theorem 2.12 (generalized Laplace expansion), which is based on multilinearity and alternating properties. We consider the expansion (2.17) along columns, where we assume $(j_1, j_2, \ldots, j_k) = (1, 2, \ldots, k)$ for notational simplicity. Define a function

$$F(\boldsymbol{a}_1, \boldsymbol{a}_2, \ldots, \boldsymbol{a}_k) = \det[\boldsymbol{a}_1, \boldsymbol{a}_2, \ldots, \boldsymbol{a}_k, \boldsymbol{a}_{k+1}, \boldsymbol{a}_{k+2}, \ldots, \boldsymbol{a}_n]$$

in k-tuple $(\boldsymbol{a}_1, \boldsymbol{a}_2, \ldots, \boldsymbol{a}_k)$ of n-dimensional vectors, where $\boldsymbol{a}_{k+1}, \boldsymbol{a}_{k+2}, \ldots, \boldsymbol{a}_n$ are kept fixed. By Theorems 2.3 and 2.4, this function is equipped with multilinearity (2.5) and alternating property (2.6). By Theorem 2.6 in Sec. 2.3.2, it can be represented as

$$F(\boldsymbol{a}_1, \boldsymbol{a}_2, \ldots, \boldsymbol{a}_k) = \sum_{i_1 < i_2 < \cdots < i_k} c_{i_1 i_2 \cdots i_k} \cdot |A|^{i_1, i_2, \ldots, i_k}_{1,2,\ldots,k},$$

where

$$c_{i_1 i_2 \cdots i_k} = F(\boldsymbol{e}_{i_1}, \boldsymbol{e}_{i_2}, \ldots, \boldsymbol{e}_{i_k}) = (-1)^{(i_1 - 1) + \cdots + (i_k - k)} \Delta^{i_1, i_2, \ldots, i_k}_{1,2,\ldots,k}.$$

Thus we obtain (2.17) for $(j_1, j_2, \ldots, j_k) = (1, 2, \ldots, k)$. The formula for other $(j_1, j_2, \ldots, j_k)$ can be proved in a similar manner. ■

2.5.3 *Binet–Cauchy Expansion*

The following expansion formula is concerned with the determinant of a product of matrices that are not necessarily square, and is called the *Binet–Cauchy expansion* or *Cauchy–Binet expansion.*

Theorem 2.13. *Let A be an $m \times n$ matrix and B an $n \times m$ matrix, where $m \leqq n$. Then*[8]

$$\det(AB) = \sum_{j_1 < j_2 < \cdots < j_m} |A|^{1,2,\ldots,m}_{j_1, j_2, \ldots, j_m} \cdot |B|^{j_1, j_2, \ldots, j_m}_{1,2,\ldots,m}, \tag{2.20}$$

where the summation is taken over all $(j_1, j_2, \ldots, j_m)$ satisfying $1 \leqq j_1 < j_2 < \cdots < j_m \leqq n$.

Proof. Two different proofs are given.

[Proof 1: Based on the generalized Laplace expansion] By Theorem 2.9(2) in Sec. 2.4 we have

$$\det \left[\begin{array}{c|c} O & A \\ \hline B & I_n \end{array}\right] = (-1)^m \det(AB).$$

To the matrix on the left-hand side we apply the generalized Laplace expansion of Theorem 2.12(1) with $k = m$ and $(i_1, i_2, \ldots, i_m) = (1, 2, \ldots, m)$, to obtain $(-1)^m$ times the right-hand side of (2.20). Thus (2.20) is proved.

[8] $|A|^{1,2,\ldots,m}_{j_1, j_2, \ldots, j_m}$ denotes the minor of A corresponding to the column indices $j_1 < j_2 < \cdots < j_m$ (cf., Definition 2.7 in Sec. 2.2). Similarly, $|B|^{j_1, j_2, \ldots, j_m}_{1,2,\ldots,m}$ denotes the minor of B corresponding to the row indices $j_1 < j_2 < \cdots < j_m$.

[Proof 2: Based on multilinearity][9] Let $A = [\boldsymbol{a}_1, \boldsymbol{a}_2, \ldots, \boldsymbol{a}_n]$ and $B = (b_{ij} \mid i = 1, 2, \ldots, n;\ j = 1, 2, \ldots, m)$. By multilinearity we have

$$\det(AB) = \det\left[\sum_{i_1=1}^{n} b_{i_1 1}\boldsymbol{a}_{i_1}, \sum_{i_2=1}^{n} b_{i_2 2}\boldsymbol{a}_{i_2}, \ldots, \sum_{i_m=1}^{n} b_{i_m m}\boldsymbol{a}_{i_m}\right]$$
$$= \sum_{i_1=1}^{n}\sum_{i_2=1}^{n}\cdots\sum_{i_m=1}^{n} b_{i_1 1}b_{i_2 2}\cdots b_{i_m m}\det[\boldsymbol{a}_{i_1}, \boldsymbol{a}_{i_2}, \ldots, \boldsymbol{a}_{i_m}].$$

If the indices $i_1, i_2, \ldots, i_m$ are not all distinct, it follows from the alternating property that $\det[\boldsymbol{a}_{i_1}, \boldsymbol{a}_{i_2}, \ldots, \boldsymbol{a}_{i_m}] = 0$. Hence we may restrict ourselves to the terms with $(i_1, i_2, \ldots, i_m) = (\tau(j_1), \tau(j_2), \ldots, \tau(j_m))$ for m distinct column indices $j_1 < j_2 < \cdots < j_m$ and a permutation τ of order m. We have

$$\det[\boldsymbol{a}_{i_1}, \boldsymbol{a}_{i_2}, \ldots, \boldsymbol{a}_{i_m}] = \det[\boldsymbol{a}_{\tau(j_1)}, \boldsymbol{a}_{\tau(j_2)}, \cdots, \boldsymbol{a}_{\tau(j_m)}]$$
$$= \operatorname{sgn}\tau \cdot \det[\boldsymbol{a}_{j_1}, \boldsymbol{a}_{j_2}, \ldots, \boldsymbol{a}_{j_m}]$$

by the alternating property. Substituting this into the above expression we obtain

$$\det(AB)$$
$$= \sum_{j_1<j_2<\cdots<j_m} \det[\boldsymbol{a}_{j_1}, \boldsymbol{a}_{j_2}, \ldots, \boldsymbol{a}_{j_m}] \cdot \sum_{\tau} \operatorname{sgn}\tau \cdot b_{\tau(j_1)1}b_{\tau(j_2)2}\cdots b_{\tau(j_m)m}$$
$$= \sum_{j_1<j_2<\cdots<j_m} |A|^{1,2,\ldots,m}_{j_1,j_2,\ldots,j_m} \cdot |B|^{j_1,j_2,\ldots,j_m}_{1,2,\ldots,m}.$$

□

Remark 2.9. Let A be an $m \times n$ matrix and B an $n \times m$ matrix. For the determinant of AB, we have the following:

If $m = n$, then $\det(AB) = \det(A) \cdot \det(B)$.

If $m < n$, then $\det(AB) = \displaystyle\sum_{j_1<j_2<\cdots<j_m} |A|^{1,2,\ldots,m}_{j_1,j_2,\ldots,j_m} \cdot |B|^{j_1,j_2,\ldots,j_m}_{1,2,\ldots,m}$.

If $m > n$, then $\det(AB) = 0$. ■

Remark 2.10. Let $\boldsymbol{a}_1, \boldsymbol{a}_2, \ldots, \boldsymbol{a}_m$ be n-dimensional vectors, where $m \leqq n$. The *Gram matrix* of $\boldsymbol{a}_1, \boldsymbol{a}_2, \ldots, \boldsymbol{a}_m$ is defined to be the square matrix of order m whose (i, j) entry is equal to $(\boldsymbol{a}_j, \boldsymbol{a}_i) = \boldsymbol{a}_i{}^*\boldsymbol{a}_j$, that is,

$$G = \begin{bmatrix} (\boldsymbol{a}_1, \boldsymbol{a}_1) & (\boldsymbol{a}_2, \boldsymbol{a}_1) & \cdots & (\boldsymbol{a}_m, \boldsymbol{a}_1) \\ (\boldsymbol{a}_1, \boldsymbol{a}_2) & (\boldsymbol{a}_2, \boldsymbol{a}_2) & \cdots & (\boldsymbol{a}_m, \boldsymbol{a}_2) \\ \vdots & \vdots & \vdots & \vdots \\ (\boldsymbol{a}_1, \boldsymbol{a}_m) & (\boldsymbol{a}_2, \boldsymbol{a}_m) & \cdots & (\boldsymbol{a}_m, \boldsymbol{a}_m) \end{bmatrix}.$$

[9] This proof is a generalization of the proof of Theorem 2.7(2) in Sec. 2.4.

The determinant of G is called the *Gramian* of $\boldsymbol{a}_1, \boldsymbol{a}_2, \ldots, \boldsymbol{a}_m$. For the matrix $A = [\boldsymbol{a}_1, \boldsymbol{a}_2, \ldots, \boldsymbol{a}_m]$ we have $G = A^*A$, for which the Binet–Cauchy expansion (2.20) gives

$$\det G = \sum_{i_1 < i_2 < \cdots < i_m} \left| |A|^{i_1, i_2, \ldots, i_m}_{1,2,\ldots,m} \right|^2. \tag{2.21}$$

Therefore, $\boldsymbol{a}_1, \boldsymbol{a}_2, \ldots, \boldsymbol{a}_m$ is linearly independent if and only if $\det G \neq 0$ (cf., Theorem 1.3 in Sec. 1.3.1). Thus the vanishing or non-vanishing of $\det G$ can be used as a criterion for linear independence. Moreover, in engineering applications, the value of $\det G$ itself can be used as a quantitative measure for linear independence.

In the particular case of $m = 2$, we have

$$\det G = \det \begin{bmatrix} (\boldsymbol{a}, \boldsymbol{a}) & (\boldsymbol{b}, \boldsymbol{a}) \\ (\boldsymbol{a}, \boldsymbol{b}) & (\boldsymbol{b}, \boldsymbol{b}) \end{bmatrix} = (\boldsymbol{a}, \boldsymbol{a}) \cdot (\boldsymbol{b}, \boldsymbol{b}) - |(\boldsymbol{a}, \boldsymbol{b})|^2.$$

Hence (2.21) implies

$$(\boldsymbol{a}, \boldsymbol{a}) \cdot (\boldsymbol{b}, \boldsymbol{b}) - |(\boldsymbol{a}, \boldsymbol{b})|^2 = \sum_{i<j} |a_i b_j - a_j b_i|^2, \tag{2.22}$$

which is called the *Cauchy–Lagrange identity.* Since the right-hand side of this identity is a sum of nonnegative real numbers, we can regard (2.22) as a refinement of the Cauchy–Schwarz inequality introduced in Remark 1.18 in Sec. 1.6. ■

2.6 Cofactors

Let $A = (a_{ij})$ be a square matrix of order n. We continue to use the notation Δ_{ij} to mean the determinant of the submatrix of A obtained by deleting row i and column j; see Sec. 2.5.1.

Definition 2.9. $\tilde{\Delta}_{ij} = (-1)^{i+j} \Delta_{ij}$ is called the (i, j) *cofactor* of A. ■

Definition 2.10. The matrix $\tilde{A} = (\tilde{a}_{ij} \mid i, j = 1, 2, \ldots, n)$ with entries

$$\tilde{a}_{ij} = \tilde{\Delta}_{ji} = (-1)^{i+j} \Delta_{ji} \tag{2.23}$$

is called the *adjoint* of A. Note that the (i, j) entry of the adjoint matrix is the (j, i) cofactor $\tilde{\Delta}_{ji}$ of A. The adjoint matrix is also called the *adjugate matrix.* ■

There is a close relationship between the cofactors and the determinant.[10]

Theorem 2.14. *Let $A = (a_{ij})$ be a square matrix of order n.*

[10] The symbol δ_{ij} is the Kronecker delta introduced in Remark 1.3 in Sec. 1.1.1.

(1) $\sum_{j=1}^{n} a_{ij}\tilde{\Delta}_{kj} = \delta_{ik} \cdot \det A \qquad (1 \leqq i, k \leqq n).$

(2) $\sum_{i=1}^{n} a_{ij}\tilde{\Delta}_{ik} = \delta_{jk} \cdot \det A \qquad (1 \leqq j, k \leqq n).$

Proof. Since $\det A = \det A^\top$, Part (1) follows from Part (2) applied to the transpose of A. Therefore we may concentrate on Part (2). If $j = k$, this identity is nothing but the Laplace expansion along the kth column stated in Theorem 2.11(2) of Sec. 2.5.1. Suppose that $j \neq k$ and let B be the matrix obtained from $A = [\boldsymbol{a}_1, \ldots, \boldsymbol{a}_k, \ldots, \boldsymbol{a}_n]$ by replacing the kth column with $\boldsymbol{a}_j$, that is, $B = [\boldsymbol{a}_1, \ldots, \overset{\overset{k}{\vee}}{\boldsymbol{a}_j}, \ldots, \boldsymbol{a}_n]$. Denote the (i, k) cofactor of B by $\tilde{\Gamma}_{ik}$. The Laplace expansion along the kth column of B shows

$$\sum_{i=1}^{n} b_{ik}\tilde{\Gamma}_{ik} = \det B,$$

in which $b_{ik} = a_{ij}$ and $\tilde{\Gamma}_{ik} = \tilde{\Delta}_{ik}$ for $i = 1, 2, \ldots, n$, and $\det B = 0$ by the definition of B. Therefore,

$$\sum_{i=1}^{n} a_{ij}\tilde{\Delta}_{ik} = 0.$$

□

Theorem 2.14 can be expressed in a matrix form as follows.

Theorem 2.15. *For a square matrix A and its adjoint matrix $\tilde{A}$, we have the following relation:*

$$A\tilde{A} = \tilde{A}A = (\det A)I.$$

Proof. Let $B = A\tilde{A}$. With the use of (2.23) the left-hand side of Theorem 2.14(1) can be rewritten as

$$\sum_{j} a_{ij}\tilde{\Delta}_{kj} = \sum_{j} a_{ij}\tilde{a}_{jk} = b_{ik}.$$

Therefore we have $B = (\det A)I$. Similarly, we can obtain $\tilde{A}A = (\det A)I$ from Theorem 2.14(2). □

The inverse matrix can be expressed in terms of the adjoint matrix.

Theorem 2.16.

(1) *A square matrix A is nonsingular if and only if* $\det A \neq 0$.

(2) *If* $\det A \neq 0$, *then*

$$A^{-1} = \frac{1}{\det A}\tilde{A},$$

where $\tilde{A}$ *denotes the adjoint of* A.

Proof. If A is nonsingular, the inverse X of A satisfies $AX = I$. From this we obtain $(\det A) \cdot (\det X) = 1$, which implies $\det A \neq 0$. Conversely, if $\det A \neq 0$, then Theorem 2.15 shows that $X = (1/\det A)\tilde{A}$ gives the inverse of A. □

Remark 2.11. Theorem 2.16 enables us to prove Theorem 1.2 in Sec. 1.3.1. (1) Given a matrix A, suppose that $XA = I$ holds for some X. Since $(\det X) \cdot (\det A) = 1$, we have $\det A \neq 0$. By Theorem 2.16, A is nonsingular and the inverse A^{-1} exists. By multiplying $XA = I$ with A^{-1} from the right we obtain $X = A^{-1}$. Part (2) of Theorem 1.2, where we have X with $AX = I$, can be proved similarly. ■

2.7 Computing Determinants

When we want to calculate a determinant by hand, we try to simplify the computation with the aid of various properties and identities such as multilinearity, alternating property, and the Laplace expansion. It is especially effective to turn a non-zero entry to zero by making use of the facts that the determinant remains unchanged when a scalar multiple of one row is added to another row, and that it remains unchanged when a scalar multiple of one column is added to another column (Proposition 2.1 and Remark 2.4 in Sec. 2.3.1). For example, when $a_{11} \neq 0$ we can annihilate the $(1, 2)$ entry by adding

$$(-a_{12}/a_{11}) \times (\text{the first column})$$

to the second column, while keeping the value of the determinant unchanged:

$$\begin{vmatrix} a_{11} & a_{12} & a_{13} \\ a_{21} & a_{22} & a_{23} \\ a_{31} & a_{32} & a_{33} \end{vmatrix} = \begin{vmatrix} a_{11} & 0 & a_{13} \\ a_{21} & a_{22} - a_{21}a_{12}/a_{11} & a_{23} \\ a_{31} & a_{32} - a_{31}a_{12}/a_{11} & a_{33} \end{vmatrix}.$$

Example 2.15. Here is an example of computing a determinant.

$$\begin{vmatrix} 5 & 5 & 0 & 3 & 1 \\ 1 & 9 & 0 & 1 & 2 \\ 4 & 6 & 5 & 3 & 3 \\ 7 & 2 & 0 & 2 & 2 \\ 0 & 8 & 0 & 0 & 0 \end{vmatrix} \overset{\substack{\text{expand}\\\text{along}\\\text{row 5}}}{=\!=\!=} -8 \cdot \begin{vmatrix} 5 & 0 & 3 & 1 \\ 1 & 0 & 1 & 2 \\ 4 & 5 & 3 & 3 \\ 7 & 0 & 2 & 2 \end{vmatrix} \overset{\substack{\text{expand}\\\text{along}\\\text{col 2}}}{=\!=\!=} (-8)\cdot(-5)\cdot \begin{vmatrix} 5 & 3 & 1 \\ 1 & 1 & 2 \\ 7 & 2 & 2 \end{vmatrix}$$

$$\overset{\text{col 1}-\text{col 2}}{=\!=\!=} 40 \cdot \begin{vmatrix} 2 & 3 & 1 \\ 0 & 1 & 2 \\ 5 & 2 & 2 \end{vmatrix} \overset{\text{col 3}-2\times\text{col 2}}{=\!=\!=} 40 \cdot \begin{vmatrix} 2 & 3 & -5 \\ 0 & 1 & 0 \\ 5 & 2 & -2 \end{vmatrix}$$

$$\overset{\substack{\text{expand}\\\text{along}\\\text{row 2}}}{=\!=\!=} 40 \cdot 1 \cdot \begin{vmatrix} 2 & -5 \\ 5 & -2 \end{vmatrix} = 40 \times (2 \times (-2) - 5 \times (-5)) = 40 \times 21 = 840.$$

■

Chapter 3

Elementary Transformations and Elimination

In this chapter we describe the method of elimination using elementary transformations as a fundamental tool for computing and transforming matrices. Elementary transformations are used effectively to derive the rank normal form and the reduced echelon form and to compute inverse matrices.

3.1 Elementary Transformations of Matrices

3.1.1 *Elementary Row Transformations*

In solving a system of linear equations (simultaneous linear equations), it is effective to eliminate variables through appropriate combination of equations. Suppose, for example, that we want to solve a system of equations in two variables:

$$\left\{ \begin{array}{rcl} a_{11}x_1 + a_{12}x_2 &=& b_1, \\ a_{21}x_1 + a_{22}x_2 &=& b_2. \end{array} \right.$$

If $a_{11} \neq 0$, we can eliminate variable x_1 from the second equation by adding $-a_{21}/a_{11}$ times the first equation to the second equation. This elimination process amounts to transforming the coefficient matrix as

$$\begin{bmatrix} a_{11} & a_{12} \\ a_{21} & a_{22} \end{bmatrix} \longrightarrow \begin{bmatrix} a'_{11} & a'_{12} \\ a'_{21} & a'_{22} \end{bmatrix} = \begin{bmatrix} a_{11} & a_{12} \\ 0 & a_{22} - \dfrac{a_{21}a_{12}}{a_{11}} \end{bmatrix}.$$

In this section we consider such operations on matrices.

Definition 3.1. For a matrix over K, the following three kinds of operations are called *elementary row transformations* (or *elementary row operations*):

(1) exchanging two rows,

(2) multiplying a certain row by a nonzero scalar ($\in K$), and
(3) adding a scalar ($\in K$) multiple of a certain row to another row. ■

The above three types of elementary row transformations are represented, respectively, by elementary matrices, defined as follows.

Definition 3.2. The following three kinds of matrices are called *elementary matrices*, where $p \neq q$; $b, c \in K$; $b \neq 0$.

(1) $E_1(p,q)$: A square matrix whose submatrix corresponding to row and column indices $\{p,q\}$ is $\begin{bmatrix} 0 & 1 \\ 1 & 0 \end{bmatrix}$, and the rest is a unit matrix.
(2) $E_2(p;b)$: A square matrix whose submatrix corresponding to row and column index $\{p\}$ is $[\,b\,]$, and the rest is a unit matrix.
(3) $E_3(p,q;c)$: A square matrix whose (p,q) entry is c, the other off-diagonal entries are all 0, and the diagonal entries are all 1. The submatrix corresponding to row and column indices $\{p,q\}$ is $\begin{bmatrix} 1 & c \\ 0 & 1 \end{bmatrix}$ if $p < q$, and $\begin{bmatrix} 1 & 0 \\ c & 1 \end{bmatrix}$ if $q < p$.

■

Example 3.1. The elementary matrices for $(p,q) = (1,2)$ are given as follows:

$$E_1(1,2) = \left[\begin{array}{cc|cccc} & 1 & & & & \\ 1 & & & & & \\ \hline & & 1 & & & \\ & & & & \ddots & \\ & & & & & 1 \end{array}\right], \quad E_2(1;b) = \left[\begin{array}{c|cccc} b & & & & \\ \hline & 1 & & & \\ & & 1 & & \\ & & & \ddots & \\ & & & & 1 \end{array}\right], \quad E_3(1,2;c) = \left[\begin{array}{cc|cccc} 1 & c & & & & \\ & 1 & & & & \\ \hline & & 1 & & & \\ & & & & \ddots & \\ & & & & & 1 \end{array}\right].$$

■

Elementary row transformations on a matrix correspond to multiplying the matrix by elementary matrices $E_1(p,q)$, $E_2(p;b)$, and $E_3(p,q;c)$ from the left.

Proposition 3.1.

(1) *If we multiply A by $E_1(p,q)$ from the left, the pth row and qth row of A are exchanged.*
(2) *If we multiply A by $E_2(p;b)$ from the left, the pth row of A is multiplied by b.*

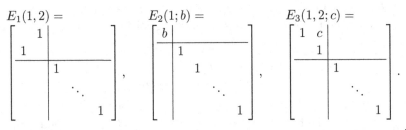

(3) *If we multiply A by $E_3(p,q;c)$ from the left, then c times the qth row of A is added to the pth row of A.*

Proposition 3.2. *An elementary matrix is nonsingular, and the inverse of an elementary matrix is an elementary matrix. More specifically, we have the following.*

(1) $\det E_1(p,q) = -1, \qquad E_1(p,q)^{-1} = E_1(p,q).$
(2) $\det E_2(p;b) = b \neq 0, \qquad E_2(p;b)^{-1} = E_2(p;1/b).$
(3) $\det E_3(p,q;c) = 1, \qquad E_3(p,q;c)^{-1} = E_3(p,q;-c).$

Example 3.2. Let

$$A = \begin{bmatrix} 1 & 1 & 2 & 3 \\ 1 & 2 & 3 & 2 \\ 0 & 3 & 3 & -2 \\ 1 & 3 & 4 & 2 \end{bmatrix}. \tag{3.1}$$

The elementary row operation that adds -1 times the first row to the second row transforms A to

$$A' = \begin{bmatrix} 1 & 1 & 2 & 3 \\ 0 & 1 & 1 & -1 \\ 0 & 3 & 3 & -2 \\ 1 & 3 & 4 & 2 \end{bmatrix}.$$

We have $A' = E_3(2,1;-1) \cdot A$. This elementary transformation has turned the $(2,1)$ entry to 0. Next, we turn the $(4,1)$ entry to 0 by applying to A' the elementary row transformation that adds -1 times the first row to the fourth row. The resulting matrix is

$$A'' = \begin{bmatrix} 1 & 1 & 2 & 3 \\ \underline{0} & 1 & 1 & -1 \\ \underline{0} & 3 & 3 & -2 \\ \underline{0} & 2 & 2 & -1 \end{bmatrix},$$

in which the entries in first column, excepting the $(1,1)$ entry, are all equal to 0. We have the relation $A'' = E_3(4,1;-1) \cdot A' = E_3(4,1;-1) \cdot E_3(2,1;-1) \cdot A$. In this way, by applying elementary row transformations repeatedly, we can transform a column of a matrix to a scalar multiple of a unit vector. ∎

The method of transformation illustrated above can be described for the general case as follows. Assume that the (k,l) entry a_{kl} of a matrix A is distinct from 0. The elementary row operation that adds $-a_{il}/a_{kl}$ times the kth row to the ith row, where $i \neq k$, transforms A to another matrix

$A' = E_3(i,k;-a_{il}/a_{kl}) \cdot A$, where the (i,l) entry of the matrix A' is equal to 0. By applying such elementary row transformation for each i ($\neq k$), we obtain a matrix A'' such that all entries in the lth column, excepting the (k,l) entry, are equal to zero. This matrix A'' is obtained from A by multiplying A by the product of the matrices $E_3(i,k;-a_{il}/a_{kl})$ over $i \neq k$ from the left.[1] The operation of transforming A to A'' is called the *clearing* of a column with *pivot* (k,l).

3.1.2 *Elementary Column Transformations*

Elementary column transformations are defined similarly as elementary row transformations.

Definition 3.3. For a matrix over K, the following three kinds of operations are called *elementary column transformations* (or *elementary column operations*):

(1) exchanging two columns,
(2) multiplying a certain column by a nonzero scalar ($\in K$), and
(3) adding a scalar ($\in K$) multiple of a certain column to another column.

■

Elementary column transformations on a matrix correspond to multiplying the matrix by elementary matrices $E_1(p,q)$, $E_2(p;b)$, and $E_3(p,q;c)$ from the right.

Proposition 3.3.

(1) *If we multiply A by $E_1(p,q)$ from the right, the pth column and qth column of A are exchanged.*
(2) *If we multiply A by $E_2(p;b)$ from the right, the pth column of A is multiplied by b.*
(3) *If we multiply A by $E_3(p,q;c)$ from the right, then c times the pth column of A is added to the qth column of A.*

By applying elementary column transformations repeatedly, we can transform a row of a matrix to a scalar multiple of a unit vector. Assume that the (k,l) entry a_{kl} of a matrix A is distinct from 0. The elementary column operation that adds $-a_{kj}/a_{kl}$ times the lth column to the jth column,

[1] Since $E_3(i,k;c)E_3(i',k;c') = E_3(i',k;c')E_3(i,k;c)$ in general, this product is not affected by the order of the elementary matrices in the product.

where $j \neq l$, transforms A to another matrix $A' = A \cdot E_3(l, j; -a_{kj}/a_{kl})$, where the (k, j) entry of the matrix A' is equal to 0. By applying such elementary column transformation for each j $(\neq l)$, we obtain a matrix A'' such that all entries in the kth row, excepting the (k, l) entry, are equal to zero. This matrix A'' is obtained from A by multiplying A by the product of the matrices $E_3(l, j; -a_{kj}/a_{kl})$ over $j \neq l$ from the right. This operation is called the *clearing* of a row with *pivot* (k, l).

Example 3.3. For the matrix A in (3.1), the clearing of the first row with pivot $(1, 1)$ consists of three elementary column operations: adding -1 times the first column to the second column, adding -2 times the first column to the third column, and adding -3 times the first column to the fourth column. This operation results in the matrix

$$A \cdot E_3(1, 2; -1) \cdot E_3(1, 3; -2) \cdot E_3(1, 4; -3) = \begin{bmatrix} 1 & \underline{0} & \underline{0} & \underline{0} \\ 1 & 1 & 1 & -1 \\ 0 & 3 & 3 & -2 \\ 1 & 2 & 2 & -1 \end{bmatrix}.$$

■

Definition 3.4. An *elementary transformation* means an elementary row or column transformation. ■

3.2 Rank Normal Form

Let us consider transforming a given matrix to a simplest possible form by repeated application of elementary row and column operations. The following theorem states that any matrix can be transformed to a simple form

$$\left[\begin{array}{c|c} I_r & O \\ \hline O & O \end{array}\right], \tag{3.2}$$

which contains the unit matrix I_r of an appropriate order r.

Theorem 3.1. *Any $m \times n$ matrix A can be brought to a matrix of the form of (3.2) by repeated application of elementary row or column transformations. In other words, there exist elementary matrices $E_1, E_2, \ldots, E_k$; $F_1, F_2, \ldots, F_l$ and a nonnegative integer r such that*

$$(E_k \cdots E_2 E_1) \cdot A \cdot (F_1 F_2 \cdots F_l) = \left[\begin{array}{c|c} I_r & O_{r,n-r} \\ \hline O_{m-r,r} & O_{m-r,n-r} \end{array}\right]. \tag{3.3}$$

Proof. If $A = O$, it is already in the form of (3.3) with $r = 0$. Suppose $A \neq O$. Then A has a nonzero entry. By appropriate permutations of rows and columns, we can move the nonzero entry to the $(1, 1)$ entry. Furthermore, we can change the $(1, 1)$ entry to 1 by an elementary row operation (a scalar multiplication) on the first row. By clearing out the first column and the first row with pivot $(1, 1)$, we can obtain

$$\left[\begin{array}{c|ccc} 1 & 0 & \cdots & 0 \\ \hline 0 & & & \\ \vdots & & A^{(1)} & \\ 0 & & & \end{array}\right].$$

By recursively applying such transformation to the remaining matrix $A^{(1)}$, we eventually arrive at a matrix of the form on the right-hand side of (3.3). □

The matrix

$$\tilde{A} = \left[\begin{array}{c|c} I_r & O \\ \hline O & O \end{array}\right]$$

on the right-hand side of (3.3) is called the *rank normal form* of the matrix A.

Remark 3.1. The integer r in Theorem 3.1 is in fact equal to the rank of matrix A and is determined by the given matrix A; see Sec. 4.1 for the definition of rank. Therefore, the rank normal form $\tilde{A}$ is uniquely determined by the given matrix A. See Theorem 4.8 in Sec. 4.2. ■

Example 3.4. We illustrate the construction of the rank normal form in the proof of Theorem 3.1:

$$A = \begin{bmatrix} 0 & -4 & -4 \\ 3 & 4 & 7 \\ 2 & -1 & 1 \end{bmatrix} \quad \xrightarrow[E_1(1,3)\times]{\text{exchange row 1 and row 3}}$$

$$\begin{bmatrix} 2 & -1 & 1 \\ 3 & 4 & 7 \\ 0 & -4 & -4 \end{bmatrix} \quad \xrightarrow[E_2(1;1/2)\times]{\text{multiply row 1 by 1/2}}$$

$$\begin{bmatrix} \boxed{1} & -1/2 & 1/2 \\ 3 & 4 & 7 \\ 0 & -4 & -4 \end{bmatrix} \quad \xrightarrow[E_3(2,1;-3)\times]{\text{clear col 1, pivot (1,1)}}$$

$$\begin{bmatrix} \boxed{1} & -1/2 & 1/2 \\ 0 & 11/2 & 11/2 \\ 0 & -4 & -4 \end{bmatrix} \xrightarrow[\times E_3(1,2;1/2)E_3(1,3;-1/2)]{\text{clear row 1, pivot } (1,1)}$$

$$\left[\begin{array}{c|cc} 1 & 0 & 0 \\ \hline 0 & 11/2 & 11/2 \\ 0 & -4 & -4 \end{array}\right] \xrightarrow[E_2(2;2/11)\times]{\text{multiply row 2 by } 2/11}$$

$$\left[\begin{array}{c|cc} 1 & 0 & 0 \\ \hline 0 & \boxed{1} & 1 \\ 0 & -4 & -4 \end{array}\right] \xrightarrow[E_3(3,2;4)\times]{\text{clear col 2, pivot } (2,2)}$$

$$\left[\begin{array}{c|cc} 1 & 0 & 0 \\ \hline 0 & \boxed{1} & 1 \\ 0 & 0 & 0 \end{array}\right] \xrightarrow[\times E_3(2,3;-1)]{\text{clear row 2, pivot } (2,2)}$$

$$\left[\begin{array}{cc|c} 1 & 0 & 0 \\ 0 & 1 & 0 \\ \hline 0 & 0 & 0 \end{array}\right] = \tilde{A} \quad \text{(rank normal form, } r = 2\text{).}$$

■

Example 3.5. For the matrix A in (3.1) we construct its rank normal form $\tilde{A}$ according to the proof of Theorem 3.1.

$$A = \begin{bmatrix} \boxed{1} & 1 & 2 & 3 \\ 1 & 2 & 3 & 2 \\ 0 & 3 & 3 & -2 \\ 1 & 3 & 4 & 2 \end{bmatrix} \xrightarrow[E_3(4,1;-1)E_3(2,1;-1)\times]{\text{clear col 1, pivot } (1,1)}$$

$$\begin{bmatrix} \boxed{1} & 1 & 2 & 3 \\ 0 & 1 & 1 & -1 \\ 0 & 3 & 3 & -2 \\ 0 & 2 & 2 & -1 \end{bmatrix} \xrightarrow[\times E_3(1,2;-1)E_3(1,3;-2)E_3(1,4;-3)]{\text{clear row 1, pivot } (1,1)}$$

$$\left[\begin{array}{c|ccc} 1 & 0 & 0 & 0 \\ \hline 0 & \boxed{1} & 1 & -1 \\ 0 & 3 & 3 & -2 \\ 0 & 2 & 2 & -1 \end{array}\right] \xrightarrow[E_3(4,2;-2)E_3(3,2;-3)\times]{\text{clear col 2, pivot } (2,2)}$$

$$\left[\begin{array}{c|ccc} 1 & 0 & 0 & 0 \\ \hline 0 & \boxed{1} & 1 & -1 \\ 0 & 0 & 0 & 1 \\ 0 & 0 & 0 & 1 \end{array}\right] \xrightarrow[\times E_3(2,3;-1)E_3(2,4;1)]{\text{clear row 2, pivot } (2,2)}$$

$$\left[\begin{array}{cc|cc} 1 & 0 & 0 & 0 \\ 0 & 1 & 0 & 0 \\ \hline 0 & 0 & 0 & 1 \\ 0 & 0 & 0 & 1 \end{array}\right] \xrightarrow[\times E_1(3,4)]{\text{exchange col 3 and col 4}}$$

$$\left[\begin{array}{cc|cc} 1 & 0 & 0 & 0 \\ 0 & 1 & 0 & 0 \\ \hline 0 & 0 & \boxed{1} & 0 \\ 0 & 0 & 1 & 0 \end{array}\right] \xrightarrow[E_3(4,3;-1)\times]{\text{clear col 3, pivot (3,3)}}$$

$$\left[\begin{array}{ccc|c} 1 & 0 & 0 & 0 \\ 0 & 1 & 0 & 0 \\ 0 & 0 & 1 & 0 \\ \hline 0 & 0 & 0 & 0 \end{array}\right] = \tilde{A} \quad \text{(rank normal form, } r = 3).$$

■

By applying Theorem 3.1 to a nonsingular matrix A, we can obtain a fundamental proposition that every nonsingular matrix can be represented as a product of elementary matrices.

Theorem 3.2.

(1) *The rank normal form of a nonsingular matrix is the unit matrix.*
(2) *Every nonsingular matrix can be represented as a product of elementary matrices.*

Proof. (1) Recall that the product of nonsingular matrices is nonsingular (Theorem 1.1 in Sec. 1.3.1). This implies that the left-hand side of (3.3) is a nonsingular matrix. On the other hand, the matrix on the right-hand side of (3.3) is nonsingular only if it is the unit matrix.

(2) It follows from (1) above that

$$(E_k \cdots E_2 E_1) \cdot A \cdot (F_1 F_2 \cdots F_l) = I,$$

which implies

$$A = E_1^{-1} E_2^{-1} \cdots E_k^{-1} \cdot F_l^{-1} \cdots F_2^{-1} F_1^{-1}.$$

Here, the inverses of elementary matrices E_i and F_j are also elementary matrices (Proposition 3.2 in Sec. 3.1.1). □

Theorem 3.1 can be recast into Theorem 3.3 below.[2] Although they are equivalent, Theorem 3.1 is more amenable to algorithm designs by featuring

[2]The implication "Theorem 3.1 $\Rightarrow$ Theorem 3.3" is obvious from Proposition 3.2, whereas the converse "Theorem 3.1 $\Leftarrow$ Theorem 3.3" follows from Theorem 3.2(2).

operational aspects in terms of elementary transformations, whereas Theorem 3.3 is more suitable for the description of intrinsic structures. In this book we make use of Theorem 3.3 in an analysis of systems of equations (Sec. 5.1).

Theorem 3.3. *For any $m \times n$ matrix A, there exist a nonsingular matrix S of order m, a nonsingular matrix T of order n, and a nonnegative integer r for which*

$$SAT = \left[\begin{array}{c|c} I_r & O_{r,n-r} \\ \hline O_{m-r,r} & O_{m-r,n-r} \end{array}\right]. \tag{3.4}$$

Example 3.6. In Example 3.5 we computed the rank normal form of the matrix A in (3.1). The transformation matrices S and T in (3.4) are given by

$$\begin{aligned} S &= E_3(4,3;-1) \times E_3(4,2;-2)E_3(3,2;-3) \times E_3(4,1;-1)E_3(2,1;-1) \\ &= \left[\begin{array}{cccc} 1 & 0 & 0 & 0 \\ -1 & 1 & 0 & 0 \\ 3 & -3 & 1 & 0 \\ \hline -2 & 1 & -1 & 1 \end{array}\right], \end{aligned}$$

$$\begin{aligned} T &= E_3(1,2;-1)E_3(1,3;-2)E_3(1,4;-3) \times E_3(2,3;-1)E_3(2,4;1) \times E_1(3,4) \\ &= \left[\begin{array}{ccc|c} 1 & -1 & -4 & -1 \\ 0 & 1 & 1 & -1 \\ 0 & 0 & 0 & 1 \\ 0 & 0 & 1 & 0 \end{array}\right]. \end{aligned}$$

■

Remark 3.2. For an $m \times n$ matrix A, a transformation of the form SAT is called an *equivalence transformation*, where S is a nonsingular matrix of order m and T is a nonsingular matrix of order n. Two matrices are said to be *equivalent* if they are connected by an equivalence transformation with some nonsingular matrices S and T. Theorem 3.3 shows that any matrix is equivalent to a matrix of the form of (3.2). ■

Remark 3.3. For a square matrix A, similarity and congruence transformations are defined in addition to equivalence transformations. We introduce these transformations here although we use them in later chapters.

- A transformation of the form SAT using nonsingular matrices S and T is called an equivalence transformation. Two matrices connected by an equivalence transformation are said to be equivalent. An equivalence transformation corresponds to a change of bases of a pair of spaces (Theorem 9.18 in Sec. 9.5.3).

- A transformation of the form $S^{-1}AS$ using a nonsingular matrix S is called a *similarity transformation.* Two matrices connected by a similarity transformation is said to be *similar.* A similarity transformation corresponds to a change of basis in a single space (Theorem 9.18 in Sec. 9.5.3), and is particularly important in the eigenvalue problem (Theorem 6.2 in Sec. 6.1.1).
- A transformation of the form S^*AS using a nonsingular matrix S is called a *congruence transformation*, where $S^*AS = S^\top AS$ if $K = \mathbb{R}$. Two matrices connected by a congruence transformation is said to be *congruent.* A congruence transformation is important in discussing quadratic forms (cf., Chapter 7).

If S is a unitary matrix, we have $S^{-1}AS = S^*AS$. We refer to such transformation as a *unitary similarity transformation* or as a *unitary congruence transformation*, depending on the context. Similarly, if S is an orthogonal matrix, we have $S^{-1}AS = S^\top AS$. We refer to such transformation as an *orthogonal similarity transformation* or as an *orthogonal congruence transformation*, depending on the context. ■

3.3 Reduced Echelon Form

While the rank normal form (Theorem 3.3) gives the simplest form under the transformation of the form SAT involving row and column operations, the normal form addressed in Theorem 3.4 of this section reveals the simplest form under the transformation of the form SA involving row operations only. Theorem 3.4 is used for the computation of inverse matrices (Sec. 3.4) and the solution of linear equations (Sec. 5.3).

Definition 3.5. An $m \times n$ matrix B is said to be in the *reduced echelon form* if there exist a nonnegative integer r and column indices

$$j_1^\circ < j_2^\circ < \cdots < j_r^\circ$$

such that

- For $i = 1, 2, \ldots, r$:
$b_{i,j_i^\circ} = 1$ and $b_{ij} = 0 \quad (j \in \{1, 2, \ldots, j_i^\circ - 1\} \cup \{j_{i+1}^\circ, j_{i+2}^\circ, \ldots, j_r^\circ\})$,
- For $i = r+1, r+2, \ldots, m : b_{ij} = 0 \quad (j = 1, 2, \ldots, n)$.

The above conditions imply that, for each $i = 1, 2, \ldots, r$, the j_i°th column of B is equal to the ith unit vector $(0, \ldots, 0, \overset{\overset{i}{\vee}}{1}, 0, \ldots, 0)^\top$. ■

Example 3.7. The following two 4×7 matrices are in the reduced echelon form:

$$\begin{bmatrix} 1 & * & 0 & * & * & 0 & * \\ 0 & 0 & 1 & * & * & 0 & * \\ 0 & 0 & 0 & 0 & 0 & 1 & * \\ 0 & 0 & 0 & 0 & 0 & 0 & 0 \end{bmatrix}, \qquad \begin{bmatrix} 0 & 1 & 0 & * & 0 & * & 0 \\ 0 & 0 & 1 & * & 0 & * & 0 \\ 0 & 0 & 0 & 0 & 1 & * & 0 \\ 0 & 0 & 0 & 0 & 0 & 0 & 1 \end{bmatrix}.$$

For the first matrix we have $r = 3$, $j_1^\circ = 1$, $j_2^\circ = 3$, and $j_3^\circ = 6$. For the second matrix we have $r = 4$, $j_1^\circ = 2$, $j_2^\circ = 3$, $j_3^\circ = 5$, and $j_4^\circ = 7$. ■

Theorem 3.4. *For any $m \times n$ matrix A, there exists a nonsingular matrix S of order m such that $\hat{A} = SA$ is a matrix in the reduced echelon form.*

Proof. If $A = O$, it is in the reduced echelon form with $r = 0$. Suppose $A \neq O$. Let j_1° be the smallest column index j such that the jth column of A is not $\mathbf{0}$. By an appropriate permutation of rows, we can move a nonzero entry to the $(1, j_1^\circ)$ entry. Furthermore, we can change the $(1, j_1^\circ)$ entry to 1 by an elementary row operation (a scalar multiplication) on the first row. By clearing out the column with pivot $(1, j_1^\circ)$, we can obtain

$$A' = \left[\begin{array}{c|c|c} \mathbf{0}^\top & 1 & * \ \cdots \ * \\ \hline & 0 & \\ O_{m-1, j_1^\circ - 1} & \vdots & A^{(1)} \\ & 0 & \end{array}\right],$$

where $A^{(1)}$ is an $(m-1) \times (n - j_1^\circ)$ matrix. If $A^{(1)} = O$, A' is a matrix in the reduced echelon form with $r = 1$. Suppose $A^{(1)} \neq O$. Let j_2° $(> j_1^\circ)$ be the smallest column index j of A' such that the $(j - j_1^\circ)$th column of $A^{(1)}$ is not $\mathbf{0}$. By an appropriate permutation of rows i of A' with $i \geq 2$, we can move a nonzero entry to the $(2, j_2^\circ)$ entry. Furthermore, we can change the $(2, j_2^\circ)$ entry to 1 by an elementary row operation (a scalar multiplication) on the second row. We clear out the column with pivot $(2, j_2^\circ)$, where the $(1, j_2^\circ)$ entry is also changed to 0. Continuing in this way, we eventually arrive at a matrix of the reduced echelon form. □

It can be shown (Theorem 4.9 in Sec. 4.2) that the matrix $\hat{A}$ in Theorem 3.4 is uniquely determined by matrix A, and it is referred to as the reduced echelon form of A.

Example 3.8. According to the procedure described in the proof of Theorem 3.4, we can transform the matrix A in (3.1) to its reduced echelon form $\hat{A}$ as follows:

$$A = \begin{bmatrix} \boxed{1} & 1 & 2 & 3 \\ 1 & 2 & 3 & 2 \\ 0 & 3 & 3 & -2 \\ 1 & 3 & 4 & 2 \end{bmatrix} \xrightarrow[E_3(4,1;-1)E_3(2,1;-1)\times]{\text{clear col 1, pivot }(1,1)}$$

$$\left[\begin{array}{c|ccc} 1 & 1 & 2 & 3 \\ \hline 0 & \boxed{1} & 1 & -1 \\ 0 & 3 & 3 & -2 \\ 0 & 2 & 2 & -1 \end{array}\right] \xrightarrow[E_3(4,2;-2)E_3(3,2;-3)E_3(1,2;-1)\times]{\text{clear col 2, pivot }(2,2)}$$

$$\left[\begin{array}{ccc|c} 1 & 0 & 1 & 4 \\ 0 & 1 & 1 & -1 \\ \hline 0 & 0 & 0 & \boxed{1} \\ 0 & 0 & 0 & 1 \end{array}\right] \xrightarrow[E_3(4,3;-1)E_3(2,3;1)E_3(1,3;-4)\times]{\text{clear col 4, pivot }(3,4)}$$

$$\begin{bmatrix} 1 & 0 & 1 & 0 \\ 0 & 1 & 1 & 0 \\ 0 & 0 & 0 & 1 \\ 0 & 0 & 0 & 0 \end{bmatrix} = \hat{A} \quad \text{(reduced echelon form)}$$
$$(r = 3,\ j_1^\circ = 1,\ j_2^\circ = 2,\ j_3^\circ = 4).$$

■

Remark 3.4. An $m \times n$ matrix B is said to be in the *echelon form* if there exist a nonnegative integer r and column indices

$$j_1^\circ < j_2^\circ < \cdots < j_r^\circ$$

such that

- For $i = 1, 2, \ldots, r : b_{i,j_i^\circ} \neq 0$ and $b_{ij} = 0$ $(j \in \{1, 2, \ldots, j_i^\circ - 1\})$,
- For $i = r+1, r+2, \ldots, m : b_{ij} = 0$ $(j = 1, 2, \ldots, n)$.

In comparison with Definition 3.5 of the reduced echelon form, the first condition is weaker here. Namely, a reduced echelon form is an echelon form with the additional property that, for each $i = 1, 2, \ldots, r$, b_{i,j_i° is normalized to one and

$$b_{ij} = 0 \ \text{ for } \ j \in \{j_{i+1}^\circ, j_{i+2}^\circ, \ldots, j_r^\circ\}.$$

■

3.4 Computing Inverse Matrices

We can compute the inverse of a matrix with the aid of elementary row transformations. For a given square matrix A of order n, we consider an $n \times (2n)$ matrix

$$\left[\, A \,\middle|\, I_n \,\right] \tag{3.5}$$

consisting of the given matrix A and the unit matrix I_n. If we can transform this matrix, by means of elementary row transformations, to a matrix of the form

$$\left[\, I_n \,\middle|\, B \,\right] \tag{3.6}$$

containing the unit matrix I_n in the left, then the given matrix A is nonsingular and $B = A^{-1}$. To see this, let $E_1, E_2, \ldots, E_k$ be the elementary matrices corresponding to the elementary row operations employed in the transformation. Then we have

$$(E_k \cdots E_2 E_1) \left[\, A \,\middle|\, I_n \,\right] = \left[\, I_n \,\middle|\, B \,\right], \tag{3.7}$$

which implies

$$(E_k \cdots E_2 E_1) \cdot A = I_n, \qquad (E_k \cdots E_2 E_1) \cdot I_n = B.$$

We can find the elementary matrices $E_1, E_2, \ldots, E_k$ in (3.7) by repeating elimination operations with elementary row transformations. In the course of this process we can also test for nonsingularity of the given matrix. For any nonsingular A, the existence of such $E_1, E_2, \ldots, E_k$ is guaranteed by Theorem 3.2(2) in Sec. 3.2.

Example 3.9. Let us compute the inverse of

$$A = \begin{bmatrix} 1 & 1 & 2 & 3 \\ 1 & 2 & 3 & 2 \\ 0 & 3 & 3 & -2 \\ 1 & 3 & 3 & 2 \end{bmatrix} \tag{3.8}$$

according to the procedure described above. A possible elimination process for $[\,A \mid I\,]$ is as follows:

$$\left[\begin{array}{cccc|cccc} \boxed{1} & 1 & 2 & 3 & 1 & 0 & 0 & 0 \\ 1 & 2 & 3 & 2 & 0 & 1 & 0 & 0 \\ 0 & 3 & 3 & -2 & 0 & 0 & 1 & 0 \\ 1 & 3 & 3 & 2 & 0 & 0 & 0 & 1 \end{array}\right] \xrightarrow[E_3(4,1;-1)E_3(2,1;-1)\times]{\text{clear col 1, pivot } (1,1)}$$

$$\left[\begin{array}{cccc|cccc} 1 & 1 & 2 & 3 & 1 & 0 & 0 & 0 \\ 0 & \boxed{1} & 1 & -1 & -1 & 1 & 0 & 0 \\ 0 & 3 & 3 & -2 & 0 & 0 & 1 & 0 \\ 0 & 2 & 1 & -1 & -1 & 0 & 0 & 1 \end{array}\right] \xrightarrow[E_3(4,2;-2)E_3(3,2;-3)E_3(1,2;-1)\times]{\text{clear col 2, pivot } (2,2)}$$

$$\left[\begin{array}{cccc|cccc} 1 & 0 & 1 & 4 & 2 & -1 & 0 & 0 \\ 0 & 1 & 1 & -1 & -1 & 1 & 0 & 0 \\ 0 & 0 & 0 & 1 & 3 & -3 & 1 & 0 \\ 0 & 0 & -1 & 1 & 1 & -2 & 0 & 1 \end{array}\right] \xrightarrow[E_1(3,4)\times]{\text{exchange row 3 and row 4}}$$

$$\left[\begin{array}{cccc|cccc} 1 & 0 & 1 & 4 & 2 & -1 & 0 & 0 \\ 0 & 1 & 1 & -1 & -1 & 1 & 0 & 0 \\ 0 & 0 & -1 & 1 & 1 & -2 & 0 & 1 \\ 0 & 0 & 0 & 1 & 3 & -3 & 1 & 0 \end{array}\right] \xrightarrow[E_2(3;-1)\times]{\text{multiply row 3 by } -1}$$

$$\left[\begin{array}{cccc|cccc} 1 & 0 & 1 & 4 & 2 & -1 & 0 & 0 \\ 0 & 1 & 1 & -1 & -1 & 1 & 0 & 0 \\ 0 & 0 & \boxed{1} & -1 & -1 & 2 & 0 & -1 \\ 0 & 0 & 0 & 1 & 3 & -3 & 1 & 0 \end{array}\right] \xrightarrow[E_3(2,3;-1)E_3(1,3;-1)\times]{\text{clear col 3, pivot } (3,3)}$$

$$\left[\begin{array}{cccc|cccc} 1 & 0 & 0 & 5 & 3 & -3 & 0 & 1 \\ 0 & 1 & 0 & 0 & 0 & -1 & 0 & 1 \\ 0 & 0 & 1 & -1 & -1 & 2 & 0 & -1 \\ 0 & 0 & 0 & \boxed{1} & 3 & -3 & 1 & 0 \end{array}\right] \xrightarrow[E_3(3,4;1)E_3(1,4;-5)\times]{\text{clear col 4, pivot } (4,4)}$$

$$\left[\begin{array}{cccc|cccc} 1 & 0 & 0 & 0 & -12 & 12 & -5 & 1 \\ 0 & 1 & 0 & 0 & 0 & -1 & 0 & 1 \\ 0 & 0 & 1 & 0 & 2 & -1 & 1 & -1 \\ 0 & 0 & 0 & 1 & 3 & -3 & 1 & 0 \end{array}\right] = [I \mid A^{-1}].$$

The inverse matrix A^{-1} is given by the right half of the above matrix as

$$A^{-1} = \begin{bmatrix} -12 & 12 & -5 & 1 \\ 0 & -1 & 0 & 1 \\ 2 & -1 & 1 & -1 \\ 3 & -3 & 1 & 0 \end{bmatrix},$$

which is equal to the product of the elementary matrices used in the process as

$$\begin{aligned}A^{-1} = {} & E_3(3,4;1)E_3(1,4;-5) \times E_3(2,3;-1)E_3(1,3;-1) \\ & \times E_2(3;-1) \times E_1(3,4) \times E_3(4,2;-2)E_3(3,2;-3)E_3(1,2;-1) \\ & \times E_3(4,1;-1)E_3(2,1;-1).\end{aligned}$$

Note that the first (left-most) factor $E_3(3,4;1)E_3(1,4;-5)$ on the right-hand side of this expression corresponds to last step of the elimination process. It follows from this expression and Proposition 3.2 (Sec. 3.1.1) that

$$\begin{aligned}A = {} & E_3(2,1;1)E_3(4,1;1) \times E_3(1,2;1)E_3(3,2;3)E_3(4,2;2) \\ & \times E_1(3,4) \times E_2(3;-1) \times E_3(1,3;1)E_3(2,3;1) \\ & \times E_3(1,4;5)E_3(3,4;-1),\end{aligned}$$

which is a representation of A as a product of elementary matrices (Theorem 3.2(2)). ■

Example 3.10. In this example we consider the (singular) matrix A in (3.1). The elimination process for $[\,A \mid I\,]$ is as follows:

$$\left[\begin{array}{cccr|cccc} \boxed{1} & 1 & 2 & 3 & 1 & 0 & 0 & 0 \\ 1 & 2 & 3 & 2 & 0 & 1 & 0 & 0 \\ 0 & 3 & 3 & -2 & 0 & 0 & 1 & 0 \\ 1 & 3 & 4 & 2 & 0 & 0 & 0 & 1 \end{array}\right] \xrightarrow[E_3(4,1;-1)E_3(2,1;-1)\times]{\text{clear col 1, pivot } (1,1)}$$

$$\left[\begin{array}{cccr|rccc} 1 & 1 & 2 & 3 & 1 & 0 & 0 & 0 \\ 0 & \boxed{1} & 1 & -1 & -1 & 1 & 0 & 0 \\ 0 & 3 & 3 & -2 & 0 & 0 & 1 & 0 \\ 0 & 2 & 2 & -1 & -1 & 0 & 0 & 1 \end{array}\right] \xrightarrow[E_3(4,2;-2)E_3(3,2;-3)E_3(1,2;-1)\times]{\text{clear col 2, pivot } (2,2)}$$

$$\left[\begin{array}{cccr|rrcc} 1 & 0 & 1 & 4 & 2 & -1 & 0 & 0 \\ 0 & 1 & 1 & -1 & -1 & 1 & 0 & 0 \\ 0 & 0 & 0 & \boxed{1} & 3 & -3 & 1 & 0 \\ 0 & 0 & 0 & 1 & 1 & -2 & 0 & 1 \end{array}\right] \xrightarrow[E_3(4,3;-1)E_3(2,3;1)E_3(1,3;-4)\times]{\text{clear col 4, pivot } (3,4)}$$

$$\left[\begin{array}{cccc|rrrc} 1 & 0 & 1 & 0 & -10 & 11 & -4 & 0 \\ 0 & 1 & 1 & 0 & 2 & -2 & 1 & 0 \\ 0 & 0 & 0 & 1 & 3 & -3 & 1 & 0 \\ 0 & 0 & 0 & 0 & -2 & 1 & -1 & 1 \end{array}\right] = [\hat{A} \mid S].$$

At this point it is revealed that the given matrix A is singular. The 4×4 submatrix in the left coincides with the reduced echelon form $\hat{A}$ computed in Example 3.8, whereas the 4×4 submatrix in the right gives the transformation matrix S in Theorem 3.4. ■

Remark 3.5. In Theorem 1.5(2) in Sec. 1.3.2, we gave the following formula for the inverse matrix:

$$\left[\begin{array}{c|c} A & B \\ \hline C & D \end{array}\right]^{-1} = \left[\begin{array}{c|c} A^{-1}+A^{-1}BS^{-1}CA^{-1} & -A^{-1}BS^{-1} \\ \hline -S^{-1}CA^{-1} & S^{-1} \end{array}\right], \tag{3.9}$$

where A and $S = D - CA^{-1}B$ are assumed to be nonsingular. This formula can be derived in a natural way by means of elimination operations. The elimination operations can be described block-wise as follows:[3]

$$\left[\begin{array}{cc|cc} A & B & I & O \\ C & D & O & I \end{array}\right] \longrightarrow \left[\begin{array}{cc|cc} \boxed{I} & A^{-1}B & A^{-1} & O \\ C & D & O & I \end{array}\right]$$

$$\Longrightarrow \left[\begin{array}{cc|cc} I & A^{-1}B & A^{-1} & O \\ O & D-CA^{-1}B & -CA^{-1} & I \end{array}\right] = \left[\begin{array}{cc|cc} I & A^{-1}B & A^{-1} & O \\ O & S & -CA^{-1} & I \end{array}\right]$$

$$\longrightarrow \left[\begin{array}{cc|cc} I & A^{-1}B & A^{-1} & O \\ O & \boxed{I} & -S^{-1}CA^{-1} & S^{-1} \end{array}\right]$$

$$\Longrightarrow \left[\begin{array}{cc|cc} I & O & A^{-1}+A^{-1}BS^{-1}CA^{-1} & -A^{-1}BS^{-1} \\ O & I & -S^{-1}CA^{-1} & S^{-1} \end{array}\right].$$

The right half of the last matrix coincides with the right-hand side of (3.9). ■

[3] $\longrightarrow$ denotes a block-wise scaling (transformation to a unit matrix) and $\Longrightarrow$ a block-wise clearing.

Chapter 4

Rank

The rank of a matrix is closely related to the degrees of freedom of engineering systems. In this chapter the rank of a matrix is defined in three different ways, by subdeterminants, by column vectors, and by row vectors, which are then shown to be equivalent. Fundamental properties of the rank are presented, and examples from different disciplines of engineering are given to demonstrate the role of this concept in engineering.

4.1 Definitions of Rank

We define the concept of rank of a matrix. Three different definitions are introduced, independently of each other, namely, the definitions by subdeterminants in Sec. 4.1.1, by column vectors in Sec. 4.1.2, and by row vectors in Sec. 4.1.3. Then in Sec. 4.1.4 we show that the three definitions are equivalent. The rank is defined for a matrix of an arbitrary type (not necessarily square). In this section we assume that A is an $m \times n$ matrix.

4.1.1 *Definition by Subdeterminants*

Definition 4.1. The maximum order of a nonzero subdeterminant of A is called[1] the *rank* of A, and is denoted as $\mathrm{rank}^{(\mathrm{det})} A$. That is,

$$\mathrm{rank}^{(\mathrm{det})} A = \max\{k \mid |A|_{j_1,j_2,\ldots,j_k}^{i_1,i_2,\ldots,i_k} \neq 0 \ \text{ for some } \ i_1 < i_2 < \cdots < i_k \text{ and } j_1 < j_2 < \cdots < j_k\}, \tag{4.1}$$

where $\mathrm{rank}^{(\mathrm{det})} A = 0$ for $A = O_{m,n}$. ■

[1]At this moment it is more precise to call it "the rank defined by nonzero subdeterminants." The notation $\mathrm{rank}^{(\mathrm{det})}$ is introduced here for the sake of argument in this chapter, and not commonly used in the literature.

By (4.1) we can also say that the rank of a matrix A ($\neq O_{m,n}$) is the maximum order of a nonsingular submatrix of A.

Example 4.1. For the matrix

$$A = \begin{bmatrix} 1 & 1 & 2 & 3 \\ 1 & 2 & 3 & 2 \\ 0 & 3 & 3 & -2 \\ 1 & 3 & 4 & 2 \end{bmatrix}$$

in (3.1), we have $|A|_{234}^{123} \neq 0$ and $|A|_{1234}^{1234} = 0$. Therefore, $\operatorname{rank}^{(\det)} A = 3$. ■

Theorem 4.1.

(1) $\operatorname{rank}^{(\det)} A$ *remains invariant under elementary row transformations of* A.
(2) $\operatorname{rank}^{(\det)} A$ *remains invariant under elementary column transformations of* A.

Proof. We prove (2) first. Let E be an elementary matrix and $B = AE$. Express A and B column-wise as $A = [\boldsymbol{a}_1, \boldsymbol{a}_2, \ldots, \boldsymbol{a}_n]$ and $B = [\boldsymbol{b}_1, \boldsymbol{b}_2, \ldots, \boldsymbol{b}_n]$. If E represents an exchange of columns ($E = E_1(p, q)$) or a scalar multiplication ($E = E_2(p; b)$), we obviously have $\operatorname{rank}^{(\det)} B = \operatorname{rank}^{(\det)} A$.

In the following we assume that $E = E_3(p, q; c)$. Then we have

$$\boldsymbol{b}_q = \boldsymbol{a}_q + c\,\boldsymbol{a}_p, \qquad \boldsymbol{b}_j = \boldsymbol{a}_j \quad (j \neq q).$$

Let $k = \operatorname{rank}^{(\det)} A$. Then there exist row indices $i_1 < i_2 < \cdots < i_k$ and column indices $j_1 < j_2 < \cdots < j_k$ for which $|A|_{j_1, j_2, \ldots, j_k}^{i_1, i_2, \ldots, i_k} \neq 0$. Let $J = \{j_1, j_2, \ldots, j_k\}$, $\alpha = |A|_{j_1, j_2, \ldots, j_k}^{i_1, i_2, \ldots, i_k}$, and $\beta = |B|_{j_1, j_2, \ldots, j_k}^{i_1, i_2, \ldots, i_k}$. We have $\alpha \neq 0$.

- If $q \notin J$, we have $\beta = \alpha \neq 0$, and hence $\operatorname{rank}^{(\det)} B \geqq k$.
- If $q \in J$, $p \in J$, we have $\beta = \alpha \neq 0$ by Proposition 2.1(3) in Sec. 2.3.1, and hence $\operatorname{rank}^{(\det)} B \geqq k$.
- If $q \in J$, $p \notin J$, let $j_1' < j_2' < \cdots < j_k'$ be the ascending sequence obtained from $j_1 < j_2 < \cdots < j_k$ by deleting q and adding p. Let $\alpha' = |A|_{j_1', j_2', \ldots, j_k'}^{i_1, i_2, \ldots, i_k}$ and $\beta' = |B|_{j_1', j_2', \ldots, j_k'}^{i_1, i_2, \ldots, i_k}$. We have $\beta' = \alpha'$. On the other hand, we have $\beta = \alpha \pm c\alpha'$ by Theorem 2.3(3) in Sec. 2.3.1. If $\alpha' \neq 0$, we have $\beta' = \alpha' \neq 0$, and hence $\operatorname{rank}^{(\det)} B \geqq k$. If $\alpha' = 0$, we have $\beta = \alpha \neq 0$, and hence $\operatorname{rank}^{(\det)} B \geqq k$.

In either case we have $\text{rank}^{(\text{det})} B \geqq \text{rank}^{(\text{det})} A$. Since $A = BE_3(p, q; -c)$, we can show $\text{rank}^{(\text{det})} A \geqq \text{rank}^{(\text{det})} B$ in a similar manner.

(1) This follows from (2) applied to $A^\top$. □

Theorem 4.2. $\text{rank}^{(\text{det})} A$ *is equal to* r *in the rank normal form* (3.3) *of* A.

Proof. It follows from (3.3) and Theorem 4.1 that

$$\text{rank}^{(\text{det})} A = \text{rank}^{(\text{det})} \begin{bmatrix} I_r & O \\ O & O \end{bmatrix},$$

where the right-hand side is obviously equal to r. □

4.1.2 *Definition by Column Vectors*

Given a set of vectors $\boldsymbol{b}_1, \boldsymbol{b}_2, \ldots, \boldsymbol{b}_k$, a vector of the form of

$$c_1 \boldsymbol{b}_1 + c_2 \boldsymbol{b}_2 + \cdots + c_k \boldsymbol{b}_k \quad (c_1, c_2, \ldots, c_k \in K)$$

is called a *linear combination* of $\boldsymbol{b}_1, \boldsymbol{b}_2, \ldots, \boldsymbol{b}_k$ (cf., Definition 9.3 in Sec. 9.2.1). Vectors $\boldsymbol{b}_1, \boldsymbol{b}_2, \ldots, \boldsymbol{b}_k$ are said to be *linearly independent* if the following implication is true:

$$c_1 \boldsymbol{b}_1 + c_2 \boldsymbol{b}_2 + \cdots + c_k \boldsymbol{b}_k = \boldsymbol{0} \ (c_1, c_2, \ldots, c_k \in K) \implies c_1 = c_2 = \cdots = c_k = 0.$$

Otherwise, they are called *linearly dependent* (cf., Definition 9.10 in Sec. 9.3).

An $m \times n$ matrix A may be regarded as a collection of n column vectors, which we denote as $A = [\boldsymbol{a}_1, \boldsymbol{a}_2, \ldots, \boldsymbol{a}_n]$.

Definition 4.2. The maximum number of linearly independent column vectors of A is called[2] the *rank* of A, and is denoted as $\text{rank}^{(\text{col})} A$. That is,

$$\text{rank}^{(\text{col})} A = \max\{k \mid \boldsymbol{a}_{j_1}, \boldsymbol{a}_{j_2}, \ldots, \boldsymbol{a}_{j_k} \text{ are linearly independent for some } j_1 < j_2 < \cdots < j_k\}. \tag{4.2}$$

■

Example 4.2. In the matrix

$$A = \left[\begin{array}{c|c|c|c} 1 & 1 & 2 & 3 \\ 1 & 2 & 3 & 2 \\ 0 & 3 & 3 & -2 \\ 1 & 3 & 4 & 2 \end{array}\right] = [\boldsymbol{a}_1, \boldsymbol{a}_2, \boldsymbol{a}_3, \boldsymbol{a}_4]$$

[2] At this moment it is more precise to call it "the rank defined by linear independence of column vectors." The notation $\text{rank}^{(\text{col})}$ is introduced here for the sake of argument in this chapter, and not commonly used in the literature.

in (3.1), the three column vectors $\boldsymbol{a}_1, \boldsymbol{a}_2, \boldsymbol{a}_4$ are linearly independent, while $\boldsymbol{a}_1, \boldsymbol{a}_2, \boldsymbol{a}_3, \boldsymbol{a}_4$ are linearly dependent.[3] Therefore, $\mathrm{rank}^{(\mathrm{col})} A = 3$. ■

Theorem 4.3.

(1) $\mathrm{rank}^{(\mathrm{col})} A$ *remains invariant under elementary row transformations of* A.

(2) $\mathrm{rank}^{(\mathrm{col})} A$ *remains invariant under elementary column transformations of* A.

Proof. (1) Let E be an elementary matrix, $B = EA$, and $B = [\boldsymbol{b}_1, \boldsymbol{b}_2, \ldots, \boldsymbol{b}_n]$. We have $\boldsymbol{b}_j = E\boldsymbol{a}_j$ $(j = 1, 2, \ldots, n)$. Since

$$c_1 \boldsymbol{b}_{j_1} + c_2 \boldsymbol{b}_{j_2} + \cdots + c_k \boldsymbol{b}_{j_k} = E(c_1 \boldsymbol{a}_{j_1} + c_2 \boldsymbol{a}_{j_2} + \cdots + c_k \boldsymbol{a}_{j_k}),$$

we have the implication:

$$\begin{aligned} &\boldsymbol{a}_{j_1}, \boldsymbol{a}_{j_2}, \ldots, \boldsymbol{a}_{j_k} \text{ are linearly dependent} \\ &\Longrightarrow \boldsymbol{b}_{j_1}, \boldsymbol{b}_{j_2}, \ldots, \boldsymbol{b}_{j_k} \text{ are linearly dependent,} \end{aligned}$$

and the converse is also true by the nonsingularity of E. That is,

$$\begin{aligned} &\boldsymbol{a}_{j_1}, \boldsymbol{a}_{j_2}, \ldots, \boldsymbol{a}_{j_k} \text{ are linearly independent} \\ &\Longleftrightarrow \boldsymbol{b}_{j_1}, \boldsymbol{b}_{j_2}, \ldots, \boldsymbol{b}_{j_k} \text{ are linearly independent.} \end{aligned}$$

Therefore, $\mathrm{rank}^{(\mathrm{col})} A = \mathrm{rank}^{(\mathrm{col})} B$.

(2) Let E be an elementary matrix, $B = AE$, and $B = [\boldsymbol{b}_1, \boldsymbol{b}_2, \ldots, \boldsymbol{b}_n]$. If E represents an exchange of columns ($E = E_1(p,q)$) or a scalar multiplication ($E = E_2(p;b)$), we obviously have $\mathrm{rank}^{(\mathrm{col})} B = \mathrm{rank}^{(\mathrm{col})} A$.

In the following we assume $E = E_3(p,q;c)$. Then we have

$$\boldsymbol{b}_q = \boldsymbol{a}_q + c\,\boldsymbol{a}_p, \qquad \boldsymbol{b}_j = \boldsymbol{a}_j \quad (j \neq q).$$

Let $k = \mathrm{rank}^{(\mathrm{col})} A$. Then there exist k linearly independent column vectors $\boldsymbol{a}_{j_1}, \boldsymbol{a}_{j_2}, \ldots, \boldsymbol{a}_{j_k}$. Let $J = \{j_1, j_2, \ldots, j_k\}$.

- If $q \notin J$, the vectors $\boldsymbol{b}_j$ $(j \in J)$ are linearly independent, and hence $\mathrm{rank}^{(\mathrm{col})} B \geqq k$.
- If $q \in J$, $p \in J$, we have

$$\sum_{j \in J} c_j \boldsymbol{b}_j = (c_p + c c_q)\boldsymbol{a}_p + c_q \boldsymbol{a}_q + \sum_{j \in J \setminus \{p,q\}} c_j \boldsymbol{a}_j,$$

$$(c_p, c_q) = (0,0) \Longleftrightarrow (c_p + c c_q, c_q) = (0,0),$$

which shows that $\boldsymbol{b}_j$ $(j \in J)$ are linearly independent. Therefore, $\mathrm{rank}^{(\mathrm{col})} B \geqq k$.

---

[3] $1 \times \boldsymbol{a}_1 + 1 \times \boldsymbol{a}_2 + (-1) \times \boldsymbol{a}_3 + 0 \times \boldsymbol{a}_4 = \boldsymbol{0}$.

- If $q \in J$, $p \notin J$, let $J' = (J \setminus \{q\}) \cup \{p\}$, which denotes the set obtained from J by deleting q and adding p. If $\boldsymbol{a}_j$ $(j \in J')$ are linearly independent, so are $\boldsymbol{b}_j$ $(j \in J')$, and hence $\operatorname{rank}^{(\mathrm{col})} B \geqq k$. If $\boldsymbol{a}_j$ $(j \in J')$ are linearly dependent, there is a set of coefficients $(c_j \mid j \in J') \neq \mathbf{0}$ satisfying

$$c_p \boldsymbol{a}_p + \sum_{j \in J \setminus \{q\}} c_j \boldsymbol{a}_j = \mathbf{0}.$$

Here we must have $c_p \neq 0$, since otherwise $(c_p = 0)$ we have a contradiction to the linear independence of $\boldsymbol{a}_j$ $(j \in J)$. Furthermore we may assume $c_p = 1$, to obtain

$$\boldsymbol{a}_p = - \sum_{j \in J \setminus \{q\}} c_j \boldsymbol{a}_j.$$

To show the linear independence of $\boldsymbol{b}_j$ $(j \in J)$, consider their linear combination $\sum_{j \in J} d_j \boldsymbol{b}_j$, which can be rewritten into a linear combination of $\boldsymbol{a}_j$ $(j \in J)$ as follows:

$$\begin{aligned}\sum_{j \in J} d_j \boldsymbol{b}_j &= d_q(\boldsymbol{a}_q + c\,\boldsymbol{a}_p) + \sum_{j \in J \setminus \{q\}} d_j \boldsymbol{a}_j \\ &= d_q \left(\boldsymbol{a}_q - c \sum_{j \in J \setminus \{q\}} c_j \boldsymbol{a}_j \right) + \sum_{j \in J \setminus \{q\}} d_j \boldsymbol{a}_j \\ &= d_q \boldsymbol{a}_q + \sum_{j \in J \setminus \{q\}} (d_j - d_q c c_j) \boldsymbol{a}_j.\end{aligned}$$

If this is equal to $\mathbf{0}$, the linear independence of $\boldsymbol{a}_j$ $(j \in J)$ implies $d_q = 0$ and $d_j - d_q c c_j = 0$ $(j \in J \setminus \{q\})$, which is equivalent to $d_j = 0$ $(j \in J)$. Therefore, $\boldsymbol{b}_j$ $(j \in J)$ are linearly independent, and hence $\operatorname{rank}^{(\mathrm{col})} B \geqq k$.

In either case above we have $\operatorname{rank}^{(\mathrm{col})} B \geqq \operatorname{rank}^{(\mathrm{col})} A$. Since $A = BE_3(p, q; -c)$, we can show $\operatorname{rank}^{(\mathrm{col})} A \geqq \operatorname{rank}^{(\mathrm{col})} B$ in a similar manner. □

Theorem 4.4. $\operatorname{rank}^{(\mathrm{col})} A$ *is equal to* r *in the rank normal form* (3.3) *of* A.

Proof. It follows from (3.3) and Theorem 4.3 that

$$\operatorname{rank}^{(\mathrm{col})} A = \operatorname{rank}^{(\mathrm{col})} \begin{bmatrix} I_r & O \\ O & O \end{bmatrix},$$

where the right-hand side is obviously equal to r. □

4.1.3 *Definition by Row Vectors*

An $m \times n$ matrix A may be regarded as a collection of m row vectors, which we denote as

$$A = \left[\begin{array}{c} \hat{\boldsymbol{a}}_1^\top \\ \hline \hat{\boldsymbol{a}}_2^\top \\ \hline \vdots \\ \hline \hat{\boldsymbol{a}}_m^\top \end{array}\right].$$

Definition 4.3. The maximum number of linearly independent row vectors of A is called[4] the *rank* of A, and is denoted as $\mathrm{rank}^{(\mathrm{row})} A$. That is,

$$\mathrm{rank}^{(\mathrm{row})} A = \max\{k \mid \hat{\boldsymbol{a}}_{i_1}, \hat{\boldsymbol{a}}_{i_2}, \ldots, \hat{\boldsymbol{a}}_{i_k} \text{ are linearly independent for some } i_1 < i_2 < \cdots < i_k\}. \tag{4.3}$$

■

Example 4.3. In the matrix

$$A = \left[\begin{array}{cccc} 1 & 1 & 2 & 3 \\ \hline 1 & 2 & 3 & 2 \\ \hline 0 & 3 & 3 & -2 \\ \hline 1 & 3 & 4 & 2 \end{array}\right] = \left[\begin{array}{c} \hat{\boldsymbol{a}}_1^\top \\ \hline \hat{\boldsymbol{a}}_2^\top \\ \hline \hat{\boldsymbol{a}}_3^\top \\ \hline \hat{\boldsymbol{a}}_4^\top \end{array}\right]$$

in (3.1), the three row vectors $\hat{\boldsymbol{a}}_1, \hat{\boldsymbol{a}}_2, \hat{\boldsymbol{a}}_3$ are linearly independent, while $\hat{\boldsymbol{a}}_1, \hat{\boldsymbol{a}}_2, \hat{\boldsymbol{a}}_3, \hat{\boldsymbol{a}}_4$ are linearly dependent.[5] Therefore, $\mathrm{rank}^{(\mathrm{row})} A = 3$. ■

Theorem 4.5.

(1) $\mathrm{rank}^{(\mathrm{row})} A$ *remains invariant under elementary row transformations of* A.
(2) $\mathrm{rank}^{(\mathrm{row})} A$ *remains invariant under elementary column transformations of* A.

Proof. (1) This follows from Theorem 4.3(2) applied to $A^\top$.
(2) This follows from Theorem 4.3(1) applied to $A^\top$. □

Theorem 4.6. $\mathrm{rank}^{(\mathrm{row})} A$ *is equal to* r *in the rank normal form* (3.3) *of* A.

[4] At this moment it is more precise to call it "the rank defined by linear independence of row vectors." The notation $\mathrm{rank}^{(\mathrm{row})}$ is introduced here for the sake of argument in this chapter, and not commonly used in the literature.

[5] $2 \times \hat{\boldsymbol{a}}_1 + (-1) \times \hat{\boldsymbol{a}}_2 + 1 \times \hat{\boldsymbol{a}}_3 + (-1) \times \hat{\boldsymbol{a}}_4 = \mathbf{0}$.

Proof. It follows from (3.3) and Theorem 4.5 that

$$\operatorname{rank}^{(\mathrm{row})} A = \operatorname{rank}^{(\mathrm{row})} \begin{bmatrix} I_r & O \\ O & O \end{bmatrix},$$

where the right-hand side is obviously equal to r. □

4.1.4 *Equivalence of the Definitions*

Theorem 4.7.

$$\operatorname{rank}^{(\mathrm{det})} A = \operatorname{rank}^{(\mathrm{col})} A = \operatorname{rank}^{(\mathrm{row})} A.$$

Proof. Theorems 4.2, 4.4, and 4.6 show that all of these numbers are equal to r characterizing the rank normal form (3.3) of A. □

Definition 4.4. The common value in Theorem 4.7 is called the *rank* of matrix A, and denoted by $\operatorname{rank} A$. That is,

$$\operatorname{rank} A = \operatorname{rank}^{(\mathrm{det})} A = \operatorname{rank}^{(\mathrm{col})} A = \operatorname{rank}^{(\mathrm{row})} A. \tag{4.4}$$

■

Note that the three numbers $\operatorname{rank}^{(\mathrm{det})} A$, $\operatorname{rank}^{(\mathrm{col})} A$, and $\operatorname{rank}^{(\mathrm{row})} A$, as defined by (4.1), (4.2), and (4.3), are uniquely determined by the matrix A.

Definition 4.5. An $m \times n$ matrix A is said to be of *full row rank*, if it has rank m. Similarly, A is said to be of *full column rank*, if it has rank n. ■

Remark 4.1. The equation (4.4) in Definition 4.4 implies the equivalence of the conditions (b), (c), (f), and (g) for nonsingularity given in Theorem 1.3 in Sec. 1.3.1. Condition (c) reads $\operatorname{rank} A = n$. Condition $\det A \neq 0$ in (b) is equivalent to $\operatorname{rank}^{(\mathrm{det})} A = n$. Linear independence of the column vectors in (f) is equivalent to $\operatorname{rank}^{(\mathrm{col})} A = n$. Linear independence of the row vectors in (g) is equivalent to $\operatorname{rank}^{(\mathrm{row})} A = n$. ■

4.2 Uniqueness of Rank Normal Form and Reduced Echelon Form

The parameter r in the rank normal form (3.3) is not yet proved to be uniquely determined by the matrix A. In the proof of Theorem 3.1 in Sec. 3.2, there remained the possibility that r depends on the elementary transformations employed in the construction of the rank normal form.

The argument in Sec. 4.1 enables us to establish the uniqueness of the rank normal form.[6]

Theorem 4.8. *In the rank normal form* (3.3) *of matrix* A, *we have* $r = \operatorname{rank} A$. *Therefore, the rank normal form of* A *is uniquely determined by* A.

Proof. We obtain $r = \operatorname{rank} A$ from Theorems 4.2, 4.4 or 4.6. □

The reduced echelon form (Sec. 3.3) is also determined uniquely by the given matrix.

Theorem 4.9. *The reduced echelon form of a matrix* A *is determined uniquely by* A. *In particular, the parameter* r *is equal to* $\operatorname{rank} A$.

Proof. Let $B = [\boldsymbol{b}_1, \boldsymbol{b}_2, \ldots, \boldsymbol{b}_n]$ be the reduced echelon form of $A = [\boldsymbol{a}_1, \boldsymbol{a}_2, \ldots, \boldsymbol{a}_n]$. We have $B = SA$ for some nonsingular matrix S, and hence $\boldsymbol{b}_j = S\boldsymbol{a}_j$ $(j = 1, 2, \ldots, n)$. Concerning their linear combinations we have the relation

$$\boldsymbol{b}_j = \sum_{k=1}^{l} c_k \boldsymbol{b}_{j_k} \iff \boldsymbol{a}_j = \sum_{k=1}^{l} c_k \boldsymbol{a}_{j_k}, \tag{4.5}$$

where $c_1, c_2, \ldots, c_l \in K$.

The column indices $j_1^\circ < j_2^\circ < \cdots < j_r^\circ$ in the reduced echelon form B are characterized as the ascending sequence of the indices j satisfying the condition that[7]

$\boldsymbol{b}_j$ cannot be represented as a linear combination of $\boldsymbol{b}_1, \boldsymbol{b}_2, \ldots, \boldsymbol{b}_{j-1}$.

This is equivalent, by (4.5), to the condition that

$\boldsymbol{a}_j$ cannot be represented as a linear combination of $\boldsymbol{a}_1, \boldsymbol{a}_2, \ldots, \boldsymbol{a}_{j-1}$.

Therefore, the column indices $j_1^\circ < j_2^\circ < \cdots < j_r^\circ$ are determined by the given matrix A. In particular, $r = \operatorname{rank} B = \operatorname{rank} A$.

Next we consider the values of the entries of $B = (b_{ij})$. First recall that, for each $i = 1, 2, \ldots, r$, we have $\boldsymbol{b}_{j_i^\circ} = (0, \ldots, 0, \overset{\overset{i}{\vee}}{1}, 0, \ldots, 0)^\top$ (the ith unit vector). For each $j \notin \{j_1^\circ, j_2^\circ, \ldots, j_r^\circ\}$, let $k(j)$ denote the maximum value

[6] The elementary matrices in (3.3) and the transformation matrices S and T in (3.4) are not unique.

[7] This condition can be expressed as $\boldsymbol{b}_j \notin \operatorname{Span}(\boldsymbol{b}_1, \boldsymbol{b}_2, \ldots, \boldsymbol{b}_{j-1})$ using the notation to be introduced in Definition 9.4 in Sec. 9.2.1.

of k with $j_k^\circ < j$. Then $\boldsymbol{b}_j$ can be represented as a linear combination of the vectors $\boldsymbol{b}_{j_1^\circ}, \boldsymbol{b}_{j_2^\circ}, \ldots, \boldsymbol{b}_{j_{k(j)}^\circ}$ as

$$\boldsymbol{b}_j = \sum_{i=1}^{k(j)} b_{ij} \boldsymbol{b}_{j_i^\circ}$$

and this representation is unique.[8] By (4.5), this implies that the numbers b_{ij} are determined uniquely from A as the coefficients in the representation

$$\boldsymbol{a}_j = \sum_{i=1}^{k(j)} b_{ij} \boldsymbol{a}_{j_i^\circ}.$$

This completes the proof of Theorem 4.9. □

4.3 Properties of Rank

4.3.1 *Basic Properties*

Theorem 4.10. $\operatorname{rank} A \leqq \min(m, n)$ *for an* $m \times n$ *matrix* A.

Theorem 4.11. $\operatorname{rank} A = \operatorname{rank}(A^\top) = \operatorname{rank}(A^*) = \operatorname{rank}(\overline{A})$.

Theorem 4.12. *If* S *and* T *are nonsingular, then*

$$\operatorname{rank}(A) = \operatorname{rank}(SA) = \operatorname{rank}(AT) = \operatorname{rank}(SAT).$$

Proof. This follows from the facts that any nonsingular matrix can be represented as a product of elementary matrices (Theorem 3.2(2) in Sec. 3.2) and that the rank is invariant under elementary transformations (Theorems 4.2, 4.4, and 4.6). □

Theorem 4.13.

(1) $\operatorname{rank} A \leqq \operatorname{rank} \left[\, A \mid B \,\right] \leqq \operatorname{rank} A + \operatorname{rank} B$.

(2) $\operatorname{rank} A \leqq \operatorname{rank} \left[\begin{array}{c} A \\ \hline B \end{array}\right] \leqq \operatorname{rank} A + \operatorname{rank} B$.

(3) $\operatorname{rank} \left[\begin{array}{c|c} A & B \\ \hline C & D \end{array}\right] \geqq \operatorname{rank} A$.

(4) $\operatorname{rank} \left[\begin{array}{c|c} A & O \\ \hline O & B \end{array}\right] = \operatorname{rank} A + \operatorname{rank} B$.

[8] Vectors $\boldsymbol{b}_{j_1^\circ}, \boldsymbol{b}_{j_2^\circ}, \ldots, \boldsymbol{b}_{j_{k(j)}^\circ}$ form a basis (Sec. 9.4.1) of $\operatorname{Span}(\boldsymbol{b}_1, \boldsymbol{b}_2, \ldots, \boldsymbol{b}_j)$.

Proof. We prove $\operatorname{rank}[A \mid B] \leqq \operatorname{rank} A + \operatorname{rank} B$ only. Let $A = [\boldsymbol{a}_1, \boldsymbol{a}_2, \ldots, \boldsymbol{a}_m]$, $B = [\boldsymbol{b}_1, \boldsymbol{b}_2, \ldots, \boldsymbol{b}_n]$, and $r = \operatorname{rank}[A \mid B]$. From among $\boldsymbol{a}_1, \boldsymbol{a}_2, \ldots, \boldsymbol{a}_m, \boldsymbol{b}_1, \boldsymbol{b}_2, \ldots, \boldsymbol{b}_n$ we can choose r linearly independent vectors $\boldsymbol{a}_{j_1}, \boldsymbol{a}_{j_2}, \ldots, \boldsymbol{a}_{j_k}, \boldsymbol{b}_{j_{k+1}}, \boldsymbol{b}_{j_{k+2}}, \ldots, \boldsymbol{b}_{j_r}$. Since $\boldsymbol{a}_{j_1}, \boldsymbol{a}_{j_2}, \ldots, \boldsymbol{a}_{j_k}$ are linearly independent, we have $\operatorname{rank} A \geqq k$. Similarly, we have $\operatorname{rank} B \geqq r - k$, since $\boldsymbol{b}_{j_{k+1}}, \boldsymbol{b}_{j_{k+2}}, \ldots, \boldsymbol{b}_{j_r}$ are linearly independent. Therefore,

$$\operatorname{rank}[A \mid B] = r = k + (r - k) \leqq \operatorname{rank} A + \operatorname{rank} B.$$

□

Theorem 4.14.

(1) $\operatorname{rank}(AB) \leqq \operatorname{rank} B$.
(2) $\operatorname{rank}(AB) \leqq \operatorname{rank} A$.

Proof. Let A be an $l \times m$ matrix and B an $m \times n$ matrix. We give three different proofs for (1).[9] Then (2) can be proved as $\operatorname{rank}(AB) = \operatorname{rank}(B^\top A^\top) \leqq \operatorname{rank}(A^\top) = \operatorname{rank} A$ using Theorem 4.11 and applying (1) to $(B^\top, A^\top)$ instead of (A, B).

[Proof 1: By rank normal form] Let $\tilde{A} = SAT$ be the rank normal form of A, where S and T are nonsingular matrices, and let $\tilde{B} = T^{-1}B$. Since

$$\operatorname{rank}(AB) = \operatorname{rank}(\tilde{A}\tilde{B}), \qquad \operatorname{rank} \tilde{B} = \operatorname{rank} B,$$

we may assume that the given matrix A is already in the rank normal form, that is, $A = \begin{bmatrix} I_r & O \\ O & O \end{bmatrix}$. Let $B = \begin{bmatrix} B_{11} & B_{12} \\ B_{21} & B_{22} \end{bmatrix}$ accordingly. Then

$$AB = \begin{bmatrix} B_{11} & B_{12} \\ O & O \end{bmatrix},$$

which implies, by Theorem 4.13(2), that $\operatorname{rank}(AB) \leqq \operatorname{rank} B$.

[Proof 2: By linear independence of column vectors] Let $B = [\boldsymbol{b}_1, \boldsymbol{b}_2, \ldots, \boldsymbol{b}_n]$. Then $AB = [A\boldsymbol{b}_1, A\boldsymbol{b}_2, \ldots, A\boldsymbol{b}_n]$. Obviously we have the implication:

$$\begin{aligned} &\boldsymbol{b}_{j_1}, \boldsymbol{b}_{j_2}, \ldots, \boldsymbol{b}_{j_k} \text{ are linearly dependent} \\ &\quad \Longrightarrow A\boldsymbol{b}_{j_1}, A\boldsymbol{b}_{j_2}, \ldots, A\boldsymbol{b}_{j_k} \text{ are linearly dependent.} \end{aligned}$$

By the contrapositive of this implication we obtain $\operatorname{rank}(AB) \leqq \operatorname{rank} B$.

[9] Yet another proof based on the correspondence to linear mappings is given in Sec. 9.8 (Theorem 9.72).

[Proof 3: By block-matrices] We consider the rank of matrix $\left[\begin{array}{c|c} A & O \\ \hline -I_m & B \end{array}\right]$, where I_m denotes the unit matrix of order m. Since

$$\left[\begin{array}{c|c} I_m & A \\ \hline O & I_m \end{array}\right] \left[\begin{array}{c|c} A & O \\ \hline -I_m & B \end{array}\right] \left[\begin{array}{c|c} I_m & B \\ \hline O & I_n \end{array}\right] = \left[\begin{array}{c|c} O & AB \\ \hline -I_m & O \end{array}\right],$$

it follows from Theorem 4.12 and Theorem 4.13(4) that

$$\operatorname{rank}\left[\begin{array}{c|c} A & O \\ \hline -I_m & B \end{array}\right] = \operatorname{rank}\left[\begin{array}{c|c} O & AB \\ \hline -I_m & O \end{array}\right] = m + \operatorname{rank}(AB). \tag{4.6}$$

On the other hand, Theorem 4.13(1) shows

$$\operatorname{rank}\left[\begin{array}{c|c} A & O \\ \hline -I_m & B \end{array}\right] \leqq \operatorname{rank}\left[\begin{array}{c} A \\ \hline -I_m \end{array}\right] + \operatorname{rank}\left[\begin{array}{c} O \\ \hline B \end{array}\right] = m + \operatorname{rank} B.$$

Therefore, $\operatorname{rank}(AB) \leqq \operatorname{rank} B$. □

Theorem 4.15. $\operatorname{rank}(A+B) \leqq \operatorname{rank} A + \operatorname{rank} B$.

Proof. We prove this using block-matrices, while an alternative proof based on the correspondence to linear mappings is given in Sec. 9.8 (Theorem 9.73). Let A and B be $m \times n$ matrices, and I_m be the unit matrix of order m. We consider the rank of matrix $\left[\begin{array}{c|c} A & I_m \\ \hline B & -I_m \end{array}\right]$. Since

$$\left[\begin{array}{c|c} I_m & I_m \\ \hline O & I_m \end{array}\right] \left[\begin{array}{c|c} A & I_m \\ \hline B & -I_m \end{array}\right] \left[\begin{array}{c|c} I_n & O \\ \hline B & I_m \end{array}\right] = \left[\begin{array}{c|c} A+B & O \\ \hline O & -I_m \end{array}\right],$$

it follows from Theorem 4.12 and Theorem 4.13(4) that

$$\operatorname{rank}\left[\begin{array}{c|c} A & I_m \\ \hline B & -I_m \end{array}\right] = \operatorname{rank}\left[\begin{array}{c|c} A+B & O \\ \hline O & -I_m \end{array}\right] = \operatorname{rank}(A+B) + m.$$

On the other hand, it follows from Theorem 4.13(1), (2) that

$$\operatorname{rank}\left[\begin{array}{c|c} A & I_m \\ \hline B & -I_m \end{array}\right] \leqq \operatorname{rank}\left[\begin{array}{c} A \\ \hline B \end{array}\right] + \operatorname{rank}\left[\begin{array}{c} I_m \\ \hline -I_m \end{array}\right] \leqq \operatorname{rank} A + \operatorname{rank} B + m.$$

Therefore, $\operatorname{rank}(A+B) \leqq \operatorname{rank} A + \operatorname{rank} B$. □

Theorem 4.16.

$$\operatorname{rank}(A^*A) = \operatorname{rank}(AA^*) = \operatorname{rank} A.$$

In particular, if A is a real matrix, then

$$\operatorname{rank}(A^\top A) = \operatorname{rank}(AA^\top) = \operatorname{rank} A.$$

Proof. Let A be an $m \times n$ matrix, and $r = \operatorname{rank} A$. We give two different proofs.

[Proof 1: By reduced echelon form] By Theorem 3.4 in Sec. 3.3, we may assume that $SAP = \begin{bmatrix} I_r & Z \\ O & O \end{bmatrix}$, where S is a nonsingular matrix and P is a permutation matrix. Then we have

$$\begin{aligned} \operatorname{rank}(AA^*) &= \operatorname{rank}((SAP)(SAP)^*) \\ &= \operatorname{rank} \begin{bmatrix} I_r + ZZ^* & O \\ O & O \end{bmatrix} \\ &= \operatorname{rank}(I_r + ZZ^*). \end{aligned}$$

The matrix $I_r + ZZ^*$ has linearly independent column vectors[10] and hence its rank is equal to r. Therefore, $\operatorname{rank}(AA^*) = r = \operatorname{rank} A$. By replacing A with A^* in this equation, we obtain $\operatorname{rank}(A^*A) = \operatorname{rank}(A^*) = \operatorname{rank} A$.

[Proof 2: By Gramian] Let $A = [\boldsymbol{a}_1, \boldsymbol{a}_2, \ldots, \boldsymbol{a}_n]$. By choosing r linearly independent column vectors $\boldsymbol{a}_{j_1}, \boldsymbol{a}_{j_2}, \ldots, \boldsymbol{a}_{j_r}$, define an $m \times r$ matrix $B = [\boldsymbol{a}_{j_1}, \boldsymbol{a}_{j_2}, \ldots, \boldsymbol{a}_{j_r}]$. We consider[11] the determinant $\det(B^*B)$. By (2.21) we have

$$\det(B^*B) = \sum_{i_1 < i_2 < \cdots < i_r} \left| |B|^{i_1, i_2, \ldots, i_r}_{1, 2, \ldots, r} \right|^2. \tag{4.7}$$

Since $\operatorname{rank} B = r$, there is a nonzero term in the summation on the right-hand side, and hence $\det(B^*B) > 0$. This means that B^*B is a nonsingular submatrix of order r of A^*A, and hence $\operatorname{rank}(A^*A) \geqq r$. The reverse inequality $\operatorname{rank}(A^*A) \leqq r$ is obvious (cf., Theorem 4.14 in Sec. 4.3.1). Therefore, $\operatorname{rank}(A^*A) = r = \operatorname{rank} A$. By replacing A with A^* in this equation, we obtain $\operatorname{rank}(AA^*) = \operatorname{rank}(A^*) = \operatorname{rank} A$. □

Theorem 4.17. $\operatorname{rank}(A \otimes B) = (\operatorname{rank} A) \cdot (\operatorname{rank} B)$.

Proof. Let $r = \operatorname{rank} A$ and $s = \operatorname{rank} B$, and consider the rank normal forms of A and B, which are given as

$$SAT = \begin{bmatrix} I_r & O \\ O & O \end{bmatrix}, \qquad UBV = \begin{bmatrix} I_s & O \\ O & O \end{bmatrix}$$

with nonsingular matrices S, T, U, and V. In the identity

$$(S \otimes U)(A \otimes B)(T \otimes V) = (SAT) \otimes (UBV) = \begin{bmatrix} I_r & O \\ O & O \end{bmatrix} \otimes \begin{bmatrix} I_s & O \\ O & O \end{bmatrix},$$

---

[10] If $(I_r + ZZ^*)\mathbf{c} = \mathbf{0}$, the multiplication by $\mathbf{c}^*$ from the left yields $\|\mathbf{c}\|_2{}^2 + \|Z^*\mathbf{c}\|_2{}^2 = 0$, which shows $\mathbf{c} = \mathbf{0}$.

[11] $\det(B^*B)$ is the Gramian of the vectors $\boldsymbol{a}_{j_1}, \boldsymbol{a}_{j_2}, \ldots, \boldsymbol{a}_{j_r}$ (cf., Remark 2.10 in Sec. 2.5.3).

the matrices $S \otimes U$ and $T \otimes V$ are nonsingular by Theorem 2.7(4) in Sec. 2.4, and the right-most matrix is of rank rs as it is a diagonal matrix with rs ones on the diagonal. Therefore, $\mathrm{rank}\,(A \otimes B) = rs = (\mathrm{rank}\,A) \cdot (\mathrm{rank}\,B)$. □

4.3.2 *Submodularity*

The ranks of submatrices of a matrix are not independent of each other but are mutually related. Let $A = [\boldsymbol{a}_1, \boldsymbol{a}_2, \ldots, \boldsymbol{a}_n]$ be an $m \times n$ matrix. For a set of row indices $I = \{i_1, i_2, \ldots, i_k\}$ and a set of column indices $J = \{j_1, j_2, \ldots, j_l\}$, the corresponding submatrix of A is denoted by $A[I, J]$. Let $\rho(J)$ denote the rank of $A[I, J]$ with $I = \{1, 2, \ldots, m\}$. That is,

$$\rho(J) = \mathrm{rank}\,A[\{1, 2, \ldots, m\}, J] = \mathrm{rank}\,[\boldsymbol{a}_{j_1}, \boldsymbol{a}_{j_2}, \ldots, \boldsymbol{a}_{j_l}].$$

This function ρ, called the *rank function*, describes the linear independence structure among column vectors.

Theorem 4.18. *The rank function ρ has properties* (a), (b), *and* (c) *below.*[12]

(a) $0 \leqq \rho(J) \leqq |J|$.
(b) $J_1 \subseteqq J_2 \implies \rho(J_1) \leqq \rho(J_2)$.
(c) $\rho(J_1) + \rho(J_2) \geqq \rho(J_1 \cup J_2) + \rho(J_1 \cap J_2)$.

Proof. (a) and (b) are obvious.

(c) Permute the columns of A so that the members of $J_1 \cap J_2$ appear first, next those of $J_1 \setminus J_2$, then those of $J_2 \setminus J_1$, and finally those not contained in $J_1 \cup J_2$. Let $\tilde{A}$ be the resulting matrix. Since the column set of $\tilde{A}$ can be identified with that of A through the permutation, we may regard ρ as the rank function of $\tilde{A}$. Let $\hat{A}$ denote the reduced echelon form of $\tilde{A}$. Since the linear independence among the column vectors are invariant under elementary row transformations (Theorem 4.3(1) in Sec. 4.1.2), the rank function of $\hat{A}$ is also given by ρ. Let

$$a = \rho(J_1 \cap J_2), \qquad b = \rho(J_1), \qquad c = \rho(J_1 \cup J_2),$$

for which we have $a \leqq b \leqq c$.

Let $j_1^\circ < j_2^\circ < \cdots < j_r^\circ$ be the characteristic column indices in the reduced echelon form $\hat{A}$. By construction we have

$$\{j_1^\circ, \ldots, j_a^\circ\} \subseteqq J_1 \cap J_2, \quad \{j_{a+1}^\circ, \ldots, j_b^\circ\} \subseteqq J_1 \setminus J_2, \quad \{j_{b+1}^\circ, \ldots, j_c^\circ\} \subseteqq J_2 \setminus J_1.$$

[12]The concept of matroid is defined by these properties [68, 73] (cf., Sec. 9.9).

Let $J_2' = \{j_1^\circ, \ldots, j_a^\circ\} \cup \{j_{b+1}^\circ, \ldots, j_c^\circ\}$. Since $J_2 \supseteqq J_2'$ and the column vectors of $\hat{A}$ in J_2', being unit vectors, are linearly independent, we obtain

$$\rho(J_2) \geqq \rho(J_2') = |J_2'| = a + (c - b) = \rho(J_1 \cap J_2) + \rho(J_1 \cup J_2) - \rho(J_1).$$

Thus we obtain the desired inequality. □

An inequality of the form in (c) above is called a *submodular inequality*, and the property expressed by this inequality is referred to as *submodularity*.

The submodularity of function ρ implies the following inequality for $\operatorname{rank} A[I, J]$ representing the rank of submatrices of A.

Theorem 4.19. *The following inequality holds for* $\operatorname{rank} A[I, J]$:

$$\begin{aligned} &\operatorname{rank} A[I_1, J_1] + \operatorname{rank} A[I_2, J_2] \\ &\geqq \operatorname{rank} A[I_1 \cap I_2, J_1 \cup J_2] + \operatorname{rank} A[I_1 \cup I_2, J_1 \cap J_2]. \end{aligned}$$

Proof. Let $R = \{1, 2, \ldots, m\}$ and $C = \{1, 2, \ldots, n\}$ denote the sets of row and column indices of A, respectively. Consider an $m \times (m + n)$ matrix

$$\tilde{A} = [I_m \mid A],$$

which contains the unit matrix I_m of order m to the left of A. The column set of $\tilde{A}$ can be identified with $R \cup C$. Let $\tilde{\rho}$ be the rank function associated with $\tilde{A}$, where $\tilde{\rho}$ is defined on $R \cup C$. The rank of a submatrix $A[I, J]$ can be expressed in terms of $\tilde{\rho}$ as

$$\operatorname{rank} A[I, J] = \tilde{\rho}((R \setminus I) \cup J) - (m - |I|)$$

(Fig. 4.1). With this relation, the submodular inequality for $\tilde{\rho}$ translates into the desired inequality. □

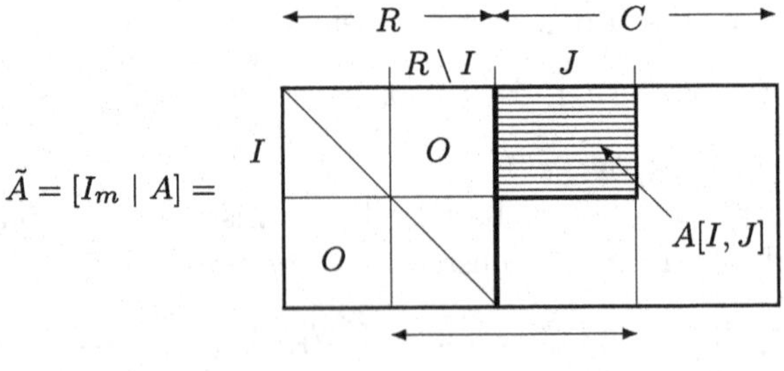

Fig. 4.1 Relation between $\operatorname{rank} A[I, J]$ and $\tilde{\rho}$ in the proof of Theorem 4.19.

As an application of Theorem 4.19 we show two inequalities.[13]

Theorem 4.20. *Let A be an $l \times m$ matrix, B an $m \times n$ matrix, and C an $n \times k$ matrix.*

(1) Sylvester inequality:

$$\operatorname{rank} A + \operatorname{rank} B \leqq \operatorname{rank}(AB) + m. \tag{4.8}$$

(2) Frobenius inequality:

$$\operatorname{rank}(AB) + \operatorname{rank}(BC) \leqq \operatorname{rank}(ABC) + \operatorname{rank} B. \tag{4.9}$$

Proof. (1) Consider

$$\begin{bmatrix} A & O \\ -I_m & B \end{bmatrix},$$

where I_m denotes the unit matrix of order m. The rank of this matrix is equal to $m + \operatorname{rank}(AB)$ by (4.6), whereas Theorem 4.19 implies

$$\operatorname{rank}\begin{bmatrix} A & O \\ -I_m & B \end{bmatrix} + \operatorname{rank}\begin{bmatrix} O \end{bmatrix} \geqq \operatorname{rank}\begin{bmatrix} A & O \end{bmatrix} + \operatorname{rank}\begin{bmatrix} O \\ B \end{bmatrix}.$$

Therefore, $m + \operatorname{rank}(AB) \geqq \operatorname{rank} A + \operatorname{rank} B$.

(2) Consider

$$\begin{bmatrix} A & O & O \\ -I_m & B & O \\ O & -I_n & C \end{bmatrix},$$

where I_m and I_n denote the unit matrices of orders m and n, respectively. Theorem 4.19 implies

$$\operatorname{rank}\begin{bmatrix} A & O & O \\ -I_m & B & O \\ O & -I_n & C \end{bmatrix} + \operatorname{rank}\begin{bmatrix} O & O \\ B & O \end{bmatrix}$$

$$\geqq \operatorname{rank}\begin{bmatrix} A & O & O \\ -I_m & B & O \end{bmatrix} + \operatorname{rank}\begin{bmatrix} O & O \\ B & O \\ -I_n & C \end{bmatrix}.$$

The left-hand side of this inequality is equal to

$$(m + n + \operatorname{rank}(ABC)) + \operatorname{rank} B,$$

and the right-hand side to

$$(m + \operatorname{rank}(AB)) + (n + \operatorname{rank}(BC)).$$

Therefore, we have (4.9). □

[13] An alternative proof of Theorem 4.20 is given in Sec. 9.8 (Theorem 9.74) on the basis of the correspondence to linear mappings.

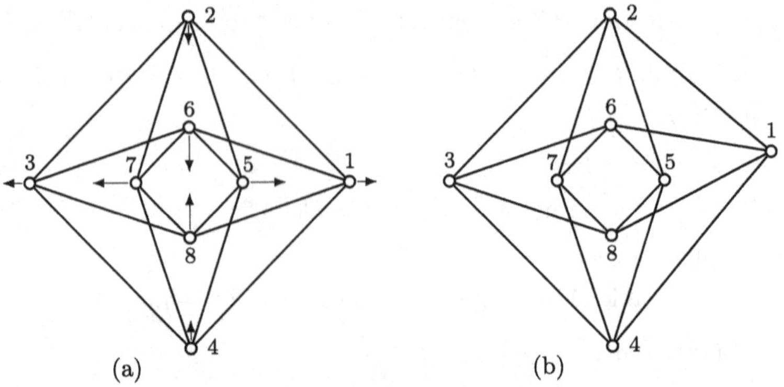

Fig. 4.2 Planar bar-joint structures: (a) with symmetry, (b) without symmetry.

4.4 Engineering Significances of Rank

Two examples are given to illustrate the significance of the concept of rank in engineering context. The rank corresponds to the degrees-of-freedom.

4.4.1 *Degrees-of-Freedom of Linkage Mechanisms*

Consider the planar bar-joint structures (consisting of bars connected by joints) shown in Fig. 4.2, where the links (bars) are assumed to be rigid. We are interested in whether these structures admit deformation to change their shapes[14] (that is, whether they are a *linkage mechanism* or not). The structure in (a) is symmetric and the structure in (b) is obtained from the symmetric structure by slightly perturbing the location of node 1. It will turn out that the symmetric structure (a) can change its shape, while the asymmetric structure (b) cannot. We can reveal this difference by investigating the rank of a certain matrix associated with the structures.

Suppose that there are n nodes (joints) and their locations are described by

$$\boldsymbol{p}_i = (x_i, y_i)^\top \qquad (i = 1, 2, \ldots, n).$$

Let L denote the set of pairs (i, j) of node i and node j $(> i)$ that are connected by a link. In Fig. 4.2 we have $n = 8$ and $|L| = 16$. If there is a link of length l_{ij} between node i and node j, we have an equation or a constraint

$$\|\boldsymbol{p}_i - \boldsymbol{p}_j\|_2{}^2 = (\boldsymbol{p}_i - \boldsymbol{p}_j, \boldsymbol{p}_i - \boldsymbol{p}_j) = l_{ij}{}^2,$$

[14]We allow the links to cross each other.

where $(\cdot,\cdot)$ means the (standard) inner product. Denoting the displacements of the nodes by

$$(\Delta\boldsymbol{p}_1, \Delta\boldsymbol{p}_2, \ldots, \Delta\boldsymbol{p}_n),$$

we can rewrite the above constraints to

$$((\boldsymbol{p}_i + \Delta\boldsymbol{p}_i) - (\boldsymbol{p}_j + \Delta\boldsymbol{p}_j), (\boldsymbol{p}_i + \Delta\boldsymbol{p}_i) - (\boldsymbol{p}_j + \Delta\boldsymbol{p}_j)) = {l_{ij}}^2 \quad ((i,j) \in L),$$

which, when linearized, gives rise to a system of linear equations

$$(\boldsymbol{p}_i - \boldsymbol{p}_j, \Delta\boldsymbol{p}_i - \Delta\boldsymbol{p}_j) = 0 \qquad ((i,j) \in L)$$

in $(\Delta\boldsymbol{p}_1, \Delta\boldsymbol{p}_2, \ldots, \Delta\boldsymbol{p}_n)$. For our structures in Fig. 4.2, this system of equations is given as[15]

$$\left[\begin{array}{cccc|cccc}
\boldsymbol{p}_1-\boldsymbol{p}_2 & \boldsymbol{p}_2-\boldsymbol{p}_1 & & & & & & \\
 & \boldsymbol{p}_2-\boldsymbol{p}_3 & \boldsymbol{p}_3-\boldsymbol{p}_2 & & & & & \\
 & & \boldsymbol{p}_3-\boldsymbol{p}_4 & \boldsymbol{p}_4-\boldsymbol{p}_3 & & & & \\
\boldsymbol{p}_1-\boldsymbol{p}_4 & & & \boldsymbol{p}_4-\boldsymbol{p}_1 & & & & \\
\hline
 & & & & \boldsymbol{p}_5-\boldsymbol{p}_6 & \boldsymbol{p}_6-\boldsymbol{p}_5 & & \\
 & & & & & \boldsymbol{p}_6-\boldsymbol{p}_7 & \boldsymbol{p}_7-\boldsymbol{p}_6 & \\
 & & & & & & \boldsymbol{p}_7-\boldsymbol{p}_8 & \boldsymbol{p}_8-\boldsymbol{p}_7 \\
 & & & & \boldsymbol{p}_5-\boldsymbol{p}_8 & & & \boldsymbol{p}_8-\boldsymbol{p}_5 \\
\hline
\boldsymbol{p}_1-\boldsymbol{p}_6 & & & & & \boldsymbol{p}_6-\boldsymbol{p}_1 & & \\
\boldsymbol{p}_1-\boldsymbol{p}_8 & & & & & & & \boldsymbol{p}_8-\boldsymbol{p}_1 \\
 & \boldsymbol{p}_2-\boldsymbol{p}_5 & & & \boldsymbol{p}_5-\boldsymbol{p}_2 & & & \\
 & \boldsymbol{p}_2-\boldsymbol{p}_7 & & & & & \boldsymbol{p}_7-\boldsymbol{p}_2 & \\
 & & \boldsymbol{p}_3-\boldsymbol{p}_6 & & & \boldsymbol{p}_6-\boldsymbol{p}_3 & & \\
 & & \boldsymbol{p}_3-\boldsymbol{p}_8 & & & & & \boldsymbol{p}_8-\boldsymbol{p}_3 \\
 & & & \boldsymbol{p}_4-\boldsymbol{p}_5 & \boldsymbol{p}_5-\boldsymbol{p}_4 & & & \\
 & & & \boldsymbol{p}_4-\boldsymbol{p}_7 & & & \boldsymbol{p}_7-\boldsymbol{p}_4 &
\end{array}\right]
\begin{bmatrix} \Delta\boldsymbol{p}_1 \\ \Delta\boldsymbol{p}_2 \\ \Delta\boldsymbol{p}_3 \\ \Delta\boldsymbol{p}_4 \\ \Delta\boldsymbol{p}_5 \\ \Delta\boldsymbol{p}_6 \\ \Delta\boldsymbol{p}_7 \\ \Delta\boldsymbol{p}_8 \end{bmatrix}
=
\left[\begin{array}{c} 0 \\ 0 \\ 0 \\ 0 \\ \hline 0 \\ 0 \\ 0 \\ 0 \\ \hline 0 \\ 0 \\ 0 \\ 0 \\ 0 \\ 0 \\ 0 \\ 0 \end{array}\right].$$

Denote the coefficient matrix above by A, which is a 16×16 matrix for our structures in Fig. 4.2 and an $|L| \times (2n)$ matrix in general.

For the symmetric structure in Fig. 4.2(a), we have

$$\begin{array}{llll}
\boldsymbol{p}_1 = (a,0)^\top, & \boldsymbol{p}_2 = (0,a)^\top, & \boldsymbol{p}_3 = (-a,0)^\top, & \boldsymbol{p}_4 = (0,-a)^\top, \\
\boldsymbol{p}_5 = (b,0)^\top, & \boldsymbol{p}_6 = (0,b)^\top, & \boldsymbol{p}_7 = (-b,0)^\top, & \boldsymbol{p}_8 = (0,-b)^\top
\end{array}$$

---

[15] In this expression, $\boldsymbol{p}_i$ should be regarded as a row vector (x_i, y_i), although it was introduced as a column vector $(x_i, y_i)^\top$. Note, however, that $\Delta\boldsymbol{p}_i$ is a column vector.

($a > b > 0$) and the matrix A is given by

$$A = \left[\begin{array}{cccccccc|cccccccc}
a & -a & -a & a & & & & & & & & & & & & \\
 & & a & a & -a & -a & & & & & & & & & & \\
 & & & & -a & a & a & -a & & & & & & & & \\
a & a & & & & & -a & -a & & & & & & & & \\
\hline
 & & & & & & & & b & -b & -b & b & & & & \\
 & & & & & & & & & & b & b & -b & -b & & \\
 & & & & & & & & & & & & -b & b & b & -b \\
 & & & & & & & & b & b & & & & & -b & -b \\
\hline
a & -b & & & & & & & & & -a & b & & & & \\
a & b & & & & & & & & & & & & & -a & -b \\
 & & -b & a & & & & & b & -a & & & & & & \\
 & & b & a & & & & & & & & & -b & -a & & \\
 & & & & -a & -b & & & & & a & b & & & & \\
 & & & & -a & b & & & & & & & & & a & -b \\
 & & & & & & -b & -a & b & a & & & & & & \\
 & & & & & & b & -a & & & & & -b & a & &
\end{array}\right].$$

We have $\operatorname{rank} A = 12$ whenever $a > b > 0$. This implies that the number of degrees of freedom of the infinitesimal displacements $\Delta \boldsymbol{p}_i$ ($i = 1, 2, \ldots, n$) is equal to $2n - \operatorname{rank} A = 16 - 12 = 4$ (see Sec. 5.2). Among the four degrees of freedom, three of them:

translation in direction x: $(1, 0 \mid 1, 0 \mid 1, 0 \mid 1, 0 \mid 1, 0 \mid 1, 0 \mid 1, 0 \mid 1, 0)^\top$,
translation in direction y: $(0, 1 \mid 0, 1 \mid 0, 1 \mid 0, 1 \mid 0, 1 \mid 0, 1 \mid 0, 1 \mid 0, 1)^\top$,
infinitesimal rotation: $(0, a \mid -a, 0 \mid 0, -a \mid a, 0 \mid 0, b \mid -b, 0 \mid 0, -b \mid b, 0)^\top$,

correspond to rigid-body motions that do not change the shape of the structure, and the remaining one:

$$(b, 0 \mid 0, -b \mid -b, 0 \mid 0, b \mid a, 0 \mid 0, -a \mid -a, 0 \mid 0, a)^\top$$

corresponds to an infinitesimal displacement[16] that changes the shape of the structure, depicted by arrows in Fig. 4.2(a).

Figure 4.2(b) shows an asymmetric structure that is obtained by changing the location of node 1 to $\boldsymbol{p}_1 = (a, c)^\top$ with $c > 0$. It can be verified that $\operatorname{rank} A = 13$ whenever $a > b > 0$ and $c > 0$. This shows that the number of degrees of freedom of the infinitesimal displacements is equal to 3. That is, only the rigid-body motions without affecting the shape of the structure is possible and there is no displacement that changes the shape of the structure.

[16] It can be shown that a finite displacement is also possible in this example.

Thus the degrees-of-freedom of the possible infinitesimal displacement can be determined by computing the rank of a matrix. Methods of analysis for systems with symmetry are explained in Chapter 7 of Volume II [9].

4.4.2 *Controllability of Control Systems*

In control theory [50, 52, 60, 61], a dynamical system (linear time-invariant system) is described by a system of differential equations

$$\frac{\mathrm{d}\boldsymbol{x}}{\mathrm{d}t}(t) = A\boldsymbol{x}(t) + B\boldsymbol{u}(t), \tag{4.10}$$

where $\boldsymbol{x}(t)$ is the state-vector at time t and $\boldsymbol{u}(t)$ is the input-vector representing the control. If $\boldsymbol{x}(t)$ is n-dimensional and $\boldsymbol{u}(t)$ is m-dimensional, A in (4.10) is a square matrix of order n and B is an $n \times m$ matrix. The equation (4.10) is called the *state–space equation.*

Controllability is one of the important properties required of a control system. This property says, roughly, that from an arbitrarily given initial state $\boldsymbol{x}(0)$ it is possible to make the state-vector the zero vector $\mathbf{0}$ within finite time by choosing appropriate input $\boldsymbol{u}(t)$. It is known that a necessary and sufficient condition for controllability is given in terms of the rank of an $n \times (mn)$ matrix $[B \mid AB \mid A^2B \mid \cdots \mid A^{n-1}B]$. That is, a system is *controllable* if and only if

$$\operatorname{rank}[B \mid AB \mid A^2B \mid \cdots \mid A^{n-1}B] = n. \tag{4.11}$$

It is noteworthy that a property of a continuous-time dynamical system can be expressed in such a concrete form using the rank of a matrix.

Chapter 5

Systems of Linear Equations

In this chapter we discuss the existence and uniqueness of solutions to a system of linear equations with the aid of the rank normal form and the reduced echelon form. A computational procedure based on elimination operations and a solution formula called Cramer's rule are presented. A system of equations arising from discretization of a differential equation (the Laplace equation) is investigated with respect to its unique solvability. We also cover the Sylvester equations and the Lyapunov equations, which are important in engineering applications.

5.1 Existence and Uniqueness of Solutions

In this chapter we discuss the existence and uniqueness of solution $\boldsymbol{x}$ to a system of linear equations

$$A\boldsymbol{x} = \boldsymbol{b}, \tag{5.1}$$

where matrix A and vector $\boldsymbol{b}$ are given and the components of vector $\boldsymbol{x}$ are unknown variables. If A is an $m \times n$ matrix and $\boldsymbol{b}$ is an m-dimensional vector, we have

$$\left\{ \begin{array}{rcl} a_{11}x_1 + a_{12}x_2 + \cdots + a_{1n}x_n & = & b_1, \\ a_{21}x_1 + a_{22}x_2 + \cdots + a_{2n}x_n & = & b_2, \\ \vdots & & \vdots \\ a_{m1}x_1 + a_{m2}x_2 + \cdots + a_{mn}x_n & = & b_m \end{array} \right.$$

consisting of m equations in n unknown variables. This is called a *system of linear equations* or *simultaneous linear equations.*

The following theorem states a most fundamental fact about the unique solvability of the equation $A\boldsymbol{x} = \boldsymbol{b}$ for varying $\boldsymbol{b}$ in the case where there are as many equations as unknowns (*i.e.*, when $m = n$).

Theorem 5.1. *The following three conditions* (a) *to* (c) *are equivalent for a square matrix A.*

(a) *For each vector $\boldsymbol{b}$, $A\boldsymbol{x} = \boldsymbol{b}$ has a unique solution $\boldsymbol{x}$.*
(b) *For each vector $\boldsymbol{b}$, $A\boldsymbol{x} = \boldsymbol{b}$ has a solution $\boldsymbol{x}$.*
(c) *A is a nonsingular matrix.*

Proof. [(a) $\Rightarrow$ (b)] This implication is obvious.

[(b) $\Rightarrow$ (c)] Assume that A is an $n \times n$ matrix. For each $i = 1, 2, \ldots, n$, let $\boldsymbol{x}_i$ be a solution $\boldsymbol{x}$ for the ith unit vector $\boldsymbol{b} = (0, \ldots, 0, \overset{\overset{i}{\vee}}{1}, 0, \ldots, 0)^\top$. The matrix $X = [\boldsymbol{x}_1, \boldsymbol{x}_2, \ldots, \boldsymbol{x}_n]$ formed by these solutions satisfies $AX = I_n$. This implies that A is nonsingular (cf., Theorem 1.2(2) in Sec. 1.3.1 and Remark 2.11 in Sec. 2.6).

[(c) $\Rightarrow$ (a)] Let $\boldsymbol{b}$ be any vector. The vector $\boldsymbol{x} = A^{-1}\boldsymbol{b}$ satisfies $A\boldsymbol{x} = \boldsymbol{b}$, that is, the system of equations $A\boldsymbol{x} = \boldsymbol{b}$ has a solution. Suppose that $\boldsymbol{y}$ is also a solution. It follows from $A\boldsymbol{y} = \boldsymbol{b}$ and $A\boldsymbol{x} = \boldsymbol{b}$ that $A\boldsymbol{y} = A\boldsymbol{x}$. Multiplication with A^{-1} from the left shows $\boldsymbol{y} = \boldsymbol{x}$. □

Next we go on to investigate the solvability of a system of equations $A\boldsymbol{x} = \boldsymbol{b}$ with an $m \times n$ matrix A, which is not necessarily square. It turns out to be convenient to introduce a terminology for the set of the vectors $\boldsymbol{b}$ for which the system of equations $A\boldsymbol{x} = \boldsymbol{b}$ admits a solution $\boldsymbol{x}$.

Definition 5.1. For an $m \times n$ matrix $A = [\boldsymbol{a}_1, \boldsymbol{a}_2, \ldots, \boldsymbol{a}_n]$, the set of m-dimensional vectors defined by

$$\operatorname{Im} A = \{\boldsymbol{b} \mid \boldsymbol{b} = A\boldsymbol{x} \ \text{ for some } \boldsymbol{x}\} \tag{5.2}$$

is called the *image* (or *range*)[1] of A. $\operatorname{Im} A$ coincides with the set of all linear combinations $\displaystyle\sum_{j=1}^{n} x_j \boldsymbol{a}_j$ of the column vectors of A. ■

Conditions for the existence of a solution $\boldsymbol{x}$ to the system of equations $A\boldsymbol{x} = \boldsymbol{b}$ are given in the following theorem.[2]

Theorem 5.2. *The following four conditions* (a) *to* (d) *are equivalent for an $m \times n$ matrix A and an m-dimensional vector $\boldsymbol{b}$:*

(a) *$A\boldsymbol{x} = \boldsymbol{b}$ has a solution $\boldsymbol{x}$.*

[1] $\operatorname{Im} A$ is also called the *image space* as it forms a linear subspace of K^m (cf., Sec. 9.5).
[2] Note that vector $\boldsymbol{b}$ is also given in Theorem 5.2, which is not the case in Theorem 5.1.

(b) *For any vector* $\boldsymbol{y}$, $\boldsymbol{y}^\top A = \boldsymbol{0}^\top$ *implies* $\boldsymbol{y}^\top \boldsymbol{b} = 0$.
(c) $\operatorname{rank} A = \operatorname{rank}[A \mid \boldsymbol{b}]$.
(d) $\boldsymbol{b} \in \operatorname{Im} A$.

Proof. For any nonsingular matrices S and T, the system of equations $A\boldsymbol{x} = \boldsymbol{b}$ can be rewritten as $(SAT)(T^{-1}\boldsymbol{x}) = S\boldsymbol{b}$. In other words, $A\boldsymbol{x} = \boldsymbol{b}$ can be transformed to an equivalent system $\tilde{A}\tilde{\boldsymbol{x}} = \tilde{\boldsymbol{b}}$ with

$$\tilde{A} = SAT, \qquad \tilde{\boldsymbol{x}} = T^{-1}\boldsymbol{x}, \qquad \tilde{\boldsymbol{b}} = S\boldsymbol{b}. \tag{5.3}$$

With a suitable choice of nonsingular matrices S and T, we can assume, by Theorem 3.3 in Sec. 3.2, that the transformed matrix $\tilde{A} = SAT$ is in the rank normal form:

$$\tilde{A} = \left[\begin{array}{c|c} I_r & O_{r,n-r} \\ \hline O_{m-r,r} & O_{m-r,n-r} \end{array}\right],$$

where $r = \operatorname{rank} A$. The transformed equation $\tilde{A}\tilde{\boldsymbol{x}} = \tilde{\boldsymbol{b}}$ means

$$\tilde{x}_i = \tilde{b}_i \quad (i = 1, 2, \ldots, r), \qquad 0 = \tilde{b}_i \quad (i = r+1, r+2, \ldots, m), \tag{5.4}$$

for which the solvability is equivalent to

$$\tilde{b}_i = 0 \quad (i = r+1, r+2, \ldots, m). \tag{5.5}$$

Since $\tilde{A}\tilde{\boldsymbol{x}} = \tilde{\boldsymbol{b}}$ and $A\boldsymbol{x} = \boldsymbol{b}$ are equivalent equations, (5.5) gives the solvability condition (a) for $A\boldsymbol{x} = \boldsymbol{b}$.

[(a) $\Leftrightarrow$ (b)] Let $\tilde{\boldsymbol{y}} = (S^{-1})^\top \boldsymbol{y} = (\tilde{y}_1, \tilde{y}_2, \ldots, \tilde{y}_m)^\top$. The condition (b) can be rephrased as

$$\tilde{y}_i = 0 \quad (i = 1, 2, \ldots, r) \implies \sum_{i=1}^{r} \tilde{y}_i \tilde{b}_i + \sum_{i=r+1}^{m} \tilde{y}_i \tilde{b}_i = 0,$$

which is equivalent to the condition in (5.5).

[(a) $\Leftrightarrow$ (c)] Since

$$\tilde{A} = SAT, \qquad [\tilde{A} \mid \tilde{\boldsymbol{b}}] = S[A \mid \boldsymbol{b}] \begin{bmatrix} T & \boldsymbol{0} \\ \boldsymbol{0}^\top & 1 \end{bmatrix}$$

and the rank of a matrix remains invariant under multiplications with nonsingular matrices (Theorem 4.12 in Sec. 4.3.1), we have

$$\operatorname{rank} \tilde{A} = \operatorname{rank} A, \qquad \operatorname{rank}[\tilde{A} \mid \tilde{\boldsymbol{b}}] = \operatorname{rank}[A \mid \boldsymbol{b}].$$

On the other hand, it follows from

$$[\tilde{A} \mid \tilde{\boldsymbol{b}}] = \left[\begin{array}{ccc|ccc|c} 1 & & & & & & \tilde{b}_1 \\ & \ddots & & & O & & \vdots \\ & & 1 & & & & \tilde{b}_r \\ \hline & & & 0 & & & \tilde{b}_{r+1} \\ & O & & & \ddots & & \vdots \\ & & & & & 0 & \tilde{b}_m \end{array}\right]$$

that the condition in (5.5) is equivalent to $\operatorname{rank} \tilde{A} = \operatorname{rank} [\tilde{A} \mid \tilde{\boldsymbol{b}}]$. Therefore, we have the equivalence between (a) and (c).

[(a) ⇔ (d)] This is obvious from the definition (5.2) of the image of A. □

Remark 5.1. The above proof of [(a) ⇔ (b)] in Theorem 5.2 is based on a general approach consisting of the following three steps:

- Express condition (a) for the normal form.
- Express condition (b) for the normal form.
- Show the equivalence [(a) ⇔ (b)] for the normal form.

In this book (including Volume II [9]) we often adopt this approach to demonstrate the use of various kinds of normal forms. ■

To investigate the structure of solutions of a system of equations $A\boldsymbol{x} = \boldsymbol{b}$, it turns out to be convenient to consider another system of equations $A\boldsymbol{x} = \boldsymbol{0}$, which is obtained from the given system by replacing the right-hand side vector to the zero vector $\boldsymbol{0}$. The latter equation, $A\boldsymbol{x} = \boldsymbol{0}$, is called the associated *system of homogeneous equations*.

The set of the solutions $\boldsymbol{x}$ to the homogeneous equations $A\boldsymbol{x} = \boldsymbol{0}$ is called the kernel of A.

Definition 5.2. For an $m \times n$ matrix $A = [\boldsymbol{a}_1, \boldsymbol{a}_2, \ldots, \boldsymbol{a}_n]$, the set of n-dimensional vectors defined by

$$\operatorname{Ker} A = \{\boldsymbol{x} \mid A\boldsymbol{x} = \boldsymbol{0}\} \tag{5.6}$$

is called the *kernel* (or *null space*)[3] of A. $\operatorname{Ker} A$ coincides with the set of all vectors $\boldsymbol{x} = (x_1, x_2, \ldots, x_n)^\top$ such that the linear combinations $\sum_{j=1}^{n} x_j \boldsymbol{a}_j$ of the column vectors of A with coefficients $x_1, x_2, \ldots, x_n$ is equal to $\boldsymbol{0}$. ■

[3] $\operatorname{Ker} A$ is also called the *kernel space* as it forms a linear subspace of K^n (cf., Sec. 9.5).

Fix an arbitrary solution $\boldsymbol{x}_{\mathrm{s}}$ to the system of equations $A\boldsymbol{x} = \boldsymbol{b}$. Then every solution $\boldsymbol{x}$ to $A\boldsymbol{x} = \boldsymbol{b}$ can be represented as the sum of $\boldsymbol{x}_{\mathrm{s}}$ and a solution $\boldsymbol{x}_0$ to the system of homogeneous equations $A\boldsymbol{x} = \boldsymbol{0}$. The arbitrarily chosen solution $\boldsymbol{x}_{\mathrm{s}}$ is often referred to as a *particular solution.*

Theorem 5.3. *Suppose that a system of equations $A\boldsymbol{x} = \boldsymbol{b}$ has a solution, and let $\boldsymbol{x} = \boldsymbol{x}_{\mathrm{s}}$ denote an arbitrary solution thereof. For every $\boldsymbol{x}_0 \in \operatorname{Ker} A$, the vector $\boldsymbol{x} = \boldsymbol{x}_{\mathrm{s}} + \boldsymbol{x}_0$ is a solution to $A\boldsymbol{x} = \boldsymbol{b}$, and conversely, any solution $\boldsymbol{x}$ to $A\boldsymbol{x} = \boldsymbol{b}$ can be represented as $\boldsymbol{x} = \boldsymbol{x}_{\mathrm{s}} + \boldsymbol{x}_0$ for some $\boldsymbol{x}_0 \in \operatorname{Ker} A$. That is,*

$$\{\boldsymbol{x} \mid A\boldsymbol{x} = \boldsymbol{b}\} = \{\boldsymbol{x}_{\mathrm{s}} + \boldsymbol{x}_0 \mid \boldsymbol{x}_0 \in \operatorname{Ker} A\}.$$

Proof. It follows from $A\boldsymbol{x}_{\mathrm{s}} = \boldsymbol{b}$ and $A\boldsymbol{x}_0 = \boldsymbol{0}$ that $A\boldsymbol{x} = A(\boldsymbol{x}_{\mathrm{s}} + \boldsymbol{x}_0) = \boldsymbol{b} + \boldsymbol{0} = \boldsymbol{b}$. For the converse, take any solution $\boldsymbol{x}$ and define $\boldsymbol{x}_0 = \boldsymbol{x} - \boldsymbol{x}_{\mathrm{s}}$. Then $A\boldsymbol{x}_0 = A(\boldsymbol{x} - \boldsymbol{x}_{\mathrm{s}}) = \boldsymbol{b} - \boldsymbol{b} = \boldsymbol{0}$ and $\boldsymbol{x} = \boldsymbol{x}_{\mathrm{s}} + \boldsymbol{x}_0$. □

Conditions for the uniqueness of the solution are given in the following theorem.

Theorem 5.4. *The following three conditions* (a) *to* (c) *are equivalent for an $m \times n$ matrix A and an m-dimensional vector $\boldsymbol{b}$:*

(a) *$A\boldsymbol{x} = \boldsymbol{b}$ has a unique solution $\boldsymbol{x}$.*
(b) $\operatorname{rank} A = \operatorname{rank}\,[A \mid \boldsymbol{b}] = n$.
(c) $\boldsymbol{b} \in \operatorname{Im} A$ *and* $\operatorname{Ker} A = \{\boldsymbol{0}\}$.

Proof. [(a) ⇔ (b)] Let $r = \operatorname{rank} A$. By Theorem 5.2, a solution $\boldsymbol{x}$ exists if and only if $\operatorname{rank}\,[A \mid \boldsymbol{b}] = r$. By (5.4) in the proof of Theorem 5.2, the $n - r$ variables $\tilde{x}_j$ $(r < j \leqq n)$, if any, are arbitrary. Therefore we must have $r = n$ for the uniqueness of the solution.

[(a) ⇔ (c)] By Theorem 5.2, the first condition $\boldsymbol{b} \in \operatorname{Im} A$ in (c) is a necessary and sufficient condition for the existence of a solution. By Theorem 5.3, the second condition $\operatorname{Ker} A = \{\boldsymbol{0}\}$ in (c) is a necessary and sufficient condition for the uniqueness. □

Remark 5.2. Here we apply Theorem 5.1 to complete the proof of Theorem 1.3 in Sec. 1.3.1 characterizing nonsingularity of a matrix. To be specific, we prove here the following two equivalences:

(d) $\operatorname{Im} A = K^n \iff$ (a) A is nonsingular (the inverse A^{-1} exists),

(e) $\operatorname{Ker} A = \{\boldsymbol{0}\} \iff$ (f) Column vectors of A are linearly independent.

By Definition 5.1 of $\operatorname{Im} A$, the condition (d) above is equivalent to (b) in Theorem 5.1, while (a) above is the same as (c) in Theorem 5.1. Therefore, [(d) $\Leftrightarrow$ (a)] in Theorem 1.3 follows from [(b) $\Leftrightarrow$ (c)] in Theorem 5.1. By Definition 5.2 of $\operatorname{Ker} A$, the condition (e) above is equivalent to the statement

$$\sum_{j=1}^{n} x_j \boldsymbol{a}_j = \boldsymbol{0} \quad \Longrightarrow \quad (x_1, x_2, \ldots, x_n) = (0, 0, \ldots, 0),$$

which is nothing but the condition (f) for the linear independence of the column vectors of A. The proof of Theorem 1.3 is thus completed. (The other equivalences have already been established in Theorem 2.16 in Sec. 2.6 and Remark 4.1 in Sec. 4.1.4.) ■

5.2 Parametric Representation of Solutions

In this section we consider a parametric representation of solutions to $A\boldsymbol{x} = \boldsymbol{b}$ when the solution is not unique. We continue to assume that A is an $m \times n$ matrix of rank r.

It follows from (5.4) in the proof of Theorem 5.2 that the solution $\tilde{\boldsymbol{x}} = (\tilde{x}_1, \tilde{x}_2, \ldots, \tilde{x}_n)^\top$ to $\tilde{A}\tilde{\boldsymbol{x}} = \tilde{\boldsymbol{b}}$ is given as

$$\tilde{\boldsymbol{x}} = (\tilde{b}_1, \tilde{b}_2, \ldots, \tilde{b}_r, \tilde{\xi}_1, \tilde{\xi}_2, \ldots, \tilde{\xi}_{n-r})^\top$$

with $n - r$ parameters $\tilde{\xi}_1, \tilde{\xi}_2, \ldots, \tilde{\xi}_{n-r}$. Hence, the solution $\boldsymbol{x}$ to $A\boldsymbol{x} = \boldsymbol{b}$ is represented as

$$\boldsymbol{x} = T\tilde{\boldsymbol{x}} = T\,(\tilde{b}_1, \tilde{b}_2, \ldots, \tilde{b}_r, \tilde{\xi}_1, \tilde{\xi}_2, \ldots, \tilde{\xi}_{n-r})^\top = \sum_{i=1}^{r} \tilde{b}_i \boldsymbol{t}_i + \sum_{j=1}^{n-r} \tilde{\xi}_j \boldsymbol{t}_{r+j}, \quad (5.7)$$

where $\boldsymbol{t}_1, \boldsymbol{t}_2, \ldots, \boldsymbol{t}_n$ denote the column vectors of T, *i.e.*, $T = [\boldsymbol{t}_1, \boldsymbol{t}_2, \ldots, \boldsymbol{t}_n]$, which are linearly independent since T is nonsingular. This expression shows that the number of degrees of freedom of the solutions is equal to $n - r$.

Theorem 5.5. *Suppose that a system of equations $A\boldsymbol{x} = \boldsymbol{b}$ has a solution $\boldsymbol{x}$, where A is an $m \times n$ matrix and $\boldsymbol{b}$ is an m-dimensional vector. Then the number of degrees of freedom of the solutions is equal to $n - \operatorname{rank} A$.*

The decomposition $\boldsymbol{x} = \boldsymbol{x}_{\mathrm{s}} + \boldsymbol{x}_0$ in Theorem 5.3 can be obtained from (5.7) by defining

$$\boldsymbol{x}_{\mathrm{s}} = \sum_{i=1}^{r} \tilde{b}_i \boldsymbol{t}_i, \qquad \boldsymbol{x}_0 = \sum_{j=1}^{n-r} \tilde{\xi}_j \boldsymbol{t}_{r+j}.$$

This expression also shows that $\operatorname{Ker} A$ is equal to the set of all linear combinations of $\boldsymbol{t}_{r+1}, \boldsymbol{t}_{r+2}, \ldots, \boldsymbol{t}_n$, that is,

$$\operatorname{Ker} A = \Big\{ \sum_{j=1}^{n-r} \tilde{\xi}_j \boldsymbol{t}_{r+j} \mid \tilde{\xi}_1, \tilde{\xi}_2, \ldots, \tilde{\xi}_{n-r} \in K \Big\}. \tag{5.8}$$

Example 5.1. Consider a system of equations $A\boldsymbol{x} = \boldsymbol{b}$ for

$$A = \begin{bmatrix} 1 & 1 & 2 & 3 \\ 1 & 2 & 3 & 2 \\ 0 & 3 & 3 & -2 \\ 1 & 3 & 4 & 2 \end{bmatrix}, \qquad \boldsymbol{b} = \begin{bmatrix} 1 \\ 1 \\ 1 \\ \beta \end{bmatrix}. \tag{5.9}$$

We use the rank normal form of A with $r = 3$ found in Example 3.5 in Sec. 3.2 and the transformation matrices S and T found in Example 3.6 in Sec. 3.2. Since

$$\tilde{\boldsymbol{b}} = S\boldsymbol{b} = \left[\begin{array}{cccc} 1 & 0 & 0 & 0 \\ -1 & 1 & 0 & 0 \\ 3 & -3 & 1 & 0 \\ \hline -2 & 1 & -1 & 1 \end{array}\right] \begin{bmatrix} 1 \\ 1 \\ 1 \\ \beta \end{bmatrix} = \left[\begin{array}{c} 1 \\ 0 \\ 1 \\ \hline \beta - 2 \end{array}\right],$$

a solution exists if and only if $\beta = 2$. When $\beta = 2$, the solution to the transformed equation $\tilde{A}\tilde{\boldsymbol{x}} = \tilde{\boldsymbol{b}}$ is given by $\tilde{\boldsymbol{x}} = (1,\ 0,\ 1,\ \tilde{\xi}_1)^\top$ with a parameter $\tilde{\xi}_1$. Accordingly, a parametric representation of the solution $\boldsymbol{x}$ to the original system $A\boldsymbol{x} = \boldsymbol{b}$ is given by

$$\boldsymbol{x} = T\tilde{\boldsymbol{x}} = \left[\begin{array}{ccc|c} 1 & -1 & -4 & -1 \\ 0 & 1 & 1 & -1 \\ 0 & 0 & 0 & 1 \\ 0 & 0 & 1 & 0 \end{array}\right] \left[\begin{array}{c} 1 \\ 0 \\ 1 \\ \hline \tilde{\xi}_1 \end{array}\right] = \begin{bmatrix} -3 \\ 1 \\ 0 \\ 1 \end{bmatrix} + \tilde{\xi}_1 \begin{bmatrix} -1 \\ -1 \\ 1 \\ 0 \end{bmatrix}.$$

The decomposition $\boldsymbol{x} = \boldsymbol{x}_\mathrm{s} + \boldsymbol{x}_0$ in Theorem 5.3 can be obtained with $\boldsymbol{x}_\mathrm{s} = (-3,\ 1,\ 0,\ 1)^\top$ and $\boldsymbol{x}_0 = \tilde{\xi}_1(-1,\ -1,\ 1,\ 0)^\top$, where $\boldsymbol{x}_0$ is a solution to the homogeneous equations $A\boldsymbol{x} = \boldsymbol{0}$. ■

5.3 Solution by Elimination

We can solve a system of equations $A\boldsymbol{x} = \boldsymbol{b}$ via *elimination* using elementary row transformations.

First we consider the (basic) case where A is a square matrix of order n. By putting $\boldsymbol{b}$ to the right of A, make an $n \times (n+1)$ matrix $[A \mid \boldsymbol{b}]$. Suppose that we can transform, by means of elementary row transformations, this matrix to a matrix of the form $[I_n \mid \boldsymbol{c}]$ that contains the unit matrix I_n in the left. This means that there is a nonsingular matrix S such that $SA = I_n$ and $S\boldsymbol{b} = \boldsymbol{c}$, which implies that A is nonsingular and $\boldsymbol{c} = A^{-1}\boldsymbol{b}$.

Next we consider the general case where A may possibly be a singular square matrix or an $m \times n$ matrix. Elimination by elementary row transformations is also effective in this general case. Let $\hat{A}$ denote the reduced echelon form of A (cf., Sec. 3.3). Then there exist a nonnegative integer r and column indices $j_1^\circ < j_2^\circ < \cdots < j_r^\circ$ such that

- For $i = 1, 2, \ldots, r$:
  $\hat{a}_{i,j_i^\circ} = 1$ and $\hat{a}_{ij} = 0 \quad (j \in \{1, 2, \ldots, j_i^\circ - 1\} \cup \{j_{i+1}^\circ, j_{i+2}^\circ, \ldots, j_r^\circ\})$,
- For $i = r+1, r+2, \ldots, m : \hat{a}_{ij} = 0 \quad (j = 1, 2, \ldots, n)$.

Define $J^\circ = \{j_1^\circ, j_2^\circ, \ldots, j_r^\circ\}$ and $J = \{1, 2, \ldots, n\} \setminus J^\circ$ and represent $\hat{A}$ as $\hat{A} = [\hat{\boldsymbol{a}}_1, \hat{\boldsymbol{a}}_2, \ldots, \hat{\boldsymbol{a}}_n]$.

The $m \times (n+1)$ matrix $[A \mid \boldsymbol{b}]$ is transformed by elementary row transformations to a matrix of the form $[\hat{A} \mid \hat{\boldsymbol{b}}]$ that contains the reduced echelon form $\hat{A}$ in the left. This means that there is a nonsingular matrix S such that $SA = \hat{A}$ and $S\boldsymbol{b} = \hat{\boldsymbol{b}}$. Therefore, the given system of equations $A\boldsymbol{x} = \boldsymbol{b}$ is equivalent to $\hat{A}\boldsymbol{x} = \hat{\boldsymbol{b}}$. The solvability condition is given by

$$\hat{b}_i = 0 \quad (i = r+1, r+2, \ldots, m), \tag{5.10}$$

and the solutions are represented as

$$x_j = \begin{cases} \hat{\xi}_j & (j \in J), \\ \hat{b}_i - \sum_{k \in J} \hat{a}_{ik}\hat{\xi}_k & (j = j_i^\circ \in J^\circ) \end{cases} \tag{5.11}$$

using $n - r$ parameters $\hat{\xi}_j \ (j \in J)$.

To carry out the solution procedure described above we only need to repeat elementary row operations to clear out columns of the coefficient matrix. The parametric representation (5.11) is illustrated in the following example.

Example 5.2. Let $A\boldsymbol{x} = \boldsymbol{b}$ be the system of equations considered in Example 5.1, where matrix A and vector $\boldsymbol{b}$ are given by (5.9). By elimination operations as in Example 3.8 in Sec. 3.3, we can transform the matrix $[A \mid \boldsymbol{b}]$ as

$$[A \mid \boldsymbol{b}] = \left[\begin{array}{cccc|c} 1 & 1 & 2 & 3 & 1 \\ 1 & 2 & 3 & 2 & 1 \\ 0 & 3 & 3 & -2 & 1 \\ 1 & 3 & 4 & 2 & \beta \end{array}\right] \longrightarrow \left[\begin{array}{cccc|c} \boxed{1} & 0 & 1 & 0 & -3 \\ 0 & \boxed{1} & 1 & 0 & 1 \\ 0 & 0 & 0 & \boxed{1} & 1 \\ 0 & 0 & 0 & 0 & \beta - 2 \end{array}\right] = [\hat{A} \mid \hat{\boldsymbol{b}}].$$

This shows that $r = 3$, $J^\circ = \{1, 2, 4\}$, and $J = \{3\}$. The condition (5.10) for solvability is given by $\beta = 2$, and the parametric representation (5.11) is given by

$$\boldsymbol{x} = \begin{bmatrix} -3 \\ 1 \\ 0 \\ 1 \end{bmatrix} + \hat{\xi}_3 \begin{bmatrix} -1 \\ -1 \\ 1 \\ 0 \end{bmatrix},$$

which coincides with the expression obtained in Example 5.1 (up to the notations $\hat{\xi}_3$ and $\tilde{\xi}_1$ of the parameter). ■

Remark 5.3. When we want to solve $A\boldsymbol{x} = \boldsymbol{b}$ for different vectors $\boldsymbol{b} = \boldsymbol{b}_1, \boldsymbol{b}_2, \ldots, \boldsymbol{b}_k$, where A is assumed to be nonsingular, we apply elementary row transformations to the matrix

$$\left[\, A \,\middle|\, \boldsymbol{b}_1 \,\middle|\, \boldsymbol{b}_2 \,\middle|\, \cdots \,\middle|\, \boldsymbol{b}_k \,\right]$$

to transform the left-most submatrix corresponding to A to the unit matrix. If $k = n$ and $\boldsymbol{b}_i = (0, \ldots, 0, \overset{\overset{i}{\vee}}{1}, 0, \ldots, 0)^\top$ for $i = 1, 2, \ldots, n$, this method coincides with the procedure of computing the inverse matrix presented in Sec. 3.4. ■

Remark 5.4. The elimination method is suitable for the solution of small-sized systems of equations by hand. A slightly different method, called the Gaussian elimination, is employed for practical numerical computations in engineering. Stationary iterative methods and conjugate gradient methods are also used for numerical solution of systems of linear equations [37, 43]. ■

5.4 Cramer's Rule

There is a formula for the solution of $A\boldsymbol{x} = \boldsymbol{b}$, known as *Cramer's rule.*[4]

Theorem 5.6. *Let $A = [\boldsymbol{a}_1, \boldsymbol{a}_2, \ldots, \boldsymbol{a}_n]$ be a nonsingular matrix of order n and $\boldsymbol{b}$ an n-dimensional vector. The solution $\boldsymbol{x} = (x_1, x_2, \ldots, x_n)^\top$ to the system of equations $A\boldsymbol{x} = \boldsymbol{b}$ is given by*

$$x_j = \frac{\det[\boldsymbol{a}_1, \boldsymbol{a}_2, \ldots, \boldsymbol{a}_{j-1}, \boldsymbol{b}, \boldsymbol{a}_{j+1}, \ldots, \boldsymbol{a}_n]}{\det[\boldsymbol{a}_1, \boldsymbol{a}_2, \ldots, \boldsymbol{a}_{j-1}, \boldsymbol{a}_j, \boldsymbol{a}_{j+1}, \ldots, \boldsymbol{a}_n]} \qquad (j = 1, 2, \ldots, n). \quad (5.12)$$

[4] Cramer's rule is a useful formula, but it is never used for numerical computations as it requires too much amount of computation and is weak against rounding errors. See also Remark 5.4.

Proof. We prove (5.12) for $j = 1$. It follows from the equation $A\boldsymbol{x} = \boldsymbol{b}$ that

$$\boldsymbol{b} = x_1\boldsymbol{a}_1 + x_2\boldsymbol{a}_2 + \cdots + x_n\boldsymbol{a}_n.$$

Using this expression as well as multilinearity and alternating properties of determinants (Sec. 2.3) we obtain

$$\begin{aligned}\det[\boldsymbol{b}, \boldsymbol{a}_2, \ldots, \boldsymbol{a}_n] &= \det[x_1\boldsymbol{a}_1 + x_2\boldsymbol{a}_2 + \cdots + x_n\boldsymbol{a}_n, \boldsymbol{a}_2, \ldots, \boldsymbol{a}_n] \\ &= \sum_{j=1}^{n} x_j \det[\boldsymbol{a}_j, \boldsymbol{a}_2, \ldots, \boldsymbol{a}_n] \\ &= x_1 \det[\boldsymbol{a}_1, \boldsymbol{a}_2, \ldots, \boldsymbol{a}_n].\end{aligned}$$

Therefore,

$$x_1 = \frac{\det[\boldsymbol{b}, \boldsymbol{a}_2, \ldots, \boldsymbol{a}_n]}{\det[\boldsymbol{a}_1, \boldsymbol{a}_2, \ldots, \boldsymbol{a}_n]}$$

as desired. □

Example 5.3. In case of $n = 3$, the formula in (5.12) reads as follows:

$$x_1 = \frac{\begin{vmatrix} b_1 & a_{12} & a_{13} \\ b_2 & a_{22} & a_{23} \\ b_3 & a_{32} & a_{33} \end{vmatrix}}{\begin{vmatrix} a_{11} & a_{12} & a_{13} \\ a_{21} & a_{22} & a_{23} \\ a_{31} & a_{32} & a_{33} \end{vmatrix}}, \quad x_2 = \frac{\begin{vmatrix} a_{11} & b_1 & a_{13} \\ a_{21} & b_2 & a_{23} \\ a_{31} & b_3 & a_{33} \end{vmatrix}}{\begin{vmatrix} a_{11} & a_{12} & a_{13} \\ a_{21} & a_{22} & a_{23} \\ a_{31} & a_{32} & a_{33} \end{vmatrix}}, \quad x_3 = \frac{\begin{vmatrix} a_{11} & a_{12} & b_1 \\ a_{21} & a_{22} & b_2 \\ a_{31} & a_{32} & b_3 \end{vmatrix}}{\begin{vmatrix} a_{11} & a_{12} & a_{13} \\ a_{21} & a_{22} & a_{23} \\ a_{31} & a_{32} & a_{33} \end{vmatrix}}.$$

■

5.5 Discretization of Differential Equations

In this section we investigate the solvability of a system of linear equations arising from discretization of a differential equation. Specifically, we consider the *Dirichlet problem*[5] on the unit square

$$\Omega = [0, 1] \times [0, 1] = \{(x, y) \mid 0 \leqq x \leqq 1, 0 \leqq y \leqq 1\}.$$

The problem is to find a function $u(x, y)$ that satisfies the *Laplace equation*

$$\frac{\partial^2 u}{\partial x^2} + \frac{\partial^2 u}{\partial y^2} = 0 \qquad ((x, y) \in (0, 1) \times (0, 1)) \tag{5.13}$$

[5] A Dirichlet problem sometimes means, more generally, the problem of finding a function satisfying an elliptic partial differential equation (not necessarily the Laplace equation) under a Dirichlet boundary condition.

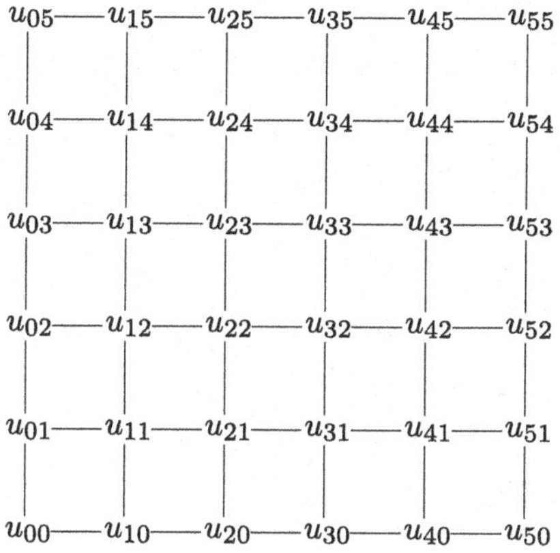

Fig. 5.1 Discretized spatial domain ($N = 5$) for the unit square $\Omega = [0, 1] \times [0, 1]$.

in the interior of Ω and the *Dirichlet boundary condition*

$$u(x, y) = g(x, y) \qquad ((x, y) \in \Gamma) \tag{5.14}$$

on the boundary Γ of Ω, where $g(x, y)$ is a given function defined on Γ.

We can obtain an approximate solution $\boldsymbol{u}$ to this problem via discretization (*finite difference method*) as follows. As in Fig. 5.1, we discretize the spatial domain Ω of unit square with equally-spaced mesh of width $h = 1/N$. The approximate solution to $u(ih, jh)$ will be denoted by u_{ij}. Through the (standard) approximations of the second partial derivatives

$$\frac{\partial^2 u}{\partial x^2} \approx \frac{u_{i-1,j} - 2u_{ij} + u_{i+1,j}}{h^2}, \qquad \frac{\partial^2 u}{\partial y^2} \approx \frac{u_{i,j-1} - 2u_{ij} + u_{i,j+1}}{h^2}, \tag{5.15}$$

the differential equation in (5.13) is replaced by

$$4u_{ij} - u_{i-1,j} - u_{i+1,j} - u_{i,j-1} - u_{i,j+1} = 0 \qquad (1 \leqq i, j \leqq N - 1), \tag{5.16}$$

which is a system of linear equations in u_{ij} $(1 \leqq i, j \leqq N - 1)$. Note that

$$u_{0j} = g(0, jh), \qquad u_{Nj} = g(1, jh) \qquad (1 \leqq j \leqq N - 1),$$
$$u_{i0} = g(ih, 0), \qquad u_{iN} = g(ih, 1) \qquad (1 \leqq i \leqq N - 1)$$

are known variables (constants) corresponding to the boundary condition (5.14).

By introducing an $(N - 1)^2$-dimensional vector

$$\boldsymbol{u} = (u_{11}, u_{12}, \ldots, u_{1,N-1}; u_{21}, u_{22}, \ldots, u_{2,N-1}; \ldots; u_{N-1,1}, \ldots, u_{N-1,N-1})^\top$$

for the unknown variables u_{ij} $(1 \leqq i, j \leqq N - 1)$, the system (5.16) of linear equations can be expressed as

$$A\boldsymbol{u} = \boldsymbol{g}, \tag{5.17}$$

where A is a square matrix of order $(N-1)^2$ and $\boldsymbol{g}$ is an $(N-1)^2$-dimensional vector determined by the boundary condition. For $N = 5$, for instance, A and $\boldsymbol{g}$ are given by

$$A = \left[\begin{array}{cccc|cccc|cccc|cccc}
4 & -1 & & & -1 & & & & & & & & & & & \\
-1 & 4 & -1 & & & -1 & & & & & & & & & & \\
 & -1 & 4 & -1 & & & -1 & & & & & & & & & \\
 & & -1 & 4 & & & & -1 & & & & & & & & \\
\hline
-1 & & & & 4 & -1 & & & -1 & & & & & & & \\
 & -1 & & & -1 & 4 & -1 & & & -1 & & & & & & \\
 & & -1 & & & -1 & 4 & -1 & & & -1 & & & & & \\
 & & & -1 & & & -1 & 4 & & & & -1 & & & & \\
\hline
 & & & & -1 & & & & 4 & -1 & & & -1 & & & \\
 & & & & & -1 & & & -1 & 4 & -1 & & & -1 & & \\
 & & & & & & -1 & & & -1 & 4 & -1 & & & -1 & \\
 & & & & & & & -1 & & & -1 & 4 & & & & -1 \\
\hline
 & & & & & & & & -1 & & & & 4 & -1 & & \\
 & & & & & & & & & -1 & & & -1 & 4 & -1 & \\
 & & & & & & & & & & -1 & & & -1 & 4 & -1 \\
 & & & & & & & & & & & -1 & & & -1 & 4
\end{array}\right] \tag{5.18}$$

and

$$\boldsymbol{g} = (g_{01} + g_{10},\ g_{02},\ g_{03},\ g_{04} + g_{15};\ g_{20},\ 0,\ 0,\ g_{25};\ g_{30},\ 0,\ 0,\ g_{35};\ g_{51} + g_{40},\ g_{52},\ g_{53},\ g_{54} + g_{45})^{\top},$$

where

$$g_{0j} = g(0, jh), \quad g_{Nj} = g(1, jh) \qquad (1 \leqq j \leqq N-1),$$
$$g_{i0} = g(ih, 0), \quad g_{iN} = g(ih, 1) \qquad (1 \leqq i \leqq N-1)$$

with $N = 5$.

The approximate solution $\boldsymbol{u}$ by the finite difference method is determined uniquely if the coefficient matrix A in (5.17) is nonsingular. The nonsingularity of A can be shown on the basis of its diagonal dominance (Sec. 1.7.1) as follows.

Theorem 1.15 says that a strictly diagonally dominant matrix is nonsingular. This theorem, however, does not apply to the matrix A here, as this matrix is not strictly diagonally dominant though it is diagonally dominant; see the example in (5.18). Instead of Theorem 1.15 itself, we apply its corollary (Theorem 1.16), stating that a generalized strictly diagonally dominant matrix is nonsingular. Let $d_1, d_2, \ldots, d_{N-1}$ be positive numbers satisfying

$$2d_j > d_{j-1} + d_{j+1} \quad (j = 1, 2, \ldots, N-1), \tag{5.19}$$

where $d_0 = d_N = 0$. For example, we may take $d_j = j(N-j)$. Define a diagonal matrix

$$D = \mathrm{diag}\,(\overbrace{d_1,\ldots,d_1}^{(N-1)}\ ;\ \overbrace{d_2,\ldots,d_2}^{(N-1)}\ ;\ \ldots;\overbrace{d_{N-1},\ldots,d_{N-1}}^{(N-1)}\ ) \tag{5.20}$$

of order $(N-1)^2$. It is easy to verify that AD is a strictly diagonally dominant matrix (cf., Definition 1.46). In other words, A is a generalized strictly diagonally dominant matrix, to which Theorem 1.16 applies to show its nonsingularity.

In this way we can guarantee that, for each N, the approximate solution $\boldsymbol{u}$ by the finite difference method is determined uniquely.

Remark 5.5. By introducing an $(N-1)\times(N-1)$ matrix

$$B = \begin{bmatrix} 0 & 1 & & \\ 1 & \ddots & \ddots & \\ & \ddots & \ddots & 1 \\ & & 1 & 0 \end{bmatrix} \tag{5.21}$$

and using the Kronecker product $\otimes$, we can represent the matrix A concisely as

$$A = 4\,I_{N-1}\otimes I_{N-1} - B\otimes I_{N-1} - I_{N-1}\otimes B. \tag{5.22}$$

The matrix D in (5.20) can be expressed similarly as $D = \mathrm{diag}\,(d_1, d_2, \ldots, d_{N-1})\otimes I_{N-1}$. ■

5.6 Sylvester Equation

An equation of the form

$$AX - XB = C \tag{5.23}$$

in a matrix variable X is called a *Sylvester equation*, where A, B, and C are given matrices, and A and B are assumed to be square.[6] If A is an $m\times m$ matrix and B is an $n\times n$ matrix, then C and X must be $m\times n$ matrices.

Since the left-hand side of (5.23) is linear in the entries x_{ij} of the matrix X, this is a system of linear equations in mn unknown variables x_{ij} $(i =$

[6] We may also consider a more general case where A and B are not necessarily square. In this book, however, we assume that A and B are square.

$1, 2, \ldots, m; j = 1, 2, \ldots, n$). This system of equations can be put in the familiar form

$$[\text{coefficient matrix}] \times [\text{unknown vector}] = [\text{given vector}]$$

as follows. Let $\boldsymbol{x}$ and $\boldsymbol{c}$ be mn-dimensional column vectors consisting of the entries of the matrices X and C, respectively, arranged as

$$\boldsymbol{x} = (x_{11}, x_{21}, \ldots, x_{m1}; x_{12}, x_{22}, \ldots, x_{m2}; \ldots; x_{1n}, x_{2n}, \ldots, x_{mn})^\top,$$
$$\boldsymbol{c} = (c_{11}, c_{21}, \ldots, c_{m1}; c_{12}, c_{22}, \ldots, c_{m2}; \ldots; c_{1n}, c_{2n}, \ldots, c_{mn})^\top.$$

Then the equation (5.23) can be rewritten as

$$(I_n \otimes A - B^\top \otimes I_m)\boldsymbol{x} = \boldsymbol{c}, \tag{5.24}$$

where

$$I_n \otimes A = \begin{bmatrix} A & & & \\ & A & & \\ & & \ddots & \\ & & & A \end{bmatrix}, \quad B^\top \otimes I_m = \begin{bmatrix} b_{11}I_m & b_{21}I_m & \cdots & b_{n1}I_m \\ b_{12}I_m & b_{22}I_m & \cdots & b_{n2}I_m \\ \vdots & \vdots & \ddots & \vdots \\ b_{1n}I_m & b_{2n}I_m & \cdots & b_{nn}I_m \end{bmatrix}.$$

A necessary and sufficient condition for unique solvability of the Sylvester equation (5.23) is given by the condition of nonsingularity of the coefficient matrix $(I_n \otimes A - B^\top \otimes I_m)$ in (5.24). The latter can be expressed in terms of the eigenvalues[7] of A and B.

Theorem 5.7. *The Sylvester equation* (5.23) *has a unique solution for each C if and only if the matrices A and B have no common eigenvalues.*

Proof. First recall that a matrix is nonsingular if and only if zero is not an eigenvalue of the matrix (Theorem 1.3 in Sec. 1.3.1). Let α be an eigenvalue of A with eigenvector $\boldsymbol{u}$, and similarly, let β be an eigenvalue of $B^\top$ with eigenvector $\boldsymbol{v}$. That is, $A\boldsymbol{u} = \alpha\boldsymbol{u}$ and $B^\top\boldsymbol{v} = \beta\boldsymbol{v}$. Then

$$(I_n \otimes A)(\boldsymbol{v} \otimes \boldsymbol{u}) = \alpha(\boldsymbol{v} \otimes \boldsymbol{u}), \qquad (B^\top \otimes I_m)(\boldsymbol{v} \otimes \boldsymbol{u}) = \beta(\boldsymbol{v} \otimes \boldsymbol{u})$$

by Proposition 1.4(4) in Sec. 1.2.4, and hence

$$(I_n \otimes A - B^\top \otimes I_m)(\boldsymbol{v} \otimes \boldsymbol{u}) = (\alpha - \beta)(\boldsymbol{v} \otimes \boldsymbol{u}).$$

This shows that $\alpha - \beta$ is an eigenvalue of matrix $(I_n \otimes A - B^\top \otimes I_m)$. Therefore, if A and B have a common eigenvalue, the coefficient matrix $(I_n \otimes A - B^\top \otimes I_m)$ is singular.

[7]Eigenvalues are treated in Chapter 6. In particular, the definition of eigenvalues can be found in Sec. 6.1.1.

To prove the converse, we make use of the fact that there exist nonsingular matrices S and T such that $\tilde{A} = S^{-1}AS$ is upper triangular and $\tilde{B} = T^{-1}BT$ is lower triangular; this fact follows form the Jordan normal form in Sec. 6.8 or the Schur decomposition in Sec. 6.5. The diagonal entries of $\tilde{A}$, denoted by $\alpha_1, \alpha_2, \ldots, \alpha_m$, are the eigenvalues of A, and similarly, the diagonal entries of $\tilde{B}$, denoted by $\beta_1, \beta_2, \ldots, \beta_n$, are the eigenvalues of B. Using the matrices S and T we also transform the matrices C and X to $\tilde{C} = S^{-1}CT$ and $\tilde{X} = S^{-1}XT$. Then the equation (5.23) is rewritten to

$$\tilde{A}\tilde{X} - \tilde{X}\tilde{B} = \tilde{C}. \tag{5.25}$$

Now consider the corresponding equation of the form of (5.24). It is not difficult to see (cf., Example 5.4 below) that the coefficient matrix $(I_n \otimes \tilde{A} - \tilde{B}^\top \otimes I_m)$ is an upper triangular matrix whose diagonal entries are $\alpha_i - \beta_j$ $(i = 1, 2, \ldots, m; j = 1, 2, \ldots, n)$. This matrix is nonsingular if $\alpha_i - \beta_j \neq 0$ for all i and j. □

Example 5.4. We verify, in terms of a concrete example, that the matrix $(I_n \otimes \tilde{A} - \tilde{B}^\top \otimes I_m)$ is upper triangular if $\tilde{A}$ is upper triangular and $\tilde{B}$ is lower triangular. For

$$\tilde{A} = \begin{bmatrix} \tilde{a}_{11} & \tilde{a}_{12} \\ 0 & \tilde{a}_{22} \end{bmatrix}, \qquad \tilde{B} = \begin{bmatrix} \tilde{b}_{11} & 0 & 0 \\ \tilde{b}_{21} & \tilde{b}_{22} & 0 \\ \tilde{b}_{31} & \tilde{b}_{32} & \tilde{b}_{33} \end{bmatrix}$$

with $m = 2$ and $n = 3$, we have

$$I_3 \otimes \tilde{A} - \tilde{B}^\top \otimes I_2 =$$

$$\left[\begin{array}{cc|cc|cc} \tilde{a}_{11} - \tilde{b}_{11} & \tilde{a}_{12} & -\tilde{b}_{21} & 0 & -\tilde{b}_{31} & 0 \\ 0 & \tilde{a}_{22} - \tilde{b}_{11} & 0 & -\tilde{b}_{21} & 0 & -\tilde{b}_{31} \\ \hline 0 & 0 & \tilde{a}_{11} - \tilde{b}_{22} & \tilde{a}_{12} & -\tilde{b}_{32} & 0 \\ 0 & 0 & 0 & \tilde{a}_{22} - \tilde{b}_{22} & 0 & -\tilde{b}_{32} \\ \hline 0 & 0 & 0 & 0 & \tilde{a}_{11} - \tilde{b}_{33} & \tilde{a}_{12} \\ 0 & 0 & 0 & 0 & 0 & \tilde{a}_{22} - \tilde{b}_{33} \end{array}\right],$$

which is an upper triangular matrix. ■

5.7 Lyapunov Equation

An equation of the form

$$AX + XA^\top = -Q \tag{5.26}$$

in a real matrix variable X is called a *Lyapunov equation*, where A and Q are given square real matrices, and Q is assumed to be symmetric ($Q = Q^\top$).

The following theorem holds concerning the solvability of this equation.

Theorem 5.8. *The Lyapunov equation* (5.26) *has a unique solution for each symmetric matrix Q if and only if, for each complex number λ, not both of λ and $-\lambda$ are eigenvalues of A. If this is the case, the solution X is a symmetric matrix, i.e., $X = X^\top$.*

Proof. Equation (5.26) is a special case of the Sylvester equation (5.23) where $B = -A^\top$ and $C = -Q$, and the coefficient matrix in (5.24) is given by $(I_n \otimes A + A \otimes I_n)$. By Theorem 5.7 and its proof, the matrix $(I_n \otimes A + A \otimes I_n)$ is nonsingular if and only if A and $-A^\top$ have no common eigenvalues. If these (equivalent) conditions hold, the equation (5.26) has a unique solution for every symmetric matrix Q. Conversely, suppose that the equation (5.26) has a unique solution for every symmetric matrix Q. Then, in particular, $X = O$ is the unique solution for $Q = O$, which implies that the matrix $(I_n \otimes A + A \otimes I_n)$ is nonsingular (cf., Theorem 1.3 in Sec. 1.3.1). The first statement of the theorem follows from the above argument with an additional observation that the eigenvalues of $-A^\top$ are precisely the negative of the eigenvalues of A.

To prove the second statement, consider the transpose of the equation (5.26), which reads

$$-Q = (-Q)^\top = X^\top A^\top + AX^\top.$$

This shows that $X^\top$ is also a solution to the equation (5.26). Then the uniqueness of the solution implies $X = X^\top$. □

In control theory [50, 52, 56, 60, 61], the Lyapunov equation is used for stability analysis of dynamical systems. Suppose that a dynamical system is described by a differential equation

$$\frac{\mathrm{d}\boldsymbol{x}}{\mathrm{d}t}(t) = A\boldsymbol{x}(t) \quad (t > 0), \qquad \boldsymbol{x}(0) = \mathbf{c}, \tag{5.27}$$

where t denotes the independent variable representing time, $\boldsymbol{x} = \boldsymbol{x}(t)$ is the dependent variable (vector) representing the state of the system, A is a square real matrix, and $\mathbf{c}$ is a vector representing the initial state. The solution of this differential equation is given by[8]

$$\boldsymbol{x}(t) = \mathrm{e}^{tA}\mathbf{c}. \tag{5.28}$$

[8]The notation e^{tA} means a matrix defined by $\mathrm{e}^{tA} = \sum_{k=0}^{\infty} \frac{1}{k!}(tA)^k$ (cf., Sec. 6.8.6). In engineering, e^{tA} is often written as e^{At}.

The system is said to be (asymptotically) stable if $\lim_{t \to +\infty} \boldsymbol{x}(t) = \mathbf{0}$ holds for any initial state $\boldsymbol{c}$, and a necessary and sufficient condition for this property is given by

$$\lim_{t \to +\infty} \mathrm{e}^{tA} = O. \tag{5.29}$$

The following theorem connects this property to the eigenvalues of A.

Theorem 5.9. *For a square real matrix A,* (5.29) *holds if and only if every eigenvalue of A has a negative real part.*

Proof. This is the case (a) in Theorem 6.42 of Sec. 6.8.6. □

Definition 5.3. A square real matrix is called a *stable matrix* if every eigenvalue has a negative real part. ■

The stability of a matrix A corresponds to the positive definiteness[9] of the solution X to the Lyapunov equation defined by A.

Theorem 5.10. *Assume that Q is a positive definite symmetric matrix in* (5.26).

(1) *If A is a stable matrix, then the solution X to* (5.26) *is determined uniquely and is a positive definite symmetric matrix.*
(2) *If the Lyapunov equation* (5.26) *has a positive definite symmetric solution X, then A is a stable matrix.*

Proof. (1) Consider an identity

$$\frac{\mathrm{d}}{\mathrm{d}t}(\mathrm{e}^{tA} Q \mathrm{e}^{tA^\top}) = A\mathrm{e}^{tA} Q \mathrm{e}^{tA^\top} + \mathrm{e}^{tA} Q \mathrm{e}^{tA^\top} A^\top.$$

Integration of this expression gives

$$\int_0^\infty \frac{\mathrm{d}}{\mathrm{d}t}(\mathrm{e}^{tA} Q \mathrm{e}^{tA^\top})\mathrm{d}t = A\left(\int_0^\infty \mathrm{e}^{tA} Q \mathrm{e}^{tA^\top} \mathrm{d}t\right) + \left(\int_0^\infty \mathrm{e}^{tA} Q \mathrm{e}^{tA^\top} \mathrm{d}t\right) A^\top. \tag{5.30}$$

By defining a matrix X by[10]

$$X = \int_0^\infty \mathrm{e}^{tA} Q \mathrm{e}^{tA^\top} \mathrm{d}t, \tag{5.31}$$

we see that the right-hand side of (5.30) is equal to $AX + XA^\top$, whereas, by Theorem 5.9, the left-hand side of (5.30) is given as

$$\int_0^\infty \frac{\mathrm{d}}{\mathrm{d}t}(\mathrm{e}^{tA} Q \mathrm{e}^{tA^\top})\mathrm{d}t = \left[\mathrm{e}^{tA} Q \mathrm{e}^{tA^\top}\right]_{t=0}^{\infty} = -Q.$$

---

[9]See Sec. 7.2 for the definition of positive definiteness.

[10]The integral on the right-hand side of (5.31) exists by Theorem 5.9 and its proof.

Therefore, the matrix X defined by (5.31) is a solution to the Lyapunov equation (5.26). In addition, X is a positive definite symmetric matrix by the expression (5.31). Finally, the uniqueness of the solution follows from Theorem 5.8, because an eigenvalue λ of A has a negative real part, which implies that $-\lambda$, having a positive real part, is not an eigenvalue of A.

(2) Let λ be any eigenvalue of A with $A^\top \boldsymbol{v} = \lambda \boldsymbol{v}$, where $\boldsymbol{v}$ is a nonzero complex vector. Let $\boldsymbol{v}^*$ denote the conjugate transpose of $\boldsymbol{v}$. Since A is a real matrix, we have $\boldsymbol{v}^* A = \overline{\lambda} \boldsymbol{v}^*$. Then it follows from (5.26) and the positive definiteness of Q that

$$0 > -\boldsymbol{v}^* Q \boldsymbol{v} = \boldsymbol{v}^*(AX + XA^\top)\boldsymbol{v} = (\lambda + \overline{\lambda})(\boldsymbol{v}^* X \boldsymbol{v}),$$

in which $\boldsymbol{v}^* X \boldsymbol{v} > 0$ because X is positive definite. Therefore we have $\lambda + \overline{\lambda} < 0$, that is, the real part of λ is negative. We have shown that every eigenvalue has a negative real part, which means that A is a stable matrix. □

Chapter 6

Eigenvalues

In this chapter we present fundamental facts about eigenvalues and eigenvectors, as well as their significances in engineering. For symmetric (resp., Hermitian) matrices, diagonalization by orthogonal (resp., unitary) transformations, minimax theorems, and perturbation theorems are shown. For general (non-symmetric) matrices, the Schur decomposition and the concept of normal matrices are explained in relation to orthogonal (resp., unitary) transformations. Finally, the Jordan normal form is presented including its construction, computation, and use.

6.1 Eigenvalues and Eigenvectors

6.1.1 *Definition*

For a square matrix $A = (a_{ij})$ we consider a pair $(\lambda, \boldsymbol{x})$ of a complex number λ and a complex vector $\boldsymbol{x} \neq \boldsymbol{0}$ satisfying

$$A\boldsymbol{x} = \lambda \boldsymbol{x}. \tag{6.1}$$

Such λ is called an *eigenvalue* of A and $\boldsymbol{x}$ is an *eigenvector* corresponding to λ. The problem of finding (all or some of) such $(\lambda, \boldsymbol{x})$ for a given matrix A is called the *eigenvalue problem*. Note that an eigenvalue of a real matrix is not necessarily a real number.

Theorem 6.1. *The following four conditions* (a) *to* (d) *are equivalent for a square matrix A and a complex number λ:*

(a) *λ is an eigenvalue of A.*
(b) $\mathrm{Ker}\,(A - \lambda I) \neq \{\boldsymbol{0}\}$.
(c) *$A - \lambda I$ is a singular matrix.*
(d) $\det(A - \lambda I) = 0$.

Proof. The defining condition in (6.1) can be rewritten as $\boldsymbol{x} \in \mathrm{Ker}\,(A - \lambda I)$. This shows the equivalence between (a) and (b). The equivalence of (b), (c), and (d) follows from Theorem 1.3 in Sec. 1.3.1. □

For a square matrix A, a transformation of the form of $S^{-1}AS$ with a nonsingular matrix S is called a *similarity transformation.* Two matrices are said to be *similar* if they are connected by a similarity transformation with some S.

Theorem 6.2. *Eigenvalues of a matrix remain invariant under a similarity transformation. To be more precise, if* $A\boldsymbol{x} = \lambda \boldsymbol{x},\ \ \boldsymbol{x} \neq \boldsymbol{0}$, *and* $B = S^{-1}AS$ *for a nonsingular matrix* S, *then* $B\boldsymbol{y} = \lambda \boldsymbol{y},\ \ \boldsymbol{y} \neq \boldsymbol{0}$ *for* $\boldsymbol{y} = S^{-1}\boldsymbol{x}$.

Proof. This follows from the rewriting of $A\boldsymbol{x} = \lambda \boldsymbol{x}$ to $(S^{-1}AS)(S^{-1}\boldsymbol{x}) = \lambda(S^{-1}\boldsymbol{x})$. □

6.1.2 *Characteristic Equation*

We assume that A is a square matrix of order n. When we regard λ as an independent variable, the determinant of the matrix $\lambda I - A$,

$$\varPhi_A(\lambda) = \det(\lambda I - A), \tag{6.2}$$

is a polynomial of degree n in λ. This is called the *characteristic polynomial* of matrix A. In the expanded form

$$\varPhi_A(\lambda) = \det(\lambda I - A) = \lambda^n + c_{n-1}\lambda^{n-1} + \cdots + c_1\lambda + c_0, \tag{6.3}$$

the coefficients are given by

$$c_{n-k} = (-1)^k \sum_{i_1 < i_2 < \cdots < i_k} |A|^{i_1, i_2, \ldots, i_k}_{i_1, i_2, \ldots, i_k}, \tag{6.4}$$

where the right-hand side means $(-1)^k$ times the sum of all principal minors of order k; see Definition 2.7 in Sec. 2.2 for notation $|A|^{i_1, i_2, \ldots, i_k}_{i_1, i_2, \ldots, i_k}$. In particular, we have

$$c_{n-1} = -\mathrm{Tr}\,A, \qquad c_0 = (-1)^n \det A. \tag{6.5}$$

Since every non-constant polynomial with complex coefficients has at least one complex root (the fundamental theorem of algebra), the characteristic polynomial $\varPhi_A(\lambda)$ can be represented in a factored form

$$\varPhi_A(\lambda) = (\lambda - \lambda_1)(\lambda - \lambda_2) \cdots (\lambda - \lambda_n) \tag{6.6}$$

using n complex numbers $\lambda_1, \lambda_2, \ldots, \lambda_n$. On the other hand, by the equivalence (a) $\Leftrightarrow$ (d) in Theorem 6.1, the eigenvalues of A are characterized as the roots of the equation

$$\Phi_A(\lambda) = 0, \tag{6.7}$$

which is called the *characteristic equation* for A. Therefore, the eigenvalues of A are precisely those numbers $\lambda_1, \lambda_2, \ldots, \lambda_n$ which appear in (6.6). It is noted, however, that not all $\lambda_1, \lambda_2, \ldots, \lambda_n$ are distinct, in general. The *multiplicity*[1] of an eigenvalue λ means the number of indices i with $\lambda = \lambda_i$. A matrix of order n has n eigenvalues counting multiplicity.

It follows from (6.5) and Vieta's formulas (the relations between roots and coefficients) that

$$\lambda_1 + \lambda_2 + \cdots + \lambda_n = \operatorname{Tr} A, \qquad \lambda_1 \lambda_2 \cdots \lambda_n = \det A. \tag{6.8}$$

In the characteristic polynomial $\Phi_A(\lambda)$ of matrix A represented in the form of (6.3), we may substitute the matrix A for the variable λ. The *Cayley–Hamilton theorem* states that the resulting matrix is equal to the zero matrix.

Theorem 6.3 (Cayley–Hamilton theorem). *For the characteristic polynomial $\Phi_A(\lambda)$ of A in (6.3) we have*

$$\Phi_A(A) = A^n + c_{n-1}A^{n-1} + \cdots + c_1 A + c_0 I = O. \tag{6.9}$$

Proof. Let B denote the adjoint matrix of $\lambda I - A$. By Theorem 2.15 in Sec. 2.6 we have

$$(\lambda I - A)B = \Phi_A(\lambda)I. \tag{6.10}$$

Each entry of B is a polynomial in λ whose degree is at most $n - 1$, and therefore we can express B as

$$B = \lambda^{n-1}B_{n-1} + \lambda^{n-2}B_{n-2} + \cdots + \lambda B_1 + B_0, \tag{6.11}$$

where $B_{n-1}, B_{n-2}, \ldots, B_1, B_0$ are constant matrices free from λ. By substituting (6.11) into (6.10) and comparing the coefficients appearing on both sides we obtain

$$\begin{aligned}
B_{n-1} \qquad\quad &= I, \\
B_{n-2} - AB_{n-1} &= c_{n-1}I, \\
&\;\vdots \\
B_0 - AB_1 \quad &= c_1 I, \\
-AB_0 \quad &= c_0 I.
\end{aligned}$$

[1]The multiplicity defined here is called the *algebraic multiplicity*. In Sec. 6.8 we shall define "geometric multiplicity."

By multiplying the above equations with $A^n, A^{n-1}, \ldots, A, I$ from the left and adding them all, we obtain (6.9). □

Remark 6.1. The *minimal polynomial* of a matrix A means the monic polynomial $\Psi(\lambda)$ of the lowest degree that satisfies $\Psi(A) = O$, where a polynomial is called *monic* when the leading coefficient is equal to 1. Theorem 6.3 shows the existence of a polynomial $\Psi(\lambda)$ that satisfies $\Psi(A) = O$. The minimal polynomial is determined uniquely (Proof: If $\Psi_1(\lambda)$ and $\Psi_2(\lambda)$ are monic polynomials of the minimum degree that satisfy $\Psi_1(A) = O$ and $\Psi_2(A) = O$, then $\Psi_3(\lambda) = \Psi_1(\lambda) - \Psi_2(\lambda)$ is a polynomial of a lower degree that satisfies $\Psi_3(A) = O$, which implies that $\Psi_3(\lambda) = 0$). The characteristic polynomial $\Phi_A(\lambda)$ of a matrix A is divisible by the minimal polynomial $\Psi(\lambda)$ of A. The minimal polynomial plays important roles in investigating polynomials and functions in a matrix A [22, 26]. ■

6.1.3 *Gershgorin Theorem*

We continue to assume that $A = (a_{ij})$ is a square matrix of order n. The maximum absolute value of an eigenvalue of matrix A is called the *spectral radius* of A. We denote this by $\rho(A)$.

Theorem 6.4. *For $p \geqq 1$ or $p = \infty$, we have*[2]

$$\rho(A) \leqq \|A\|_p, \tag{6.12}$$

where $\|A\|_p$ denotes the p-norm of A defined in (1.24).

Proof. Let λ be an eigenvalue of A and $A\boldsymbol{x} = \lambda\boldsymbol{x}$ for $\boldsymbol{x} \neq \boldsymbol{0}$. By the definition (1.24) of p-norm we obtain

$$\|A\|_p = \sup_{\boldsymbol{y}\neq\boldsymbol{0}} \frac{\|A\boldsymbol{y}\|_p}{\|\boldsymbol{y}\|_p} \geqq \frac{\|A\boldsymbol{x}\|_p}{\|\boldsymbol{x}\|_p} = \frac{\|\lambda\boldsymbol{x}\|_p}{\|\boldsymbol{x}\|_p} = |\lambda|.$$

Since this is true for any eigenvalue λ, we have $\|A\|_p \geqq \rho(A)$. □

It follows from the inequality (6.12) for $p = \infty$ and the expression $\|A\|_\infty = \max_{1\leqq i\leqq n} \sum_{j=1}^{n} |a_{ij}|$ in Theorem 1.12(2) in Sec. 1.6 that all eigenvalues of A are contained in the disk

$$\hat{C} = \{\lambda \in \mathbb{C} \mid |\lambda| \leqq \max_{1\leqq i\leqq n} \sum_{j=1}^{n} |a_{ij}|\} \tag{6.13}$$

on the complex plane.

[2] A similar inequality $\rho(A) \leqq \|A\|$ holds for any norm $\|A\|$ defined by (1.28) for $\|\cdot\|_X = \|\cdot\|_Y$. The proof is similar.

The following theorem, known as the *Gershgorin theorem*, is a refinement of this statement. For each $i = 1, 2, \ldots, n$, we define a disk[3]

$$C_i = \{\lambda \in \mathbb{C} \mid |\lambda - a_{ii}| \leqq \sum_{j \neq i} |a_{ij}|\}, \tag{6.14}$$

of which the center is located at the ith diagonal entry a_{ii} and the radius is equal to the sum $\sum_{j \neq i} |a_{ij}|$ of the absolute values of the off-diagonal entries in the ith row.

Theorem 6.5 (Gershgorin theorem). *Let $A = (a_{ij})$ be an $n \times n$ complex matrix. All eigenvalues of A are contained in the union $C_1 \cup C_2 \cup \cdots \cup C_n$ of n disks $C_1, C_2, \ldots, C_n$ defined by (6.14). More precisely, each connected component of $C_1 \cup C_2 \cup \cdots \cup C_n$ contains as many eigenvalues as the disks that constitute the connected component.*[4]

Proof. Let λ be an eigenvalue of A. Then $\lambda I - A$ is a singular matrix. This implies that $\lambda I - A$ is not a strictly diagonally dominant matrix by (the contrapositive of) Theorem 1.15 in Sec. 1.7.1, stating that a strictly diagonally dominant matrix is nonsingular. Therefore, there exists some i such that

$$|\lambda - a_{ii}| \leqq \sum_{j \neq i} |a_{ij}|,$$

which shows $\lambda \in C_i$. To prove the second statement, let $D = \mathrm{diag}\,(a_{11}, a_{22}, \ldots, a_{nn})$ and consider a parametric family of matrices

$$A(t) = D + t(A - D) \qquad (0 \leqq t \leqq 1)$$

connecting $D = A(0)$ and $A = A(1)$. When the real-valued parameter t changes continuously from 0 to 1, the coefficients of the characteristic polynomial of $A(t)$ and the eigenvalues of $A(t)$ change continuously and each disk defined from $A(t)$ grows monotonically (see Example 6.1). The second statement follows from this observation. □

Example 6.1. Let $A = \begin{bmatrix} 0 & 3 \\ -4 & 6 \end{bmatrix}$. The eigenvalues, λ_1 and λ_2, and the disks in (6.14) are given by

$$\lambda_1 = 3 - \sqrt{3}\,\mathrm{i}, \qquad \lambda_2 = 3 + \sqrt{3}\,\mathrm{i},$$
$$C_1 = \{\lambda \in \mathbb{C} \mid |\lambda| \leqq 3\}, \qquad C_2 = \{\lambda \in \mathbb{C} \mid |\lambda - 6| \leqq 4\}.$$

---

[3]We have $C_i \subseteqq \hat{C}$ for each i, and therefore $C_1 \cup C_2 \cup \cdots \cup C_n \subseteqq \hat{C}$.

[4]If k disks $C_{i_1}, C_{i_2}, \ldots, C_{i_k}$ make up a connected component, k is equal to the number of indices j $(1 \leqq j \leqq n)$ such that $\lambda_j \in C_{i_1} \cup C_{i_2} \cup \cdots \cup C_{i_k}$.

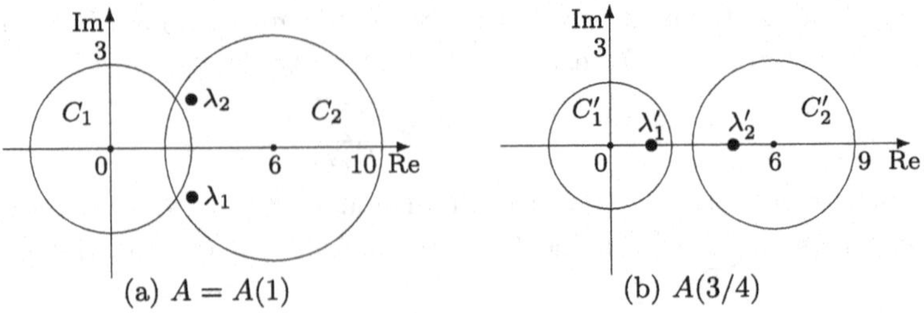

Fig. 6.1 Gershgorin theorem (Example 6.1).

The two eigenvalues λ_1 and λ_2 are both contained in C_2, and the other disk C_1 contains none of them (Fig. 6.1(a)). The disk $\hat{C}$ in (6.13), with center at the origin and radius of 10, contains both C_1 and C_2.

The matrix $A(t)$ parametrized by t ($0 \leqq t \leqq 1$) is given by $A(t) = \begin{bmatrix} 0 & 3t \\ -4t & 6 \end{bmatrix}$. The eigenvalues and the disks are

$$\lambda_1(t) = 3 - \sqrt{9 - 12t^2}, \qquad \lambda_2(t) = 3 + \sqrt{9 - 12t^2},$$
$$C_1(t) = \{\lambda \in \mathbb{C} \mid |\lambda| \leqq 3t\}, \qquad C_2(t) = \{\lambda \in \mathbb{C} \mid |\lambda - 6| \leqq 4t\}.$$

- If t is in the range of $0 \leqq t < 6/7$, the two disks $C_1(t)$ and $C_2(t)$ are disjoint ($C_1(t) \cap C_2(t) = \emptyset$) and each of them forms a connected component. We have $\lambda_1(t) \in C_1(t)$ and $\lambda_2(t) \in C_2(t)$. The configuration at $t = 3/4$ ($< 6/7$) is depicted in Fig. 6.1(b), where $\lambda_1' = \lambda_1(3/4) = 3/2$, $\lambda_2' = \lambda_2(3/4) = 9/2$, and $C_i' = C_i(3/4)$ for $i = 1, 2$.
- If $t = 6/7$, the two disks $C_1(t)$ and $C_2(t)$ touch at a point $\lambda_1(t) = 18/7$. Hence $C_1(t) \cup C_2(t)$ forms a single connected component, and we have $\lambda_1(t) \in C_1(t) \cap C_2(t)$ and $\lambda_2(t) = 24/7 \in C_2(t) \setminus C_1(t)$.
- If t is in the range of $6/7 < t \leqq 1$, $C_1(t) \cup C_2(t)$ forms a single connected component. We have $\{\lambda_1(t), \lambda_2(t)\} \subseteqq C_2(t) \setminus C_1(t)$, whereas $C_1(t)$ contains neither eigenvalue. It is noted that $\lambda_1(t)$ and $\lambda_2(t)$ are real if $t \leqq \sqrt{3}/2$, and they are complex conjugate if $t > \sqrt{3}/2$.

■

Eigenvalues remain invariant under a similarity transformation $S^{-1}AS$ (Theorem 6.2), whereas the disks in (6.14) for the Gershgorin theorem vary with the matrix S. In some cases we can take advantage of this fact to obtain a sharper bound for eigenvalues through a judicious choice of a nonsingular matrix S. This is illustrated in the following example.

Example 6.2. In Sec. 5.5 we have treated a finite difference approximation for the Dirichlet problem (Laplace equation under a Dirichlet boundary condition). A system of linear equations (5.16), or $A\boldsymbol{u} = \boldsymbol{g}$ in (5.17), has been obtained through a natural finite difference approximation (5.15) to the second partial derivatives. As an example, the coefficient matrix A for $N = 5$ is given in (5.18). Since the matrix A is symmetric, all eigenvalues of A are real numbers by Theorem 6.6 in Sec. 6.3.

Let us apply the Gershgorin theorem to this matrix A. The diagonal entries are all equal to 4, which implies that every disk C_i in (6.14) is centered at 4. The radius of C_i is equal to 2, 3, or 4. Therefore, the union of these disks is a disk of radius 4 centered at 4. It then follows from the Gershgorin theorem that all the eigenvalues of A are contained in the closed interval $[0, 8]$ of real numbers.

Next we obtain a sharper estimate with the aid of a similarity transformation $D^{-1}AD$ using the diagonal matrix

$$D = \mathrm{diag}\,(\overbrace{d_1, \ldots, d_1}^{(N-1)}\ ;\ \overbrace{d_2, \ldots, d_2}^{(N-1)}\ ;\ \ldots ; \overbrace{d_{N-1}, \ldots, d_{N-1}}^{(N-1)}\ )$$

introduced in (5.20). The diagonal entries of $D^{-1}AD$ are all equal to 4, and the union of disks C_i in (6.14) is a disk centered at 4 of radius equal to

$$r = \max_{1 \leqq j \leqq N-1} \frac{d_{j-1} + 2d_j + d_{j+1}}{d_j},$$

where $d_0 = d_N = 0$. We may use any positive numbers $d_1, d_2, \ldots, d_{N-1}$ to obtain an interval $[4 - r, 4 + r]$ to embrace all eigenvalues. With the choice of $d_j = j(N - j)$ $(j = 1, 2, \ldots, N - 1)$ we obtain

$$r = 4 - \min_{1 \leqq j \leqq N-1} \frac{2}{j(N-j)} \leqq 4 - \frac{8}{N^2},$$

which gives us a narrower interval $[8/N^2, 8 - 8/N^2]$ to embrace all eigenvalues. This sharper bound reveals, in particular, that the smallest eigenvalue is a positive number of the order $\mathrm{O}(1/N^2)$, which is an important and useful fact in applications.

It is worth mentioning that explicit formulas of the eigenvalues and eigenvectors of this matrix A are known. Define

$$\lambda_{kl} = 4 - 2\cos\left(\frac{k\pi}{N}\right) - 2\cos\left(\frac{l\pi}{N}\right) \qquad (1 \leqq k, l \leqq N - 1), \tag{6.15}$$

$$\boldsymbol{v}_k = \left(\sin\left(\frac{k\pi}{N}\right), \sin\left(\frac{2k\pi}{N}\right), \ldots, \sin\left(\frac{(N-1)k\pi}{N}\right)\right)^{\top} \quad (1 \leqq k \leqq N - 1). \tag{6.16}$$

Then we have

$$A(\boldsymbol{v}_k \otimes \boldsymbol{v}_l) = \lambda_{kl}(\boldsymbol{v}_k \otimes \boldsymbol{v}_l) \qquad (1 \leqq k, l \leqq N-1), \tag{6.17}$$

which can be verified by straightforward calculation[5] using the expression $A = 4\, I \otimes I - B \otimes I - I \otimes B$ in (5.22), where B denotes the matrix defined in (5.21). The smallest eigenvalue is

$$\lambda_{11} = 4\left(1 - \cos\left(\frac{\pi}{N}\right)\right) = 8\sin^2\left(\frac{\pi}{2N}\right) \approx \frac{2\pi^2}{N^2},$$

which is a positive number of the order $\mathrm{O}(1/N^2)$. ■

6.1.4 *Generalized Eigenvalue Problems*

Eigenvalue problems in engineering applications are not always given in the form of $A\boldsymbol{x} = \lambda\boldsymbol{x}$ in (6.1), but often take a more general form

$$A\boldsymbol{x} = \lambda B\boldsymbol{x} \tag{6.18}$$

in terms of a pair (A, B) of square matrices A and B. An eigenvalue problem of this form is called a *generalized eigenvalue problem.* In contrast, the problem of the form of (6.1) is referred to as a *standard eigenvalue problem.*

If the matrix B is nonsingular, the problem in (6.18) can be transformed to

$$B^{-1}A\boldsymbol{x} = \lambda\boldsymbol{x},$$

which is an eigenvalue problem in the standard form (6.1). Even when B is singular, we can carry out a similar reduction as long as $\det(A - \lambda B)$ is not zero as a polynomial in λ. If $\det(A - \lambda B) \neq 0$, we can choose a real (or complex) number c for which $A - cB$ is nonsingular. Then (6.18) is transformed equivalently to

$$(A - cB)^{-1}B\boldsymbol{x} = \lambda'\boldsymbol{x},$$

where $\lambda' = 1/(\lambda - c)$.

Consider a generalized eigenvalue problem (6.18) defined by a symmetric matrix A and a positive definite symmetric matrix[6] B. As will be explained in Sec. 6.3, the eigenvalue problem for a symmetric matrix is much easier to handle, and hence it is desirable to exploit the symmetry as follows, whereas $B^{-1}A$ is not symmetric in general. Since B is positive definite, we can

---

[5]To derive (6.17), we use the angle addition formula for the sine, which reads: $\sin\left(\frac{(i-1)k\pi}{N}\right) + \sin\left(\frac{(i+1)k\pi}{N}\right) = 2\cos\left(\frac{k\pi}{N}\right) \cdot \sin\left(\frac{ik\pi}{N}\right)$.

[6]See Sec. 7.2 for positive definite symmetric matrices.

decompose it as $B = SS^\top$ using a nonsingular matrix S (cf., Theorem 7.1(d) in Sec. 7.2.1), and then (6.18) is transformed to

$$S^{-1}A(S^{-1})^\top \boldsymbol{y} = \lambda \boldsymbol{y}, \qquad \boldsymbol{y} = S^\top \boldsymbol{x},$$

which is a standard eigenvalue problem for a symmetric matrix. Note that $S^{-1}A(S^{-1})^\top$ is a symmetric matrix that has the same eigenvalues as $B^{-1}A$.

6.2 Engineering Significances of Eigenvalues

Eigenvalues appear in many different contexts of engineering. In this section we show some typical examples.

6.2.1 *Linear Dynamical Systems*

Suppose that a dynamical system is described by a system of differential equations of the form

$$\frac{\mathrm{d}\boldsymbol{x}}{\mathrm{d}t}(t) = A\boldsymbol{x}(t) \qquad (t > 0), \tag{6.19}$$

where $\boldsymbol{x}(t)$ is a vector representing the state of the system at time t and A is a square matrix. It is known that each component of the solution $\boldsymbol{x}(t)$ is a linear combination of functions of the form $t^k \mathrm{e}^{\lambda t}$, where the constant λ in the exponential part is an eigenvalue of matrix A and the exponent k in the polynomial part is determined from the multiplicity of the eigenvalue λ; see the Jordan normal form in Sec. 6.8 for details. In addition, as is discussed in Sec. 5.7, the (asymptotic) stability of a dynamical system is equivalent to the condition that every eigenvalue of A has a negative real part.

In control theory [50, 52, 60, 61], a system (more precisely, a time-invariant linear dynamical system) is described by a *state–space equation*:

$$\frac{\mathrm{d}\boldsymbol{x}}{\mathrm{d}t}(t) = A\boldsymbol{x}(t) + B\boldsymbol{u}(t) \tag{6.20}$$

that involves a control input vector $\boldsymbol{u}(t)$. If the input vector is determined from the state vector as[7] $\boldsymbol{u}(t) = K\boldsymbol{x}(t)$ with a constant matrix K, the state $\boldsymbol{x}(t)$ of the system is determined by the differential equation:

$$\frac{\mathrm{d}\boldsymbol{x}}{\mathrm{d}t}(t) = (A + BK)\boldsymbol{x}(t). \tag{6.21}$$

By taking advantage of the fact that the eigenvalues of the matrix $A + BK$ vary with the matrix K, which we can choose, we can design a control

[7] This is called *state feedback*.

system having desired characteristics. It is known that the eigenvalues of the matrix $A + BK$ can be changed to an arbitrarily specified set of n complex numbers by choosing an appropriate matrix K if and only if the condition of controllability in (4.11) is satisfied by (A, B).

6.2.2 *Principal Component Analysis*

Let $X_1, X_2, \ldots, X_n$ be mutually correlated n random variables (observed or measured data). We denote their expected values (means) by $m_1, m_2, \ldots, m_n$, and their covariance by[8]

$$c_{ij} = \mathrm{E}[(X_i - m_i)(X_j - m_j)] \qquad (i, j = 1, 2, \ldots, n).$$

The matrix $C = (c_{ij})$ is a symmetric matrix, called the *covariance matrix* or *variance–covariance matrix*. The correlation structure of the random variables $X_1, X_2, \ldots, X_n$ can be revealed by investigating the eigenvalues and eigenvectors of the (symmetric) matrix C. *Principal component analysis* is such a statistical method to extract information from the covariance matrix. Eigenvalues and eigenvectors of symmetric matrices play important roles also for other statistical methods [45, 63].

6.2.3 *Vibration Analysis of Structures*

As is well known, the oscillation of a *spring–mass system* is described by a differential equation of the form

$$m \frac{\mathrm{d}^2 x}{\mathrm{d}t^2}(t) = -kx(t). \tag{6.22}$$

Here m denotes the mass of a particle, k is a spring constant, and $x(t)$ denotes the displacement at time t of the particle from the equilibrium position. A simple harmonic motion $x(t) = a \sin(\omega t + \phi)$ gives the solution to (6.22), where the angular frequency ω is determined from the relation $\omega^2 m = k$.

For a multi-degree-of-freedom system involving many springs and particles, the equation in (6.22) is generalized to a differential equation

$$M \frac{\mathrm{d}^2 \boldsymbol{x}}{\mathrm{d}t^2}(t) = -K\boldsymbol{x}(t) \tag{6.23}$$

in a vector $\boldsymbol{x}(t)$ of displacements of the particles, where M is a symmetric matrix representing masses and K is also a symmetric matrix representing spring constants. The dynamical property of a structure consisting of

[8] Notation $\mathrm{E}[\,\cdot\,]$ means the expected value of a random variable. We assume that every c_{ij} is finite. See, *e.g.*, [47, 49, 51, 58, 66, 79] for background on random variables.

trusses and beams can also be described by an equation of the form of (6.23); see Sec. 7.4.2. Substituting $\boldsymbol{x}(t) = \boldsymbol{a}\sin(\omega t + \phi)$ into (6.23) we obtain

$$\omega^2 M\boldsymbol{a} = K\boldsymbol{a}. \tag{6.24}$$

That is, a function $\boldsymbol{x}(t) = \boldsymbol{a}\sin(\omega t + \phi)$ is a solution to (6.23) if and only if ω and $\boldsymbol{a}$ satisfy the equation in (6.24). Such ω is called a *natural angular frequency* and $\boldsymbol{a}$ is called an *eigenmode.* The equation (6.24) is a generalized eigenvalue problem introduced in Sec. 6.1.4.

6.2.4 *Stationary Distribution of Markov Chains*

Suppose that there is a system with a finite number of states indexed by $\{1, 2, \ldots, n\}$ and the system stochastically changes its state at discrete time $k = 0, 1, 2, \ldots$. The probability of a transition from state i to state j, denoted by p_{ij}, is assumed to be independent of the time k at which the transition occurs. Such a system is called a *Markov chain.* The matrix $P = (p_{ij})$ consisting of the transition probabilities p_{ij} is called the *transition probability matrix.*

Let $p_i^{(k)}$ denote the probability that the system is in state i at time k, and define a row vector $\boldsymbol{p}^{(k)} = (p_1^{(k)}, p_2^{(k)}, \ldots, p_n^{(k)})$. By the definition of transition probability we have the following equation

$$\boldsymbol{p}^{(k+1)} = \boldsymbol{p}^{(k)} P \qquad (k = 0, 1, 2, \ldots). \tag{6.25}$$

If a *stationary distribution* (probability distribution that is invariant under state transitions) exists, it is represented by a row vector $\boldsymbol{\pi}$ satisfying $\boldsymbol{\pi} = \boldsymbol{\pi} P$. That is, a stationary distribution is an eigenvector of $P^\top$ for eigenvalue 1.

We can capture the properties of a Markov chain by investigating eigenvalues and eigenvectors of the transition probability matrix P. In this analysis we have to take into account that the entries of matrix P and the components of vector $\boldsymbol{\pi}$ are nonnegative numbers representing probabilities. Nonnegative matrices are discussed in detail in Chapter 2 of Volume II [9].

6.3 Eigenvalues of Symmetric Matrices

In this section we assume that A is a symmetric (real) matrix of order n.

6.3.1 *Diagonalization by Orthogonal Transformation*

Theorem 6.6. *Eigenvalues of a symmetric matrix A are all real numbers.*

Proof. Assume $A\boldsymbol{x} = \lambda\boldsymbol{x}$ for a complex number λ and complex vector $\boldsymbol{x}$ ($\neq \mathbf{0}$). Then we have $\boldsymbol{x}^* A\boldsymbol{x} = \lambda\boldsymbol{x}^*\boldsymbol{x} = \lambda\|\boldsymbol{x}\|_2{}^2$. Since A is symmetric, we have $(\boldsymbol{x}^* A\boldsymbol{x})^* = \boldsymbol{x}^* A^* \boldsymbol{x} = \boldsymbol{x}^* A\boldsymbol{x}$, which shows that $\boldsymbol{x}^* A\boldsymbol{x}$ is a real number. Therefore, λ is real. □

For a real eigenvalue λ of a real matrix, we can take an eigenvector $\boldsymbol{x}$ that has real components.[9] In this book, accordingly, we always take a real vector as an eigenvector for a symmetric matrix.

Theorem 6.7. *Eigenvectors of a symmetric matrix A corresponding to distinct eigenvalues are orthogonal to each other. That is, if $A\boldsymbol{x} = \lambda\boldsymbol{x}$, $A\boldsymbol{y} = \mu\boldsymbol{y}$, and $\lambda \neq \mu$, then $\boldsymbol{x}^\top \boldsymbol{y} = 0$.*

Proof. It follows from $A^\top = A$ that

$$\lambda(\boldsymbol{x}^\top \boldsymbol{y}) = (\lambda\boldsymbol{x})^\top \boldsymbol{y} = (A\boldsymbol{x})^\top \boldsymbol{y} = \boldsymbol{x}^\top A^\top \boldsymbol{y} = \boldsymbol{x}^\top (A\boldsymbol{y}) = \boldsymbol{x}^\top (\mu\boldsymbol{y}) = \mu(\boldsymbol{x}^\top \boldsymbol{y}).$$

If $\lambda \neq \mu$, we must have $\boldsymbol{x}^\top \boldsymbol{y} = 0$. □

In Theorem 6.2 we have observed that the eigenvalues of a matrix A are invariant under a similarity transformation $S^{-1}AS$ with a nonsingular matrix S. When A is symmetric, it is natural to restrict S to an orthogonal matrix so that the transformed matrix $S^{-1}AS$ is also symmetric. That is, for a symmetric matrix A, we consider an orthogonal similarity transformation $Q^\top AQ$ with an orthogonal matrix Q.

Theorem 6.8. *Eigenvalues of a symmetric matrix remain invariant under an orthogonal similarity transformation. To be more precise, if $A\boldsymbol{x} = \lambda\boldsymbol{x}$, $\boldsymbol{x} \neq \mathbf{0}$, and $B = Q^\top AQ$ for an orthogonal matrix Q, then $B\boldsymbol{y} = \lambda\boldsymbol{y}$, $\boldsymbol{y} \neq \mathbf{0}$ for $\boldsymbol{y} = Q^\top \boldsymbol{x}$.*

Proof. This follows from the rewriting of $A\boldsymbol{x} = \lambda\boldsymbol{x}$ to $(Q^\top AQ)(Q^\top \boldsymbol{x}) = \lambda(Q^\top \boldsymbol{x})$. □

Any symmetric matrix A can be brought to a diagonal matrix by an orthogonal similarity transformation. The matrix for this transformation can be obtained as a collection of eigenvectors of A.

[9] This is because the eigenvector $\boldsymbol{x}$ is determined as a solution to $(A - \lambda I)\boldsymbol{x} = \mathbf{0}$, which is a system of equations with real coefficients.

Theorem 6.9. *For a symmetric matrix A, there exists an orthogonal matrix Q for which $Q^\top AQ$ is a diagonal matrix, i.e.,*

$$Q^\top AQ = \mathrm{diag}\,(\lambda_1, \lambda_2, \ldots, \lambda_n) \tag{6.26}$$

for some $\lambda_1, \lambda_2, \ldots, \lambda_n$. With the notation $\boldsymbol{q}_1, \boldsymbol{q}_2, \ldots, \boldsymbol{q}_n$ for the column vectors of Q, i.e., $Q = [\boldsymbol{q}_1, \boldsymbol{q}_2, \ldots, \boldsymbol{q}_n]$, this implies that

$$A\boldsymbol{q}_j = \lambda_j \boldsymbol{q}_j \qquad (j = 1, 2, \ldots, n), \tag{6.27}$$

showing that $\lambda_1, \lambda_2, \ldots, \lambda_n$ are eigenvalues of A and $\boldsymbol{q}_1, \boldsymbol{q}_2, \ldots, \boldsymbol{q}_n$ are the corresponding eigenvectors.

Proof. Take an eigenvalue λ_1 of A and a corresponding eigenvector $\boldsymbol{q}_1$. By Theorem 6.6, λ_1 is a real number, which implies that we can choose a real vector for $\boldsymbol{q}_1$. Normalize $\boldsymbol{q}_1$ to unit length ($\boldsymbol{q}_1^\top \boldsymbol{q}_1 = 1$). Let Q_1 be an orthogonal matrix of order n that has $\boldsymbol{q}_1$ in the first column. It follows from $A\boldsymbol{q}_1 = \lambda_1 \boldsymbol{q}_1$ and the symmetry that $Q_1^\top AQ_1$ takes the form of

$$Q_1^\top AQ_1 = \left[\begin{array}{c|ccc} \lambda_1 & 0 & \cdots & 0 \\ \hline 0 & & & \\ \vdots & & A_2 & \\ 0 & & & \end{array}\right].$$

We apply a similar transformation to the symmetric matrix A_2 of order $n-1$ using its eigenvalue λ_2. That is, for some orthogonal matrix P_2 of order $n-1$, we have

$$P_2^\top A_2 P_2 = \left[\begin{array}{c|ccc} \lambda_2 & 0 & \cdots & 0 \\ \hline 0 & & & \\ \vdots & & A_3 & \\ 0 & & & \end{array}\right].$$

By defining $Q_2 = \left[\begin{array}{c|c} 1 & \boldsymbol{0}^\top \\ \hline \boldsymbol{0} & P_2 \end{array}\right]$, which is an orthogonal matrix of order n, we obtain

$$(Q_1Q_2)^\top A(Q_1Q_2) = \left[\begin{array}{c|c|cccc} \lambda_1 & 0 & \cdots & & \cdots & 0 \\ \hline 0 & \lambda_2 & 0 & & \cdots & 0 \\ \hline \vdots & 0 & & & & \\ \vdots & \vdots & & & A_3 & \\ 0 & 0 & & & & \end{array}\right].$$

Continuing this process we can construct an orthogonal matrix Q such that $Q^\top AQ$ is a diagonal matrix as in (6.26).

By multiplying (6.26) with Q from the left, we obtain

$$A[\boldsymbol{q}_1, \boldsymbol{q}_2, \ldots, \boldsymbol{q}_n] = [\boldsymbol{q}_1, \boldsymbol{q}_2, \ldots, \boldsymbol{q}_n] \cdot \mathrm{diag}\,(\lambda_1, \lambda_2, \ldots, \lambda_n),$$

which shows (6.27). The expression (6.27) means that $\boldsymbol{q}_j$ is an eigenvector of A corresponding to an eigenvalue λ_j of A. □

The matrix Q for diagonalization (6.26) is not uniquely determined, but any matrix Q valid for diagonalization must consist of eigenvectors of A, as shown in (6.27). The following proposition shows a description of all eigenvectors in terms of a particular choice of Q.

Proposition 6.1. *Let $B(\lambda) = \{j \mid \lambda_j = \lambda,\ 1 \leqq j \leqq n\}$ for any real number λ. A real vector $\boldsymbol{x}$ satisfies $A\boldsymbol{x} = \lambda\boldsymbol{x}$ if and only if $\boldsymbol{x}$ can be expressed as a linear combination*[10]

$$\boldsymbol{x} = \sum_{j \in B(\lambda)} c_j \boldsymbol{q}_j \tag{6.28}$$

of $\boldsymbol{q}_j$ $(j \in B(\lambda))$ with real coefficients c_j for $j \in B(\lambda)$.

Proof. Since $\{\boldsymbol{q}_1, \boldsymbol{q}_2, \ldots, \boldsymbol{q}_n\}$ is a basis, we can represent an arbitrary vector $\boldsymbol{x}$ as $\boldsymbol{x} = \sum_{j=1}^{n} c_j \boldsymbol{q}_j$. By (6.27) we have $A\boldsymbol{x} = \sum_{j=1}^{n} c_j \lambda_j \boldsymbol{q}_j$. Then it follows from the linear independence of $\{\boldsymbol{q}_1, \boldsymbol{q}_2, \ldots, \boldsymbol{q}_n\}$ that

$$\begin{aligned} A\boldsymbol{x} = \lambda\boldsymbol{x} &\iff \sum_{j=1}^{n} c_j(\lambda_j - \lambda)\boldsymbol{q}_j = \boldsymbol{0} \\ &\iff c_j(\lambda_j - \lambda) = 0 \quad (j = 1, 2, \ldots, n) \\ &\iff c_j = 0 \quad (j \notin B(\lambda)). \end{aligned}$$

□

Theorem 6.9 shows the existence of an orthonormal basis $\{\boldsymbol{q}_1, \boldsymbol{q}_2, \ldots, \boldsymbol{q}_n\}$ of $\mathbb{R}^n$ consisting of eigenvectors of A. In the proof of Theorem 6.9 we have constructed $Q = [\boldsymbol{q}_1, \boldsymbol{q}_2, \ldots, \boldsymbol{q}_n]$ iteratively, but such iterative construction is not mandatory. All we need is a set of eigenvalues of A and the (unit-length) eigenvectors that are orthogonal to each other. For an

[10]If λ is not an eigenvalue of A, we have $B(\lambda) = \emptyset$ and the right-hand side of (6.28) is equal to $\boldsymbol{0}$. If λ is an eigenvalue of A, the number of elements of $B(\lambda)$ is the multiplicity of λ.

eigenvalue λ of multiplicity two or larger we choose an arbitrary orthonormal basis of $\mathrm{Ker}\,(A - \lambda I)$. Recall that the matrix Q in Theorem 6.9 is not uniquely determined.

Example 6.3. For a symmetric matrix

$$A = \begin{bmatrix} 0 & 1 & -2 \\ 1 & 0 & 2 \\ -2 & 2 & -3 \end{bmatrix}$$

we carry out the diagonalization in Theorem 6.9. The characteristic polynomial is given by

$$\det(\lambda I - A) = \begin{vmatrix} \lambda & -1 & 2 \\ -1 & \lambda & -2 \\ 2 & -2 & \lambda + 3 \end{vmatrix} = (\lambda + 5)(\lambda - 1)^2,$$

from which we obtain eigenvalues $\lambda_1 = -5$ and $\lambda_2 = \lambda_3 = 1$ (multiplicity 2).

For eigenvalue $\lambda_1 = -5$, the corresponding eigenvector $\boldsymbol{x} = (x_1, x_2, x_3)^\top$ is determined as a solution to the equation

$$(A + 5I)\boldsymbol{x} = \begin{bmatrix} 5 & 1 & -2 \\ 1 & 5 & 2 \\ -2 & 2 & 2 \end{bmatrix} \begin{bmatrix} x_1 \\ x_2 \\ x_3 \end{bmatrix} = \begin{bmatrix} 0 \\ 0 \\ 0 \end{bmatrix}.$$

By the elimination method (Sec. 5.3) we transform the coefficient matrix to its reduced echelon form as

$$\begin{bmatrix} 5 & 1 & -2 \\ 1 & 5 & 2 \\ -2 & 2 & 2 \end{bmatrix} \longrightarrow \begin{bmatrix} 1 & 5 & 2 \\ 0 & -24 & -12 \\ 0 & 12 & 6 \end{bmatrix} \longrightarrow \begin{bmatrix} 1 & 0 & -1/2 \\ 0 & 1 & 1/2 \\ 0 & 0 & 0 \end{bmatrix}.$$

According to (5.11), we obtain $\boldsymbol{x} = \alpha(1/2,\ -1/2,\ 1)^\top$ (α: arbitrary real number). Hence we can take $\boldsymbol{q}_1 = (1/\sqrt{6},\ -1/\sqrt{6},\ 2/\sqrt{6})^\top$ as a unit-length eigenvector of A.

For the eigenvalue $\lambda_2 = \lambda_3 = 1$, the corresponding eigenvectors are determined from the equation

$$(A - I)\boldsymbol{x} = \begin{bmatrix} -1 & 1 & -2 \\ 1 & -1 & 2 \\ -2 & 2 & -4 \end{bmatrix} \begin{bmatrix} x_1 \\ x_2 \\ x_3 \end{bmatrix} = \begin{bmatrix} 0 \\ 0 \\ 0 \end{bmatrix}.$$

This system of equations is in fact equivalent to a single equation $-x_1 + x_2 - 2x_3 = 0$, and the solution is represented as

$$\begin{bmatrix} x_1 \\ x_2 \\ x_3 \end{bmatrix} = \alpha \begin{bmatrix} 1 \\ 1 \\ 0 \end{bmatrix} + \beta \begin{bmatrix} -2 \\ 0 \\ 1 \end{bmatrix}$$

with two real parameters α and β. By applying the Gram–Schmidt orthogonalization method (cf., Theorem 9.36 in Sec. 9.6.3) to the two vectors $(1,\,1,\,0)^\top$ and $(-2,\,0,\,1)^\top$ on the right-hand side, we can obtain mutually orthogonal unit-length eigenvectors

$$\boldsymbol{q}_2 = (1/\sqrt{2},\, 1/\sqrt{2},\, 0)^\top, \qquad \boldsymbol{q}_3 = (-1/\sqrt{3},\, 1/\sqrt{3},\, 1/\sqrt{3})^\top.$$

Thus we obtain an orthogonal matrix

$$Q = [\boldsymbol{q}_1, \boldsymbol{q}_2, \boldsymbol{q}_3] = \begin{bmatrix} 1/\sqrt{6} & 1/\sqrt{2} & -1/\sqrt{3} \\ -1/\sqrt{6} & 1/\sqrt{2} & 1/\sqrt{3} \\ 2/\sqrt{6} & 0 & 1/\sqrt{3} \end{bmatrix},$$

for which $Q^\top AQ = \mathrm{diag}\,(-5,1,1)$. ■

Theorem 6.10. *For a symmetric matrix A, the sum of squares of its eigenvalues is equal to the square of the Frobenius norm $\|A\|_{\mathrm{F}}$. That is,*

$$\sum_{i=1}^{n} \lambda_i{}^2 = \|A\|_{\mathrm{F}}{}^2. \tag{6.29}$$

Proof. It follows from (1.27) and (6.26) that $\|A\|_{\mathrm{F}}{}^2 = \|Q^\top AQ\|_{\mathrm{F}}{}^2 = \sum_{i=1}^{n} \lambda_i{}^2$. □

6.3.2 *Minimax Theorems*

Eigenvalues of a symmetric matrix can be characterized as extremal values of a quadratic form. We first present a theorem for the largest and smallest eigenvalues, and then go on to general eigenvalues.

Let $\lambda_1 \geqq \lambda_2 \geqq \cdots \geqq \lambda_n$ denote the eigenvalues of a symmetric matrix A of order n arranged in a descending order. The corresponding eigenvectors are denoted by $\boldsymbol{q}_1, \boldsymbol{q}_2, \ldots, \boldsymbol{q}_n$, where they are assumed to form an orthonormal system. For any $\boldsymbol{x} \neq \boldsymbol{0}$, the expression

$$\rho(\boldsymbol{x}) = \frac{\boldsymbol{x}^\top A\boldsymbol{x}}{\boldsymbol{x}^\top \boldsymbol{x}} \tag{6.30}$$

is called the *Rayleigh quotient*. If $\boldsymbol{x}$ is represented as a linear combination of the eigenvectors as $\boldsymbol{x} = c_1\boldsymbol{q}_1 + c_2\boldsymbol{q}_2 + \cdots + c_n\boldsymbol{q}_n$, it follows from $A\boldsymbol{q}_j = \lambda_j\boldsymbol{q}_j$ $(j = 1, 2, \ldots, n)$ and the orthonormality of $\boldsymbol{q}_1, \boldsymbol{q}_2, \ldots, \boldsymbol{q}_n$ that

$$\rho(\boldsymbol{x}) = \frac{\lambda_1 c_1{}^2 + \lambda_2 c_2{}^2 + \cdots + \lambda_n c_n{}^2}{c_1{}^2 + c_2{}^2 + \cdots + c_n{}^2}. \tag{6.31}$$

Theorem 6.11. *For the largest eigenvalue* λ_1 *and the smallest eigenvalue* λ_n *of a symmetric matrix* A *of order* n*, we have the following formulas:*

$$\lambda_1 = \max_{\boldsymbol{x} \neq \boldsymbol{0}} \frac{\boldsymbol{x}^\top A\boldsymbol{x}}{\boldsymbol{x}^\top \boldsymbol{x}} = \max_{\boldsymbol{x}^\top \boldsymbol{x}=1} \boldsymbol{x}^\top A\boldsymbol{x}, \tag{6.32}$$

$$\lambda_n = \min_{\boldsymbol{x} \neq \boldsymbol{0}} \frac{\boldsymbol{x}^\top A\boldsymbol{x}}{\boldsymbol{x}^\top \boldsymbol{x}} = \min_{\boldsymbol{x}^\top \boldsymbol{x}=1} \boldsymbol{x}^\top A\boldsymbol{x}. \tag{6.33}$$

Proof. The right-hand side of (6.31) is a weighted mean of $\lambda_1, \lambda_2, \ldots, \lambda_n$. It takes the maximum value λ_1 when $c_1 = 1$ and $c_2 = c_3 = \cdots = c_n = 0$, and the minimum value λ_n when $c_1 = c_2 = \cdots = c_{n-1} = 0$ and $c_n = 1$. □

Not only the largest and smallest eigenvalues, but also other intermediate eigenvalues admit similar expressions, known as the *Courant–Fischer theorem*. As a preparation for the proof, we note the following technical fact.

Proposition 6.2. *Let* U_k *denote the subspace*[11] *spanned by* $\{\boldsymbol{q}_1, \boldsymbol{q}_2, \ldots, \boldsymbol{q}_k\}$*, and* W_k *the subspace spanned by* $\{\boldsymbol{q}_k, \boldsymbol{q}_{k+1}, \ldots, \boldsymbol{q}_n\}$*. If* $\boldsymbol{x} \in U_k$*, then* $\rho(\boldsymbol{x}) \geqq \lambda_k$*. If* $\boldsymbol{x} \in W_k$*, then* $\rho(\boldsymbol{x}) \leqq \lambda_k$*.*

Proof. If $\boldsymbol{x} \in U_k$, we have $c_{k+1} = c_{k+2} = \cdots = c_n = 0$ in (6.31), and hence

$$\rho(\boldsymbol{x}) = \frac{\lambda_1 {c_1}^2 + \lambda_2 {c_2}^2 + \cdots + \lambda_k {c_k}^2}{{c_1}^2 + {c_2}^2 + \cdots + {c_k}^2} \geqq \frac{\lambda_k ({c_1}^2 + {c_2}^2 + \cdots + {c_k}^2)}{{c_1}^2 + {c_2}^2 + \cdots + {c_k}^2} = \lambda_k.$$

Similarly, if $\boldsymbol{x} \in W_k$, we have $c_1 = c_2 = \cdots = c_{k-1} = 0$, and hence $\rho(\boldsymbol{x}) \leqq \lambda_k$. □

Theorem 6.12 (Courant–Fischer theorem). *For the eigenvalues* $\lambda_1 \geqq \lambda_2 \geqq \cdots \geqq \lambda_n$ *of a symmetric matrix* A *of order* n*, we have the following formulas:*

$$\lambda_k = \max_{S:\, \dim S = k} \ \min_{\boldsymbol{x} \in S \setminus \{\boldsymbol{0}\}} \frac{\boldsymbol{x}^\top A\boldsymbol{x}}{\boldsymbol{x}^\top \boldsymbol{x}} \quad (k = 1, 2, \ldots, n), \tag{6.34}$$

$$\lambda_k = \min_{S:\, \dim S = n-k+1} \ \max_{\boldsymbol{x} \in S \setminus \{\boldsymbol{0}\}} \frac{\boldsymbol{x}^\top A\boldsymbol{x}}{\boldsymbol{x}^\top \boldsymbol{x}} \quad (k = 1, 2, \ldots, n), \tag{6.35}$$

where S $(\subseteqq \mathbb{R}^n)$ *runs over all subspaces of the specified dimension.*

Proof. We prove (6.34) on the basis of Proposition 6.2. We have $\rho(\boldsymbol{x}) \geqq \lambda_k$ for any $\boldsymbol{x} \in U_k$, whereas $\boldsymbol{q}_k \in U_k$ and $\rho(\boldsymbol{q}_k) = \lambda_k$ by (6.31). Therefore,

$$\min_{\boldsymbol{x} \in S \setminus \{\boldsymbol{0}\}} \rho(\boldsymbol{x}) = \lambda_k \quad \text{for } S = U_k, \tag{6.36}$$

[11] For subspaces, see Sec. 9.2 (Definition 9.4 in particular).

where $\dim U_k = k$. Next, take any S with $\dim S = k$. By the inequality in Remark 9.11 in Sec. 9.4.3, we have

$$\dim(S \cap W_k) \geqq \dim S + \dim W_k - n = k + (n - k + 1) - n = 1,$$

which implies that $S \cap W_k$ contains a nonzero vector, say, $\boldsymbol{y}$. By Proposition 6.2 we have $\rho(\boldsymbol{y}) \leqq \lambda_k$, and hence

$$\min_{\boldsymbol{x} \in S \setminus \{\boldsymbol{0}\}} \rho(\boldsymbol{x}) \leqq \min_{\boldsymbol{x} \in (S \cap W_k) \setminus \{\boldsymbol{0}\}} \rho(\boldsymbol{x}) \leqq \rho(\boldsymbol{y}) \leqq \lambda_k.$$

Therefore,

$$\max_{S:\, \dim S = k} \min_{\boldsymbol{x} \in S \setminus \{\boldsymbol{0}\}} \rho(\boldsymbol{x}) \leqq \lambda_k. \tag{6.37}$$

The desired identity (6.34) follows from (6.36) and (6.37).

To prove the second identity (6.35), observe that the $(n-k+1)$th largest eigenvalue of the matrix $-A$ is equal to $-\lambda_k$. By applying (6.34) to $-A$ we obtain

$$\begin{aligned} -\lambda_k &= \max_{S:\, \dim S = n-k+1} \min_{\boldsymbol{x} \in S \setminus \{\boldsymbol{0}\}} \frac{\boldsymbol{x}^\top (-A) \boldsymbol{x}}{\boldsymbol{x}^\top \boldsymbol{x}} \\ &= - \min_{S:\, \dim S = n-k+1} \max_{\boldsymbol{x} \in S \setminus \{\boldsymbol{0}\}} \frac{\boldsymbol{x}^\top A \boldsymbol{x}}{\boldsymbol{x}^\top \boldsymbol{x}}, \end{aligned}$$

which establishes (6.35). □

The following theorem relates the eigenvalues of A to those of a smaller matrix $V^\top AV$ derived from A. This fact is used in numerical computation of eigenvalues.

Theorem 6.13. *Let A be an $n \times n$ symmetric matrix and V an $n \times m$ matrix satisfying $V^\top V = I_m$, where $m \leqq n$. The following inequalities hold between the eigenvalues $\lambda_1 \geqq \lambda_2 \geqq \cdots \geqq \lambda_n$ of A and the eigenvalues $\mu_1 \geqq \mu_2 \geqq \cdots \geqq \mu_m$ of $V^\top AV$:*

$$\lambda_k \geqq \mu_k \geqq \lambda_{k+n-m} \qquad (k = 1, 2, \ldots, m). \tag{6.38}$$

Proof. By applying Theorem 6.12 to $V^\top AV$, we obtain

$$\mu_k = \max_{\dim S' = k} \min_{\boldsymbol{y} \in S' \setminus \{\boldsymbol{0}\}} \frac{\boldsymbol{y}^\top V^\top AV \boldsymbol{y}}{\boldsymbol{y}^\top \boldsymbol{y}},$$

where the maximum is taken over all k-dimensional subspaces S' of $\mathbb{R}^m$. For such subspace S', the set $S = \{\boldsymbol{x} \mid \boldsymbol{x} = V\boldsymbol{y},\ \boldsymbol{y} \in S'\}$ is a k-dimensional subspace of $\mathbb{R}^n$. If $\boldsymbol{x} = V\boldsymbol{y}$, we have $\boldsymbol{x}^\top \boldsymbol{x} = \boldsymbol{y}^\top V^\top V \boldsymbol{y} = \boldsymbol{y}^\top \boldsymbol{y}$, and hence

$$\max_{\dim S' = k} \min_{\boldsymbol{y} \in S' \setminus \{\boldsymbol{0}\}} \frac{\boldsymbol{y}^\top V^\top AV \boldsymbol{y}}{\boldsymbol{y}^\top \boldsymbol{y}} \leqq \max_{\dim S = k} \min_{\boldsymbol{x} \in S \setminus \{\boldsymbol{0}\}} \frac{\boldsymbol{x}^\top A \boldsymbol{x}}{\boldsymbol{x}^\top \boldsymbol{x}},$$

where the maximum on the right-hand side is taken over all k-dimensional subspaces S of $\mathbb{R}^n$. Since the right-hand side above is equal to λ_k by Theorem 6.12, we obtain $\mu_k \leqq \lambda_k$, which is the first inequality in (6.38). The other inequality $\mu_k \geqq \lambda_{k+n-m}$ in (6.38) follows from this inequality applied to $-A$. □

Theorem 6.14. *The inequalities in* (6.38) *hold between the eigenvalues* $\lambda_1 \geqq \lambda_2 \geqq \cdots \geqq \lambda_n$ *of a symmetric matrix* A *of order* n *and the eigenvalues* $\mu_1 \geqq \mu_2 \geqq \cdots \geqq \mu_m$ *of the leading principal submatrix* B *of order* m *of* A.

Proof. This follows from Theorem 6.13 for $V = \begin{bmatrix} I_m \\ O_{n-m,m} \end{bmatrix}$. □

In the special case of Theorem 6.14 where $m = n-1$, the relation (6.38) shows

$$\lambda_1 \geqq \mu_1 \geqq \lambda_2 \geqq \mu_2 \geqq \lambda_3 \geqq \cdots \geqq \lambda_{n-1} \geqq \mu_{n-1} \geqq \lambda_n. \tag{6.39}$$

This is called the *separation theorem* or *Cauchy's interlace theorem.*

6.3.3 *Perturbation Theorems*

For any symmetric matrix A of order n, we denote by $\lambda_1(A) \geqq \lambda_2(A) \geqq \cdots \geqq \lambda_n(A)$ the eigenvalues of A in a descending order. The following theorem gives upper and lower bounds on the eigenvalues of $A+B$ in terms of the eigenvalues of A and B. By regarding B as a small perturbation to the given matrix A, we can use this theorem to estimate the variation in the eigenvalues under perturbation.

Theorem 6.15 (Perturbation theorem). *For symmetric matrices* A *and* B *of order* n, *we have*

$$\max_{k \leqq i \leqq n} (\lambda_i(A) + \lambda_{k-i+n}(B)) \leqq \lambda_k(A+B) \leqq \min_{1 \leqq j \leqq k} (\lambda_j(A) + \lambda_{k-j+1}(B))$$
$$(k = 1, 2, \ldots, n). \tag{6.40}$$

Proof. Let $C = A + B$ and assume $1 \leqq j \leqq k \leqq n$. Following notations in Proposition 6.2, let U_k^C be the subspace spanned by k eigenvectors of C corresponding to the k largest eigenvalues $\lambda_i(C)$ $(1 \leqq i \leqq k)$. Similarly, let W_j^A and W_{k-j+1}^B be the subspaces defined from A and B, respectively, using their eigenvectors corresponding to the tailing eigenvalues $\lambda_i(A)$ $(j \leqq i \leqq n)$ and $\lambda_i(B)$ $(k-j+1 \leqq i \leqq n)$. We have

$$\dim U_k^C = k, \qquad \dim W_j^A = n - j + 1, \qquad \dim W_{k-j+1}^B = n - k + j.$$

By the inequality in Remark 9.11 in Sec. 9.4.3, we have

$$\dim(U_k^C \cap W_j^A \cap W_{k-j+1}^B) \geqq k + (n-j+1) + (n-k+j) - 2n = 1,$$

which implies that $U_k^C \cap W_j^A \cap W_{k-j+1}^B$ contains a unit-length vector, say, $\boldsymbol{x}$. Then it follows from Proposition 6.2 that

$$\lambda_k(C) \leqq \boldsymbol{x}^\top C\boldsymbol{x} = \boldsymbol{x}^\top (A+B)\boldsymbol{x} = \boldsymbol{x}^\top A\boldsymbol{x} + \boldsymbol{x}^\top B\boldsymbol{x} \leqq \lambda_j(A) + \lambda_{k-j+1}(B).$$

This shows the second inequality in (6.40).

The first inequality in (6.40) follows from this inequality for $(-A)+(-B)$ with the observation that $\lambda_{n-k+1}((-A)+(-B)) = -\lambda_k(A+B)$, *etc.* □

The following theorem is called the *Hoffman–Wielandt theorem.*

Theorem 6.16 (Hoffman–Wielandt theorem). *For symmetric matrices A and B of order n, we have*

$$|\lambda_k(A) - \lambda_k(B)| \leqq \|A - B\|_{\mathrm{F}} \qquad (k = 1, 2, \ldots, n). \tag{6.41}$$

Proof. We use the inequality (6.40) in Theorem 6.15. By replacing $A+B$ to C and considering the terms of $i = k$ and $j = k$, we obtain

$$\lambda_k(A) + \lambda_n(C - A) \leqq \lambda_k(C) \leqq \lambda_k(A) + \lambda_1(C - A).$$

This implies

$$|\lambda_k(C) - \lambda_k(A)| \leqq \max(|\lambda_1(C - A)|, |\lambda_n(C - A)|) \leqq \|C - A\|_{\mathrm{F}},$$

where the last inequality is due to Theorem 6.10 in Sec. 6.3.1. □

6.4 Eigenvalues of Hermitian Matrices

In this section we assume that A is a Hermitian matrix of order n.

6.4.1 *Diagonalization by Unitary Transformation*

Just as a symmetric matrix, a Hermitian matrix has the property that eigenvalues are real and eigenvectors are mutually orthogonal.

Theorem 6.17. *Let A be a Hermitian matrix.*

(1) *All eigenvalues are real numbers.*
(2) *Eigenvectors corresponding to distinct eigenvalues are orthogonal to each other. That is, if $A\boldsymbol{x} = \lambda\boldsymbol{x}$, $A\boldsymbol{y} = \mu\boldsymbol{y}$, and $\lambda \neq \mu$, then $\boldsymbol{x}^*\boldsymbol{y} = 0$.*

Theorem 6.18. *Eigenvalues of a Hermitian matrix remain invariant under a unitary similarity transformation. To be more precise, if* $A\boldsymbol{x} = \lambda\boldsymbol{x}$, $\boldsymbol{x} \neq \boldsymbol{0}$, *and* $B = U^*AU$ *for a unitary matrix* U, *then* $B\boldsymbol{y} = \lambda\boldsymbol{y}$, $\boldsymbol{y} \neq \boldsymbol{0}$ *for* $\boldsymbol{y} = U^*\boldsymbol{x}$.

Diagonalization of a symmetric matrix by an orthogonal similarity transformation (Theorem 6.9 in Sec. 6.3.1) can be extended to a Hermitian matrix using a unitary similarity transformation as follows (the proof is similar).

Theorem 6.19. *For a Hermitian matrix* A, *there exists a unitary matrix* U *for which* U^*AU *is a diagonal matrix, i.e.,*

$$U^*AU = \mathrm{diag}\,(\lambda_1, \lambda_2, \ldots, \lambda_n) \tag{6.42}$$

for some $\lambda_1, \lambda_2, \ldots, \lambda_n$. *With the notations* $\boldsymbol{u}_1, \boldsymbol{u}_2, \ldots, \boldsymbol{u}_n$ *for the column vectors of* U, *i.e.,* $U = [\boldsymbol{u}_1, \boldsymbol{u}_2, \ldots, \boldsymbol{u}_n]$, *this implies that*

$$A\boldsymbol{u}_j = \lambda_j \boldsymbol{u}_j \qquad (j = 1, 2, \ldots, n), \tag{6.43}$$

showing that $\lambda_1, \lambda_2, \ldots, \lambda_n$ *are eigenvalues of* A *and* $\boldsymbol{u}_1, \boldsymbol{u}_2, \ldots, \boldsymbol{u}_n$ *are the corresponding eigenvectors. Therefore,* $\{\boldsymbol{u}_1, \boldsymbol{u}_2, \ldots, \boldsymbol{u}_n\}$ *is an orthonormal basis of* $\mathbb{C}^n$ *consisting of eigenvectors of* A.

The following are extensions to Hermitian matrices of Proposition 6.1 and Theorem 6.10 for symmetric matrices given in Sec. 6.3.1.

Proposition 6.3. *Let* $B(\lambda) = \{j \mid \lambda_j = \lambda,\ 1 \leqq j \leqq n\}$ *for any real number* λ. *A complex vector* $\boldsymbol{x}$ *satisfies* $A\boldsymbol{x} = \lambda\boldsymbol{x}$ *if and only if* $\boldsymbol{x}$ *can be expressed as a linear combination of* $\boldsymbol{u}_j$ $(j \in B(\lambda))$ *with complex coefficients.*

Theorem 6.20. *For a Hermitian matrix* A, *the sum of squares of its eigenvalues is equal to the square of the Frobenius norm* $\|A\|_{\mathrm{F}}$; *cf.,* (6.29).

6.4.2 *Minimax Theorems*

Minimax theorems for symmetric matrices (Theorems 6.11 and 6.12 in Sec. 6.3.2) can be extended to Hermitian matrices as follows (the proofs are similar).

Theorem 6.21. *For the largest eigenvalue* λ_1 *and the smallest eigenvalue* λ_n *of a Hermitian matrix* A *of order* n, *we have the following formulas:*

$$\lambda_1 = \max_{\boldsymbol{x} \neq \boldsymbol{0}} \frac{\boldsymbol{x}^* A \boldsymbol{x}}{\boldsymbol{x}^* \boldsymbol{x}} = \max_{\boldsymbol{x}^* \boldsymbol{x} = 1} \boldsymbol{x}^* A \boldsymbol{x}, \tag{6.44}$$

$$\lambda_n = \min_{\boldsymbol{x} \neq \boldsymbol{0}} \frac{\boldsymbol{x}^* A \boldsymbol{x}}{\boldsymbol{x}^* \boldsymbol{x}} = \min_{\boldsymbol{x}^* \boldsymbol{x} = 1} \boldsymbol{x}^* A \boldsymbol{x}. \tag{6.45}$$

Theorem 6.22 (Courant–Fischer theorem). *For the eigenvalues $\lambda_1 \geqq \lambda_2 \geqq \cdots \geqq \lambda_n$ of a Hermitian matrix A of order n, we have the following formulas:*

$$\lambda_k = \max_{S:\, \dim S = k} \ \min_{\boldsymbol{x} \in S \setminus \{\boldsymbol{0}\}} \frac{\boldsymbol{x}^* A \boldsymbol{x}}{\boldsymbol{x}^* \boldsymbol{x}} \qquad (k = 1, 2, \ldots, n), \tag{6.46}$$

$$\lambda_k = \min_{S:\, \dim S = n-k+1} \ \max_{\boldsymbol{x} \in S \setminus \{\boldsymbol{0}\}} \frac{\boldsymbol{x}^* A \boldsymbol{x}}{\boldsymbol{x}^* \boldsymbol{x}} \qquad (k = 1, 2, \ldots, n), \tag{6.47}$$

where S ($\subseteqq \mathbb{C}^n$) runs over all subspaces of the specified dimension.

Theorems 6.13 and 6.14 for symmetric matrices in Sec. 6.3.2 can also be extended to Hermitian matrices.

Theorem 6.23. *Let A be an $n \times n$ Hermitian matrix and V an $n \times m$ complex matrix satisfying $V^* V = I_m$, where $m \leqq n$. The following inequalities hold between the eigenvalues $\lambda_1 \geqq \lambda_2 \geqq \cdots \geqq \lambda_n$ of A and the eigenvalues $\mu_1 \geqq \mu_2 \geqq \cdots \geqq \mu_m$ of $V^* A V$:*

$$\lambda_k \geqq \mu_k \geqq \lambda_{k+n-m} \qquad (k = 1, 2, \ldots, m). \tag{6.48}$$

Theorem 6.24. *The inequalities in (6.48) hold between the eigenvalues $\lambda_1 \geqq \lambda_2 \geqq \cdots \geqq \lambda_n$ of a Hermitian matrix A of order n and the eigenvalues $\mu_1 \geqq \mu_2 \geqq \cdots \geqq \mu_m$ of the leading principal submatrix B of order m of A.*

6.4.3 *Perturbation Theorems*

Theorems 6.15 and 6.16 for symmetric matrices in Sec. 6.3.3 can also be extended to Hermitian matrices.

Theorem 6.25 (Perturbation theorem). *For Hermitian matrices A and B of order n, we have*

$$\max_{k \leqq i \leqq n} (\lambda_i(A) + \lambda_{k-i+n}(B)) \leqq \lambda_k(A+B) \leqq \min_{1 \leqq j \leqq k} (\lambda_j(A) + \lambda_{k-j+1}(B)) \qquad (k = 1, 2, \ldots, n). \tag{6.49}$$

Theorem 6.26 (Hoffman–Wielandt theorem). *For Hermitian matrices A and B of order n, we have*

$$|\lambda_k(A) - \lambda_k(B)| \leqq \|A - B\|_{\mathrm{F}} \qquad (k = 1, 2, \ldots, n). \tag{6.50}$$

6.5 Schur Decomposition

In this section A will denote a general square matrix of order n (not assumed to be symmetric or Hermitian).

Theorem 6.27.

(1) *For any complex square matrix A, there exist a unitary matrix U and an upper triangular matrix S such that*

$$A = U\, S\, U^{*}. \tag{6.51}$$

(2) *For a real square matrix A whose eigenvalues are all real, there exist an orthogonal matrix Q and a real upper triangular matrix S such that*

$$A = Q\, S\, Q^{\top}. \tag{6.52}$$

Proof. (1) Take an eigenvalue λ_1 of A and a corresponding eigenvector $\boldsymbol{u}_1$ with $\|\boldsymbol{u}_1\|_2 = 1$ (*i.e.*, $\boldsymbol{u}_1^{*}\boldsymbol{u}_1 = 1$). Let U_1 be a unitary matrix of order n that has $\boldsymbol{u}_1$ in the first column. Then we have

$$U_1^{*} A U_1 = \left[\begin{array}{c|ccc} \lambda_1 & * & \cdots & * \\ \hline 0 & & & \\ \vdots & & A_2 & \\ 0 & & & \end{array}\right].$$

We apply a similar transformation to the matrix A_2 of order $n-1$ using its eigenvalue λ_2. That is, for some unitary matrix V_2 of order $n-1$, we have

$$V_2^{*} A_2 V_2 = \left[\begin{array}{c|ccc} \lambda_2 & * & \cdots & * \\ \hline 0 & & & \\ \vdots & & A_3 & \\ 0 & & & \end{array}\right].$$

By defining $U_2 = \left[\begin{array}{c|c} 1 & \mathbf{0}^{\top} \\ \hline \mathbf{0} & V_2 \end{array}\right]$, which is a unitary matrix of order n, we obtain

$$(U_1 U_2)^{*} A (U_1 U_2) = \left[\begin{array}{c|c|cccc} \lambda_1 & * & \cdots & \cdots & * \\ \hline 0 & \lambda_2 & * & \cdots & * \\ \cline{2-5} \vdots & 0 & & & \\ \vdots & \vdots & & A_3 & \\ 0 & 0 & & & \end{array}\right].$$

Continuing this process we can construct a unitary matrix U such that $S = U^*AU$ is an upper triangular matrix. Thus we obtain $A = USU^*$ in (6.51).

(2) In the above proof, λ_1 is a real number by assumption. Since A is a real matrix, we may assume that $\boldsymbol{u}_1$ is a real vector and U_1 is an orthogonal matrix. By Theorem 6.18, the transformed matrix $U_1^*AU_1$ has the same eigenvalues as A, and hence all the eigenvalues of A_2 are real. This guarantees that we can continue the subsequent argument with real numbers without involving complex numbers. □

The decomposition given in (6.51) is called the *Schur decomposition* of A, and the column vectors $\boldsymbol{u}_1, \boldsymbol{u}_2, \dots, \boldsymbol{u}_n$ of the matrix $U = [\boldsymbol{u}_1, \boldsymbol{u}_2, \dots, \boldsymbol{u}_n]$ is called the *Schur vector* of A. The Schur decomposition is not uniquely determined by A, but it is an important and convenient concept in discussing eigenvalues and eigenvectors.

Proposition 6.4. *The Schur decomposition* $A = USU^*$ *in* (6.51) *has the following properties.*

(1) *The eigenvalues of A coincide with the diagonal entries of the upper triangular matrix S.*
(2) *For each $k = 1, 2, \dots, n$, the subspace spanned by $\{\boldsymbol{u}_j \mid j = 1, 2, \dots, k\}$ is an invariant subspace*[12] *of A.*

Proof. (1) Let $S = (s_{ij})$, which is an upper triangular matrix. Then we have

$$\begin{aligned}\det(\lambda I - A) &= \det(\lambda I - USU^*) = \det[U(\lambda I - S)U^*] \\ &= \det(\lambda I - S) = \prod_{i=1}^{n}(\lambda - s_{ii}).\end{aligned}$$

Hence the eigenvalues of A are given by $s_{11}, s_{22}, \dots, s_{nn}$.

(2) Let W_k denote the subspace spanned by $\{\boldsymbol{u}_j \mid j = 1, 2, \dots, k\}$. Since $AU = US$ by (6.51), we have

$$A\boldsymbol{u}_j = s_{1j}\boldsymbol{u}_1 + s_{2j}\boldsymbol{u}_2 + \cdots + s_{jj}\boldsymbol{u}_j \in W_k \qquad (j = 1, 2, \dots, k).$$

This shows that $A\boldsymbol{x} \in W_k$ for any $\boldsymbol{x} \in W_k$. Hence W_k is an invariant subspace of A. □

[12]See Definition 9.19 in Sec. 9.5.5 for the definition of an invariant subspace.

Theorem 6.28. *The sum of squares of the absolute values of the eigenvalues* $\lambda_1, \lambda_2, \ldots, \lambda_n$ *of* A *is bounded by the squared Frobenius norm* $\|A\|_F$. *That is,*

$$\sum_{i=1}^{n} |\lambda_i|^2 \leqq \|A\|_F{}^2.$$

Proof. Consider the Schur decomposition $A = USU^*$. It follows from (1.27) and Proposition 6.4(1) that

$$\|A\|_F{}^2 = \|S\|_F{}^2 = \sum_{i \leqq j} |s_{ij}|^2 \geqq \sum_{i=1}^{n} |s_{ii}|^2 = \sum_{i=1}^{n} |\lambda_i|^2.$$

□

The Schur decomposition (6.51) can be rewritten as

$$U^* AU = S, \tag{6.53}$$

which shows that a matrix A can be brought to an upper triangular matrix S by means of a unitary similarity transformation using U. This generalizes the diagonalization of a Hermitian matrix by a unitary similarity transformation given in Theorem 6.19 in Sec. 6.4.1. Similarly, the Schur decomposition (6.52) for a real matrix A with real eigenvalues can be rewritten as

$$Q^\top AQ = S, \tag{6.54}$$

which shows that such a matrix A can be brought to an upper triangular matrix S by means of an orthogonal similarity transformation using Q. This generalizes the diagonalization of a symmetric matrix by an orthogonal similarity transformation given in Theorem 6.9 in Sec. 6.3.1.

6.6 Normal Matrices

The Schur decomposition treated in the previous section shows that every matrix can be brought to an upper triangular matrix by a unitary similarity transformation. In this section we are concerned with diagonalization, which is a stronger requirement than upper triangularization. Our main interest lies in a necessary and sufficient condition for a matrix to be diagonalizable by a unitary similarity transformation.

A square matrix A is called a *normal matrix* if

$$A^* A = AA^*. \tag{6.55}$$

Theorem 6.29. *A square matrix A of order n is a normal matrix if and only if it can be brought to a diagonal matrix with a unitary matrix U as*

$$U^*AU = \operatorname{diag}(\lambda_1, \lambda_2, \ldots, \lambda_n). \tag{6.56}$$

Proof. Obviously, (6.56) implies $A^*A = AA^*$. Conversely, let A be a normal matrix and consider its Schur decomposition $A = USU^*$ in (6.51). It follows from the assumed normality $A^*A = AA^*$ that $S^*S = SS^*$. On the other hand, since $S = (s_{ij})$ is an upper triangular matrix, we have

$$(1,1) \text{ entry of } S^*S = |s_{11}|^2, \qquad (1,1) \text{ entry of } SS^* = \sum_{j=1}^{n} |s_{1j}|^2.$$

Therefore, $s_{1j} = 0$ ($j = 2, 3, \ldots, n$), that is, S is of the form

$$S = \begin{bmatrix} s_{11} & \mathbf{0}^\top \\ \mathbf{0} & S_2 \end{bmatrix}.$$

Here the submatrix S_2 is an upper triangular matrix satisfying ${S_2}^*S_2 = S_2{S_2}^*$. By repeating the same argument we can conclude that S is a diagonal matrix. □

Remark 6.2. For a real normal matrix A it is not always possible to choose U in (6.56) to be a real matrix (orthogonal matrix). For example, we cannot choose a real matrix U for

$$A = \begin{bmatrix} \cos\theta & -\sin\theta \\ \sin\theta & \cos\theta \end{bmatrix}$$

with $0 < \theta < \pi$, since this matrix has complex eigenvalues $\cos\theta \pm \mathrm{i}\sin\theta$, which appear as the diagonal entries on the right-hand side of (6.56). ■

Example 6.4. A Hermitian matrix (a symmetric matrix, in particular) is a normal matrix. A matrix A is Hermitian if and only if it admits a decomposition of the form of (6.56) with real $\lambda_1, \lambda_2, \ldots, \lambda_n$. This is discussed in Secs. 6.3 and 6.4. ■

Example 6.5. A skew-Hermitian matrix (an anti-symmetric matrix, in particular) is a normal matrix. A matrix A is skew-Hermitian if and only if it admits a decomposition of the form of (6.56) where $\lambda_1, \lambda_2, \ldots, \lambda_n$ are pure imaginary or zero. ■

Example 6.6. A unitary matrix (an orthogonal matrix, in particular) is a normal matrix. A matrix A is unitary if and only if it admits a decomposition of the form of (6.56) where $\lambda_1, \lambda_2, \ldots, \lambda_n$ have modulus 1 (*i.e.*, $|\lambda_i| = 1$ for $i = 1, 2, \ldots, n$). ■

Example 6.7. A circulant matrix A is a normal matrix. Denote the first row of A by $(a_0, a_1, \ldots, a_{n-1})$ and let $\zeta = \exp(2\pi \mathrm{i}/n)$. The eigenvalues and the corresponding eigenvectors of A are given by[13]

$$\lambda_i = \sum_{j=0}^{n-1} a_j \zeta^{j(i-1)}, \quad \boldsymbol{u}_i = \frac{1}{\sqrt{n}}\left(1, \zeta^{i-1}, \zeta^{2(i-1)}, \ldots, \zeta^{(n-1)(i-1)}\right)^\top$$

$(i = 1, 2, \ldots, n)$. Then $U = [\boldsymbol{u}_1, \boldsymbol{u}_2, \ldots, \boldsymbol{u}_n]$ is a unitary matrix and (6.56) holds. ■

Theorem 6.30.

(1) *A complex matrix A is a normal matrix if and only if $A^* = q(A)$ for some polynomial $q(\lambda)$ with complex coefficients.*
(2) *A real matrix A is a normal matrix if and only if $A^\top = q(A)$ for some polynomial $q(\lambda)$ with real coefficients.*

Proof. (1) Let A be a normal matrix. We have the decomposition in (6.56) with a unitary matrix U and complex numbers λ_i $(i = 1, 2, \ldots, n)$. By a standard result about polynomial interpolation [76, 77], there is a polynomial $q(\lambda)$ (Lagrange interpolating polynomial) that satisfies the condition

$$q(\lambda_i) = \overline{\lambda_i} \qquad (i = 1, 2, \ldots, n). \tag{6.57}$$

Let $D = \mathrm{diag}\,(\lambda_1, \lambda_2, \ldots, \lambda_n)$. Then we have $D^* = \overline{D} = q(D)$ and hence

$$A^* = UD^*U^* = Uq(D)U^* = q(UDU^*) = q(A).$$

Conversely, assume that $A^* = q(A)$ for some polynomial $q(\lambda)$. Then

$$A^*A = q(A)A = Aq(A) = AA^*,$$

which shows (6.55).

(2) The if-part is obvious as above. The only-if-part can be shown as follows. By the proof of (1), there is a polynomial $q(\lambda) = \sum_k c_k \lambda^k$ with $c_k \in \mathbb{C}$ such that $A^\top = A^* = q(A)$. Denoting the real part of c_k by $\hat{c}_k$ we define $\hat{q}(\lambda) = \sum_k \hat{c}_k \lambda^k$, which is a polynomial with real coefficients that satisfies $A^\top = \hat{q}(A)$. □

For a normal matrix A, the *normal degree* of A, denoted by $\nu(A)$, is defined as the lowest degree of a polynomial $q(\lambda)$ with complex coefficients that satisfies $A^* = q(A)$. The normal degree is bounded by $n - 1$, *i.e.*,

[13] With straightforward calculation we can verify $A\boldsymbol{u}_i = \lambda_i \boldsymbol{u}_i$ for $i = 1, 2, \ldots, n$.

$\nu(A) \leqq n-1$ for all A, since there exists a polynomial of degree at most $n-1$ that satisfies the interpolation condition (6.57) at n points. The normal degree of a Hermitian or symmetric matrix is equal to 1. For a real normal matrix A, there exists a polynomial $q(\lambda)$ of degree $\nu(A)$ with real coefficients that satisfies $A^\top = q(A)$.

6.7 Eigenvalue Problem for General Matrices

In this section A will denote a square complex matrix of order n (not assumed to be symmetric or Hermitian). We denote the distinct eigenvalues of A by $\lambda_1, \lambda_2, \ldots, \lambda_s$ and their multiplicities by $n_1, n_2, \ldots, n_s$. We have

$$n_1 + n_2 + \cdots + n_s = n \tag{6.58}$$

(cf., Sec. 6.1.2).

6.7.1 *Eigenspaces*

For $i = 1, 2, \ldots, s$, define a set of vectors

$$E_A(\lambda_i) = \{\boldsymbol{x} \mid A\boldsymbol{x} = \lambda_i \boldsymbol{x}\} = \mathrm{Ker}(A - \lambda_i I) \tag{6.59}$$

consisting of all eigenvectors for eigenvalue λ_i and the zero vector. This is called the *eigenspace* of A corresponding to eigenvalue λ_i. The set $E_A(\lambda_i)$ forms a subspace.[14]

Theorem 6.31. *If $\boldsymbol{x}_1, \boldsymbol{x}_2, \ldots, \boldsymbol{x}_s$ are eigenvectors corresponding, respectively, to distinct eigenvalues $\lambda_1, \lambda_2, \ldots, \lambda_s$, then $\boldsymbol{x}_1, \boldsymbol{x}_2, \ldots, \boldsymbol{x}_s$ are linearly independent.*

Proof. Let r denote the maximum number of linearly independent vectors among $\boldsymbol{x}_1, \boldsymbol{x}_2, \ldots, \boldsymbol{x}_s$. To prove by contradiction, assume $r < s$. Choose r linearly independent vectors $\boldsymbol{x}_{i_1}, \boldsymbol{x}_{i_2}, \ldots, \boldsymbol{x}_{i_r}$. By the assumption of $r < s$, there is an index $j \in \{1, 2, \ldots, s\} \setminus \{i_1, i_2, \ldots, i_r\}$. Since $\boldsymbol{x}_j$ is linearly dependent on $\boldsymbol{x}_{i_1}, \boldsymbol{x}_{i_2}, \ldots, \boldsymbol{x}_{i_r}$, it can be represented as their linear combination

$$\boldsymbol{x}_j = c_1 \boldsymbol{x}_{i_1} + c_2 \boldsymbol{x}_{i_2} + \cdots + c_r \boldsymbol{x}_{i_r}. \tag{6.60}$$

By multiplying (6.60) with A we obtain

$$\lambda_j \boldsymbol{x}_j = c_1 \lambda_{i_1} \boldsymbol{x}_{i_1} + c_2 \lambda_{i_2} \boldsymbol{x}_{i_2} + \cdots + c_r \lambda_{i_r} \boldsymbol{x}_{i_r},$$

[14]In Secs. 6.7 and 6.8, frequent use is made of basic concepts related to vector spaces such as subspace and basis, for which the reader is referred to Chapter 9.

whereas by multiplying (6.60) with λ_j we obtain

$$\lambda_j \boldsymbol{x}_j = c_1 \lambda_j \boldsymbol{x}_{i_1} + c_2 \lambda_j \boldsymbol{x}_{i_2} + \cdots + c_r \lambda_j \boldsymbol{x}_{i_r}.$$

The subtraction between these two gives

$$\mathbf{0} = c_1(\lambda_{i_1} - \lambda_j)\boldsymbol{x}_{i_1} + c_2(\lambda_{i_2} - \lambda_j)\boldsymbol{x}_{i_2} + \cdots + c_r(\lambda_{i_r} - \lambda_j)\boldsymbol{x}_{i_r},$$

which, by the linear independence of $\boldsymbol{x}_{i_1}, \boldsymbol{x}_{i_2}, \ldots, \boldsymbol{x}_{i_r}$, implies $c_k(\lambda_{i_k} - \lambda_j) = 0$ $(k = 1, 2, , \ldots, r)$. Since $\lambda_{i_k} \neq \lambda_j$ $(k = 1, 2, , \ldots, r)$, we obtain $c_k = 0$ $(k = 1, 2, , \ldots, r)$ and hence $\boldsymbol{x}_j = \mathbf{0}$ by (6.60). This is a contradiction, and therefore we must have $r = s$. □

A matrix A is said to be *diagonalizable* if there exists a nonsingular matrix S such that $S^{-1}AS$ is a diagonal matrix. The above theorem implies that a matrix with simple eigenvalues (*i.e.*, of multiplicity 1) is diagonalizable, as follows.

Theorem 6.32. *Suppose that the eigenvalues of A are all distinct, and let $\boldsymbol{x}_1, \boldsymbol{x}_2, \ldots, \boldsymbol{x}_n$ be (any) eigenvectors corresponding, respectively, to the distinct eigenvalues $\lambda_1, \lambda_2, \ldots, \lambda_n$. Then the matrix $X = [\boldsymbol{x}_1, \boldsymbol{x}_2, \ldots, \boldsymbol{x}_n]$ is nonsingular and*

$$X^{-1}AX = \mathrm{diag}\,(\lambda_1, \lambda_2, \ldots, \lambda_n). \tag{6.61}$$

In particular, A is diagonalizable.

Proof. It follows from $A\boldsymbol{x}_i = \lambda_i \boldsymbol{x}_i$ $(i = 1, 2, \ldots, n)$ that

$$A\,[\boldsymbol{x}_1, \boldsymbol{x}_2, \ldots, \boldsymbol{x}_n] = [\boldsymbol{x}_1, \boldsymbol{x}_2, \ldots, \boldsymbol{x}_n] \cdot \mathrm{diag}\,(\lambda_1, \lambda_2, \ldots, \lambda_n).$$

By Theorem 6.31, the matrix $X = [\boldsymbol{x}_1, \boldsymbol{x}_2, \ldots, \boldsymbol{x}_n]$ is nonsingular. By multiplying the above equation with X^{-1} from the left, we obtain (6.61). □

6.7.2 *Generalized Eigenspaces*

We continue to denote the distinct eigenvalues of A by $\lambda_1, \lambda_2, \ldots, \lambda_s$ and their multiplicities by $n_1, n_2, \ldots, n_s$. For each $i = 1, 2, \ldots, s$,

$$\hat{E}_A(\lambda_i) = \{\boldsymbol{x} \mid (A - \lambda_i I)^{n_i} \boldsymbol{x} = \mathbf{0}\} = \mathrm{Ker}((A - \lambda_i I)^{n_i}) \tag{6.62}$$

is called the *generalized eigenspace* of A corresponding to eigenvalue λ_i. $\hat{E}_A(\lambda_i)$ forms a subspace and contains the eigenspace $E_A(\lambda_i)$:

$$\hat{E}_A(\lambda_i) \supseteqq E_A(\lambda_i).$$

A nonzero vector in $\hat{E}_A(\lambda_i)$ is called a *generalized eigenvector* of A corresponding to λ_i.

Theorem 6.33. *The following hold for the generalized eigenspace $\hat{E}_A(\lambda_i)$ defined in* (6.62).

(1) $\dim \hat{E}_A(\lambda_i) = n_i$ *for* $i = 1, 2, \ldots, s$. *That is, the dimension of* $\hat{E}_A(\lambda_i)$ *is equal to the multiplicity* n_i *of* λ_i *for* $i = 1, 2, \ldots, s$.
(2) *If* $\boldsymbol{x}_i \in \hat{E}_A(\lambda_i)$ *and* $\boldsymbol{x}_i \neq \mathbf{0}$ *for* $i = 1, 2, \ldots, s$, *then* $\boldsymbol{x}_1, \boldsymbol{x}_2, \ldots, \boldsymbol{x}_s$ *are linearly independent.*[15]

Proof. (1) Let $A = USU^*$ be the Schur decomposition of A.[16] The proof of Theorem 6.27 in Sec. 6.5 shows an iterative construction of the upper triangular matrix S. By considering the multiple eigenvalues consecutively in this iterative construction, we obtain S of the form

$$S = \begin{bmatrix} R_1(\lambda_1) & & & \\ & R_2(\lambda_2) & & * \\ & O & \ddots & \\ & & & R_s(\lambda_s) \end{bmatrix}, \tag{6.63}$$

where each diagonal block $R_i(\lambda_i)$ is an upper triangular matrix of order n_i with diagonal entries all equal to λ_i, that is,

$$R_i(\lambda_i) = \begin{bmatrix} \lambda_i & & * \\ & \ddots & \\ O & & \lambda_i \end{bmatrix}.$$

In the following we fix an index i with $1 \leqq i \leqq s$. We have

$$\dim \hat{E}_A(\lambda_i) = \dim \operatorname{Ker}((A - \lambda_i I)^{n_i}) = n - \operatorname{rank}((A - \lambda_i I)^{n_i})$$

(cf., Theorem 9.69(2) in Sec. 9.8.1) and

$$(A - \lambda_i I)^{n_i} = (USU^* - \lambda_i I)^{n_i} = U(S - \lambda_i I)^{n_i} U^* \tag{6.64}$$

by $A = USU^*$. Therefore,

$$\dim \hat{E}_A(\lambda_i) = n - \operatorname{rank}((S - \lambda_i I)^{n_i}). \tag{6.65}$$

By (6.63), $S - \lambda_i I$ is an upper block-triangular matrix whose jth diagonal block $R_j(\lambda_j) - \lambda_i I$, where $1 \leqq j \leqq s$, is an upper triangular matrix with diagonal entries all equal to $\lambda_j - \lambda_i$. In particular, the ith diagonal block is a strictly upper triangular matrix order n_i and other diagonal blocks are

---

[15] Theorem 6.31 is an immediate corollary of Theorem 6.33(2), since $E_A(\lambda_i) \subseteqq \hat{E}_A(\lambda_i)$.
[16] Unitarity of U does not play essential role here. A decomposition $A = USU^{-1}$ with a nonsingular matrix U and an upper triangular matrix S suffices for the proof.

nonsingular. Therefore, $(S-\lambda_i I)^{n_i}$ is an upper block-triangular matrix of the form

$$(S-\lambda_i I)^{n_i} = \left[\begin{array}{ccc|c|ccc} \tilde{R}_1 & & * & * & * & \cdots & * \\ & \ddots & & \vdots & \vdots & & \vdots \\ O & & \tilde{R}_{i-1} & * & * & \cdots & * \\ \hline O & \cdots & O & O_{n_i} & * & \cdots & * \\ \hline O & \cdots & O & O & \tilde{R}_{i+1} & & * \\ \vdots & & \vdots & \vdots & & \ddots & \\ O & \cdots & O & O & O & & \tilde{R}_s \end{array}\right],$$

where the ith diagonal block is equal to O_{n_i} and other diagonal blocks $\tilde{R}_j$ $(j \neq i)$ are nonsingular upper triangular matrices. Since this matrix can be transformed by elementary row operations to[17]

$$\tilde{S}_i = \left[\begin{array}{ccc|c|ccc} I_{n_1} & & O & B_1^{(i)} & O & \cdots & O \\ & \ddots & & \vdots & \vdots & & \vdots \\ O & & I_{n_{i-1}} & B_{i-1}^{(i)} & O & \cdots & O \\ \hline O & \cdots & O & O_{n_i} & O & \cdots & O \\ \hline O & \cdots & O & O & I_{n_{i+1}} & & O \\ \vdots & & \vdots & \vdots & & \ddots & \\ O & \cdots & O & O & O & & I_{n_s} \end{array}\right], \tag{6.66}$$

we have $\operatorname{rank} \tilde{S}_i = n - n_i$, whereas $\operatorname{rank} \tilde{S}_i = \operatorname{rank}((S-\lambda_i I)^{n_i})$ since the rank is invariant under elementary row transformations. Combining these with (6.65) we finally obtain

$$\dim \hat{E}_A(\lambda_i) = n - \operatorname{rank}((S-\lambda_i I)^{n_i}) = n - \operatorname{rank} \tilde{S}_i = n_i.$$

(2) It suffices to show that the vectors $\boldsymbol{y}_i = U^* \boldsymbol{x}_i$ $(i = 1, 2, , \ldots, s)$ are linearly independent. By (6.62), (6.64), and the definition of $\tilde{S}_i$ in (6.66), we have equivalences:

$$\boldsymbol{x}_i \in \hat{E}_A(\lambda_i) \iff (A-\lambda_i I)^{n_i} \boldsymbol{x}_i = \boldsymbol{0} \iff (S-\lambda_i I)^{n_i} \boldsymbol{y}_i = \boldsymbol{0} \iff \tilde{S}_i \boldsymbol{y}_i = \boldsymbol{0}.$$

Let $\boldsymbol{y}_i = (\boldsymbol{y}_{i1}^\top, \boldsymbol{y}_{i2}^\top, \ldots, \boldsymbol{y}_{is}^\top)^\top$. From the structure of $\tilde{S}_i$ shown in (6.66), $\tilde{S}_i \boldsymbol{y}_i = \boldsymbol{0}$ holds if and only if

$$\boldsymbol{y}_{ii} = \boldsymbol{\xi}_i, \quad \boldsymbol{y}_{ij} = -B_j^{(i)} \boldsymbol{\xi}_i \quad (1 \leqq j \leqq i-1), \quad \boldsymbol{y}_{ij} = \boldsymbol{0} \quad (i+1 \leqq j \leqq s)$$

[17] The matrix $\tilde{S}_i$ can be obtained from $(S-\lambda_i I)^{n_i}$ by multiplying it with a nonsingular upper block-triangular matrix with diagonal blocks $\tilde{R}_1^{-1}, \ldots, \tilde{R}_{i-1}^{-1}, I_{n_i}, \tilde{R}_{i+1}^{-1}, \ldots, \tilde{R}_s^{-1}$ from the left.

for some n_i-dimensional vector $\boldsymbol{\xi}_i$. Note that $\boldsymbol{x}_i \neq \mathbf{0} \Leftrightarrow \boldsymbol{y}_i \neq \mathbf{0} \Leftrightarrow \boldsymbol{\xi}_i \neq \mathbf{0}$.

To show the linear independence of $\boldsymbol{y}_1, \boldsymbol{y}_2, \ldots, \boldsymbol{y}_s$, suppose that $\mathbf{0} = c_1\boldsymbol{y}_1 + c_2\boldsymbol{y}_2 + \cdots + c_s\boldsymbol{y}_s$, which is equivalent to

$$\mathbf{0} = \sum_{i=1}^{s} c_i \boldsymbol{y}_{ij} = c_j \boldsymbol{\xi}_j - \sum_{i=j+1}^{s} c_i B_j^{(i)} \boldsymbol{\xi}_i \qquad (j = 1, 2, \ldots, s).$$

This equation for $j = s$ reads $c_s\, \boldsymbol{\xi}_s = \mathbf{0}$, implying $c_s = 0$. Next, the equation for $j = s-1$ reads $c_{s-1}\boldsymbol{\xi}_{s-1} - c_s B_s^{(s)} \boldsymbol{\xi}_s = \mathbf{0}$, which implies $c_{s-1} = 0$. Continuing in this way we obtain $c_1 = c_2 = \cdots = c_s = 0$. Therefore, $\boldsymbol{y}_1, \boldsymbol{y}_2, \ldots, \boldsymbol{y}_s$ are linearly independent. □

It follows from Theorem 6.33 that $\mathbb{C}^n$ is decomposed into a direct sum of the generalized eigenspaces. That is,

$$\mathbb{C}^n = \hat{E}_A(\lambda_1) \oplus \hat{E}_A(\lambda_2) \oplus \cdots \oplus \hat{E}_A(\lambda_s). \tag{6.67}$$

By the inclusion relation $E_A(\lambda_i) \subseteqq \hat{E}_A(\lambda_i)$ between the eigenspace $E_A(\lambda_i)$ and the generalized eigenspace $\hat{E}_A(\lambda_i)$, Theorem 6.33(1) implies

$$\dim E_A(\lambda_i) \leqq n_i. \tag{6.68}$$

The following theorem states that if $\dim E_A(\lambda_i) = n_i$ (*i.e.*, $E_A(\lambda_i) = \hat{E}_A(\lambda_i)$) for each eigenvalue λ_i, the matrix A is diagonalizable. This generalizes Theorem 6.32 in Sec. 6.7.1.

Theorem 6.34. *Assume that* $\dim E_A(\lambda_i) = n_i$ *for all* $i = 1, 2, \ldots, s$. *Let* $\{\boldsymbol{x}_1^i, \ldots, \boldsymbol{x}_{n_i}^i\}$ *be an arbitrary basis of the eigenspace* $E_A(\lambda_i)$ *for* $i = 1, 2, \ldots, s$, *and*

$$X = \left[\boldsymbol{x}_1^1, \ldots, \boldsymbol{x}_{n_1}^1; \boldsymbol{x}_1^2, \ldots, \boldsymbol{x}_{n_2}^2;\ \ldots\ ; \boldsymbol{x}_1^s, \ldots, \boldsymbol{x}_{n_s}^s\right].$$

Then X is nonsingular and

$$X^{-1}AX = \operatorname{diag}(\,\overbrace{\lambda_1, \ldots, \lambda_1}^{n_1}\,;\ \overbrace{\lambda_2, \ldots, \lambda_2}^{n_2}\,;\ \ldots\ ;\ \overbrace{\lambda_s, \ldots, \lambda_s}^{n_s}\,).$$

In particular, A is diagonalizable.

Proof. The proof is similar to that of Theorem 6.32. □

In the general case where the assumption in Theorem 6.34 is not met, diagonalization is weakened to block-diagonalization, as follows.

Theorem 6.35. *Let* $\{\boldsymbol{x}_1^i, \ldots, \boldsymbol{x}_{n_i}^i\}$ *be an arbitrary basis of the generalized eigenspace* $\hat{E}_A(\lambda_i)$ *for* $i = 1, 2, \ldots, s$, *and*

$$X = \left[\boldsymbol{x}_1^1, \ldots, \boldsymbol{x}_{n_1}^1; \boldsymbol{x}_1^2, \ldots, \boldsymbol{x}_{n_2}^2;\ \ldots\ ; \boldsymbol{x}_1^s, \ldots, \boldsymbol{x}_{n_s}^s\right]$$

Then X is nonsingular and

$$X^{-1}AX = \text{diag}\,(B_1, B_2, \ldots, B_s),$$

where each B_i is a square matrix of order n_i with a single eigenvalue λ_i of multiplicity n_i.

Proof. This is an immediate consequence of the direct sum decomposition in (6.67). □

It will turn out in the next section that a more detailed block-diagonal structure can be identified by a suitable choice of a basis of $\hat{E}_A(\lambda_i)$.

6.8 Jordan Normal Form

6.8.1 *Theorem*

For any complex number λ and natural number k, the square matrix of order k defined by

$$J_k(\lambda) = \begin{bmatrix} \lambda & 1 & & \\ & \lambda & \ddots & \\ & & \ddots & 1 \\ & & & \lambda \end{bmatrix}$$

is called a *Jordan block of order k*. A block-diagonal matrix consisting of diagonal blocks of such matrices $J_k(\lambda)$ with varying k and λ is called a *Jordan matrix*.

Theorem 6.36. *Every square matrix is similar to a Jordan matrix. More precisely, for a square matrix A of order n with distinct eigenvalues $\lambda_1, \lambda_2, \ldots, \lambda_s$ of multiplicities $n_1, n_2, \ldots, n_s$, there exists a nonsingular matrix X such that*

$$\begin{aligned} X^{-1}AX = \text{diag}\,(&J_{n_{11}}(\lambda_1), J_{n_{12}}(\lambda_1), \ldots, J_{n_{1g_1}}(\lambda_1);\\ &J_{n_{21}}(\lambda_2), J_{n_{22}}(\lambda_2), \ldots, J_{n_{2g_2}}(\lambda_2);\\ &\cdots\cdots\cdot\\ &J_{n_{s1}}(\lambda_s), J_{n_{s2}}(\lambda_s), \ldots, J_{n_{sg_s}}(\lambda_s)), \end{aligned} \tag{6.69}$$

$$n_{i1} \geqq n_{i2} \geqq \cdots \geqq n_{ig_i} \quad (i = 1, 2, \ldots, s),$$

$$\sum_{j=1}^{g_i} n_{ij} = n_i \quad (i = 1, 2, \ldots, s), \qquad \sum_{i=1}^{s} n_i = n. \tag{6.70}$$

The numbers $g_1, g_2, \ldots, g_s$ of the Jordan blocks for each eigenvalue and the orders n_{ij} $(1 \leqq i \leqq s, 1 \leqq j \leqq g_i)$ of the Jordan blocks are determined uniquely by the matrix A.

Proof. A systematic construction of matrix X is described in Sec. 6.8.2, whereas the uniqueness of g_i $(1 \leqq i \leqq s)$ and n_{ij} $(1 \leqq i \leqq s, 1 \leqq j \leqq g_i)$ is proved in Sec. 6.8.3. □

For a square matrix A, the Jordan matrix similar to A, given in Theorem 6.36, is called the *Jordan normal form* (or *Jordan canonical form*) of A. The number g_i of Jordan blocks corresponding to eigenvalue λ_i is called the *geometric multiplicity* of λ_i. This is equal to the dimension of the eigenspace for λ_i, that is,

$$g_i = \dim E_A(\lambda_i) \qquad (i = 1, 2, \ldots, s). \tag{6.71}$$

The sum of the geometric multiplicities, $g = g_1 + g_2 + \cdots + g_s$, is equal to the total number of Jordan blocks in the Jordan normal form of A.

Theorem 6.37. *A square matrix is diagonalizable if and only if the geometric multiplicity g_i is equal to the algebraic multiplicity n_i for each eigenvalue.*

Example 6.8. The Jordan normal form is illustrated here for a 4×4 matrix

$$A = \begin{bmatrix} -3 & -5 & 3 & -2 \\ 9 & 11 & -5 & 4 \\ 3 & 3 & 0 & 1 \\ -3 & -3 & 2 & 1 \end{bmatrix}.$$

The characteristic polynomial $\Phi_A(\lambda)$ of this matrix is given as $\Phi_A(\lambda) = (\lambda - 2)^3(\lambda - 3)$, which shows eigenvalues $\lambda_1 = 2$ of multiplicity $n_1 = 3$ and $\lambda_2 = 3$ of multiplicity $n_2 = 1$. With the choice of

$$X = \begin{bmatrix} 3 & -1 & -1 & 5 \\ -5 & 0 & 0 & -9 \\ -2 & 0 & -1 & -3 \\ 2 & 1 & 1 & 3 \end{bmatrix}, \quad X^{-1} = \begin{bmatrix} 9/5 & 8/5 & 0 & 9/5 \\ 0 & 0 & 1 & 1 \\ -3/5 & -1/5 & -1 & -3/5 \\ -1 & -1 & 0 & -1 \end{bmatrix} \tag{6.72}$$

we obtain

$$X^{-1}AX = \left[\begin{array}{cc|c|c} 2 & 1 & & \\ 0 & 2 & & \\ \hline & & 2 & \\ \hline & & & 3 \end{array}\right] = \mathrm{diag}\,(J_2(\lambda_1), J_1(\lambda_1), J_1(\lambda_2)), \tag{6.73}$$

which is the Jordan normal form of A. In the notation of Theorem 6.36 we have $n_{11} = 2$, $n_{12} = 1$ ($g_1 = 2$) and $n_{21} = 1$ ($g_2 = 1$). While the Jordan normal form is uniquely determined by A, the transformation matrix X is not. For example, another choice

$$\tilde{X} = \begin{bmatrix} 3 & -1 & -1 & 5 \\ -5 & 2 & 1 & -9 \\ -2 & 2 & 0 & -3 \\ 2 & -1 & 0 & 3 \end{bmatrix}, \quad \tilde{X}^{-1} = \begin{bmatrix} 3/2 & 3/2 & 1/2 & 5/2 \\ 0 & 0 & 1 & 1 \\ -3/2 & -1/2 & 1/2 & 3/2 \\ -1 & -1 & 0 & -1 \end{bmatrix} \qquad (6.74)$$

is also valid for $\tilde{X}^{-1}A\tilde{X} = \mathrm{diag}\,(J_2(\lambda_1), J_1(\lambda_1), J_1(\lambda_2))$. ■

Remark 6.3. Even if the given matrix A is a real matrix, we may need a complex matrix for its Jordan normal form as well as for the transformation matrix X, since a real matrix may possibly have complex eigenvalues. On the other hand, if the given matrix A has real eigenvalues only, everything can be done with real numbers, as in Example 6.8. ■

Example 6.9. Consider a companion matrix

$$A = \begin{bmatrix} 0 & \cdots & \cdots & 0 & -c_n \\ 1 & 0 & \cdots & 0 & -c_{n-1} \\ 0 & \ddots & \ddots & \vdots & \vdots \\ & \ddots & 1 & 0 & -c_2 \\ 0 & \cdots & 0 & 1 & -c_1 \end{bmatrix}.$$

As is shown in Example 2.13 in Sec. 2.5.1, the characteristic polynomial $\Phi_A(\lambda)$ of this matrix is given as $\Phi_A(\lambda) = \lambda^n + c_1\lambda^{n-1} + \cdots + c_{n-1}\lambda + c_n$. Hence the eigenvalues are the roots of the equation $\lambda^n + c_1\lambda^{n-1} + \cdots + c_{n-1}\lambda + c_n = 0$. Let $\lambda_1, \lambda_2, \ldots, \lambda_s$ be the distinct roots and $n_1, n_2, \ldots, n_s$ their multiplicities. Here we have

$$\mathrm{rank}\,(A - \lambda_i I) = n - 1 \qquad (i = 1, 2, \ldots, s).$$

(Proof: Since λ_i is an eigenvalue, we have $\mathrm{rank}\,(A - \lambda_i I) \leqq n - 1$. The reverse direction $\mathrm{rank}\,(A - \lambda_i I) \geqq n - 1$ is also true, since the submatrix of $A - \lambda_i I$ obtained by deleting the first row and the last column is an upper triangular matrix of order $n - 1$ with diagonal entries all equal to 1.) Using (6.71) and Theorem 9.69(2) in Sec. 9.8.1, the geometric multiplicity g_i of λ_i is given as

$$g_i = \dim E_A(\lambda_i) = \dim \mathrm{Ker}(A - \lambda_i I) = n - \mathrm{rank}\,(A - \lambda_i I) = 1.$$

That is, there is only one Jordan block for each eigenvalue λ_i, whose order is equal to the algebraic multiplicity n_i. Thus the Jordan normal form of the companion matrix A is given by $\mathrm{diag}\,(J_{n_1}(\lambda_1), J_{n_2}(\lambda_2), \ldots, J_{n_s}(\lambda_s))$. ■

6.8.2 *Construction*

We show how to construct the transformation matrix X in Theorem 6.36. As a preparation we first consider the kernel spaces $\mathrm{Ker}((A-\lambda_i I)^k)$ for $k = 1, 2, \ldots$. Obviously we have monotonicity

$$\{\mathbf{0}\} \subseteqq \mathrm{Ker}(A-\lambda_i I) \subseteqq \mathrm{Ker}((A-\lambda_i I)^2) \subseteqq \mathrm{Ker}((A-\lambda_i I)^3) \subseteqq \cdots$$

and the following is a refinement of this monotonicity.

Proposition 6.5. *For each $i = 1, 2, \ldots, s$, there exists a positive integer m_i $(\leqq n_i)$ such that*

$$\begin{aligned}\{\mathbf{0}\} &\subsetneqq \mathrm{Ker}(A-\lambda_i I) \subsetneqq \mathrm{Ker}((A-\lambda_i I)^2) \subsetneqq \cdots \subsetneqq \mathrm{Ker}((A-\lambda_i I)^{m_i}) \\ &= \mathrm{Ker}((A-\lambda_i I)^{m_i+1}) = \mathrm{Ker}((A-\lambda_i I)^{m_i+2}) = \cdots ,\end{aligned} \tag{6.75}$$

where n_i denotes the (algebraic) multiplicity of λ_i.

Proof. First we show that, if $\mathrm{Ker}((A-\lambda_i I)^m) \subsetneqq \mathrm{Ker}((A-\lambda_i I)^{m+1})$ for a positive integer m, then $\mathrm{Ker}((A-\lambda_i I)^k) \subsetneqq \mathrm{Ker}((A-\lambda_i I)^{k+1})$ for all $k \leqq m$. To prove this claim, suppose that

$$\boldsymbol{x} \in \mathrm{Ker}((A-\lambda_i I)^{m+1}), \qquad \boldsymbol{x} \notin \mathrm{Ker}((A-\lambda_i I)^m)$$

for some $\boldsymbol{x}$. Then for $\boldsymbol{y} = (A-\lambda_i I)^{m-k}\boldsymbol{x}$ we have

$$(A-\lambda_i I)^{k+1}\boldsymbol{y} = (A-\lambda_i I)^{m+1}\boldsymbol{x} = \mathbf{0}, \quad (A-\lambda_i I)^k\boldsymbol{y} = (A-\lambda_i I)^m\boldsymbol{x} \neq \mathbf{0},$$

which show

$$\boldsymbol{y} \in \mathrm{Ker}((A-\lambda_i I)^{k+1}), \qquad \boldsymbol{y} \notin \mathrm{Ker}((A-\lambda_i I)^k)$$

and the claim above is proved. Let m_i be the smallest integer m $(\geqq 1)$ for which $\mathrm{Ker}((A-\lambda_i I)^m) = \mathrm{Ker}((A-\lambda_i I)^{m+1})$. Then (6.75) holds for this m_i. Finally, since Theorem 6.33(1) in Sec. 6.7.2 implies $\mathrm{Ker}((A-\lambda_i I)^{n_i}) = \mathrm{Ker}((A-\lambda_i I)^{n_i+1})$, we have $m_i \leqq n_i$. □

In Proposition 6.5, m_i is the smallest integer for which $\mathrm{Ker}((A-\lambda_i I)^{m_i}) = \mathrm{Ker}((A-\lambda_i I)^{m_i+1})$. Equivalently, m_i is the smallest integer for which $\mathrm{rank}\,((A-\lambda_i I)^{m_i}) = \mathrm{rank}\,((A-\lambda_i I)^{m_i+1})$.

Theorem 6.38. *For each index $i = 1, 2, \ldots, s$, there exists a basis $\{\boldsymbol{x}_1^i, \boldsymbol{x}_2^i, \ldots, \boldsymbol{x}_{n_i}^i\}$ of the generalized eigenspace $\hat{E}_A(\lambda_i)$ satisfying*

$$A\,[\boldsymbol{x}_1^i, \boldsymbol{x}_2^i, \ldots, \boldsymbol{x}_{n_i}^i] = [\boldsymbol{x}_1^i, \boldsymbol{x}_2^i, \ldots, \boldsymbol{x}_{n_i}^i]J^{(i)}, \tag{6.76}$$

where $J^{(i)}$ is a Jordan matrix of order n_i of the form

$$J^{(i)} = \mathrm{diag}\Big(\overbrace{J_{m_i}(\lambda_i),\ldots,J_{m_i}(\lambda_i)}^{r^{(i)}_{m_i}}\ ;\ \overbrace{J_{m_i-1}(\lambda_i),\ldots,J_{m_i-1}(\lambda_i)}^{r^{(i)}_{m_i-1}}\ ;\ \cdots\ ;$$

$$\overbrace{J_2(\lambda_i),\ldots,J_2(\lambda_i)}^{r^{(i)}_{2}}\ ;\ \overbrace{J_1(\lambda_i),\ldots,J_1(\lambda_i)}^{r^{(i)}_{1}}\Big) \tag{6.77}$$

consisting of $r_l^{(i)}$ Jordan blocks $J_l(\lambda_i)$ of order l for $l = 1,2,\ldots,m_i$, where the maximum order m_i of a Jordan block contained in $J^{(i)}$ is equal to m_i defined by (6.75)*, and, for $l = 1,2,\ldots,m_i$, the number $r_l^{(i)}$ of Jordan blocks $J_l(\lambda_i)$ is given by*

$$r_l^{(i)} = \mathrm{rank}\,((A-\lambda_i I)^{l+1}) - 2\ \mathrm{rank}\,((A-\lambda_i I)^{l}) + \mathrm{rank}\,((A-\lambda_i I)^{l-1}) \tag{6.78}$$

with the convention of $\mathrm{rank}\,((A-\lambda_i I)^0) = \mathrm{rank}\,(I_n) = n$.

While the proof of Theorem 6.38 is given later, we first show the derivation of (6.69) and (6.70) in Theorem 6.36 from Theorem 6.38. With the use of the basis $\{\boldsymbol{x}_1^i, \boldsymbol{x}_2^i, \ldots, \boldsymbol{x}_{n_i}^i\}$ in (6.76), define a square matrix X of order n by

$$X = [\boldsymbol{x}_1^1, \boldsymbol{x}_2^1, \ldots, \boldsymbol{x}_{n_1}^1; \boldsymbol{x}_1^2, \boldsymbol{x}_2^2, \ldots, \boldsymbol{x}_{n_2}^2;\ \cdots\ ;\ \boldsymbol{x}_1^s, \boldsymbol{x}_2^s, \ldots, \boldsymbol{x}_{n_s}^s],$$

which is nonsingular by Theorem 6.33(2) in Sec. 6.7.2. It follows from (6.76) that

$$AX = X \cdot \mathrm{diag}\,(J^{(1)}, J^{(2)}, \ldots, J^{(s)}),$$

that is,

$$X^{-1}AX = \mathrm{diag}\,(J^{(1)}, J^{(2)}, \ldots, J^{(s)}).$$

By the definition (6.77), each diagonal block $J^{(i)}$ can be represented as

$$J^{(i)} = \mathrm{diag}\,(J_{n_{i1}}(\lambda_i), J_{n_{i2}}(\lambda_i), \ldots, J_{n_{ig_i}}(\lambda_i))$$

with integers g_i and $n_{i1}, n_{i2}, \ldots, n_{ig_i}$ defined by

$$g_i = r_{m_i}^{(i)} + r_{m_i-1}^{(i)} + \cdots + r_1^{(i)},$$

$$(n_{i1}, n_{i2}, \ldots\ldots, n_{ig_i}) = (\overbrace{m_i,\ldots,m_i}^{r^{(i)}_{m_i}}\ ;\ \overbrace{m_i-1,\ldots,m_i-1}^{r^{(i)}_{m_i-1}}\ ;\ \cdots\ ;\ \overbrace{1,\ldots,1}^{r^{(i)}_{1}}). \tag{6.79}$$

Thus we obtain (6.69). Using (6.79), (6.78), Proposition 6.5, and Theorem 6.33(1) in this order, we obtain

$$\sum_{j=1}^{g_i} n_{ij} = \sum_{l=1}^{m_i} l \cdot r_l^{(i)} = n - \operatorname{rank}((A - \lambda_i I)^{m_i}) = \dim \operatorname{Ker}((A - \lambda_i I)^{m_i})$$
$$= \dim \operatorname{Ker}((A - \lambda_i I)^{n_i}) = \dim \hat{E}_A(\lambda_i) = n_i,$$

which shows the first identity in (6.70). The second identity $\sum_{i=1}^{s} n_i = n$ in (6.70) is given in (6.58).

Proof of Theorem 6.38

As a preparation we first introduce some terminology. For a linear space V and its subspace W, we say that vectors $\boldsymbol{y}_1, \boldsymbol{y}_2, \ldots, \boldsymbol{y}_k \in V$ are *linearly independent modulo* W if the following implication is true:

$$c_1\boldsymbol{y}_1 + c_2\boldsymbol{y}_2 + \cdots + c_k\boldsymbol{y}_k \in W \ (c_1, c_2, \ldots, c_k \in \mathbb{C}) \implies c_1 = c_2 = \cdots = c_k = 0.$$

Vectors that are linearly independent modulo W are linearly independent in the ordinary sense.

Proposition 6.6. *Let W be a subspace of a linear space V, and assume that vectors $\boldsymbol{y}_1, \boldsymbol{y}_2, \ldots, \boldsymbol{y}_k \in V$ are linearly independent modulo W.*

(1) *There exists a complementary subspace W' to W containing $\boldsymbol{y}_1, \boldsymbol{y}_2, \ldots, \boldsymbol{y}_k$.*
(2) *For any complement W' of W containing $\boldsymbol{y}_1, \boldsymbol{y}_2, \ldots, \boldsymbol{y}_k$, there exist vectors $\boldsymbol{z}_1, \boldsymbol{z}_2, \ldots, \boldsymbol{z}_l$ such that $\{\boldsymbol{y}_1, \boldsymbol{y}_2, \ldots, \boldsymbol{y}_k\} \cup \{\boldsymbol{z}_1, \boldsymbol{z}_2, \ldots, \boldsymbol{z}_l\}$ is a basis of W', where $l = \dim V - \dim W - k$.*

Proof. (1) Let $U = \operatorname{Span}(\boldsymbol{y}_1, \boldsymbol{y}_2, \ldots, \boldsymbol{y}_k)$.[18] Since $\boldsymbol{y}_1, \boldsymbol{y}_2, \ldots, \boldsymbol{y}_k$ are linearly independent modulo W, we have $W + U = W \oplus U$. Take any complement W'' of $W \oplus U$ in V and let $W' = U \oplus W''$. Then we have $V = W \oplus U \oplus W'' = W \oplus W'$.

(2) Take any basis $\{\boldsymbol{z}_1, \boldsymbol{z}_2, \ldots, \boldsymbol{z}_l\}$ of the complement of U in W'. Then $\{\boldsymbol{y}_1, \boldsymbol{y}_2, \ldots, \boldsymbol{y}_k\} \cup \{\boldsymbol{z}_1, \boldsymbol{z}_2, \ldots, \boldsymbol{z}_l\}$ is a basis of W'. ☐

In the following, we fix an index i and use shorter notations

$$B = A - \lambda_i I, \quad V_l = \operatorname{Ker}((A - \lambda_i I)^l) = \operatorname{Ker}(B^l) \quad (l = 0, 1, 2, \ldots),$$

[18] $\operatorname{Span}(\boldsymbol{y}_1, \boldsymbol{y}_2, \ldots, \boldsymbol{y}_k)$ denotes the subspace spanned by $\boldsymbol{y}_1, \boldsymbol{y}_2, \ldots, \boldsymbol{y}_k$. See Definition 9.4 in Sec. 9.2.1 for notation "Span."

where $V_0 = \{\mathbf{0}\}$. By Proposition 6.5, we have

$$V_0 \subsetneqq V_1 \subsetneqq V_2 \subsetneqq \cdots \subsetneqq V_{m_i}.$$

Proposition 6.7. *Assume $2 \leqq l \leqq m_i$. If $\boldsymbol{y}_1, \boldsymbol{y}_2, \ldots, \boldsymbol{y}_k \in V_l$ are linearly independent modulo V_{l-1}, then $B\boldsymbol{y}_1, B\boldsymbol{y}_2, \ldots, B\boldsymbol{y}_k \in V_{l-1}$ are linearly independent modulo V_{l-2}.*

Proof. We prove the implication: $\sum_{j=1}^{k} c_j B\boldsymbol{y}_j \in V_{l-2} \Longrightarrow c_j = 0$ $(j = 1, 2, \ldots, k)$. It follows from $\sum_{j=1}^{k} c_j B\boldsymbol{y}_j \in V_{l-2}$ that

$$B^{l-1}\Big(\sum_{j=1}^{k} c_j \boldsymbol{y}_j\Big) = B^{l-2}\Big(\sum_{j=1}^{k} c_j B\boldsymbol{y}_j\Big) = \mathbf{0},$$

which shows $\sum_{j=1}^{k} c_j \boldsymbol{y}_j \in V_{l-1}$. Then the assumed linear independence modulo V_{l-1} gives $c_j = 0$ $(j = 1, 2, \ldots, k)$. □

With the preparations above, we can construct a basis of the generalized eigenspace $\hat{E}_A(\lambda_i)$ that satisfies (6.76), as follows. We write m for m_i.

- Take any complement U_m of V_{m-1} in $\hat{E}_A(\lambda_i) = V_m$ and let $\boldsymbol{b}_1^{(m)}, \boldsymbol{b}_2^{(m)}, \ldots, \boldsymbol{b}_{r_m}^{(m)}$ be a basis of U_m.
- Since $\boldsymbol{b}_1^{(m)}, \boldsymbol{b}_2^{(m)}, \ldots, \boldsymbol{b}_{r_m}^{(m)} \in V_m$ are linearly independent modulo V_{m-1}, Proposition 6.7 implies that $B\boldsymbol{b}_1^{(m)}, B\boldsymbol{b}_2^{(m)}, \ldots, B\boldsymbol{b}_{r_m}^{(m)} \in V_{m-1}$ are linearly independent modulo V_{m-2}. By Proposition 6.6, we can choose vectors $\boldsymbol{b}_1^{(m-1)}, \boldsymbol{b}_2^{(m-1)}, \ldots, \boldsymbol{b}_{r_{m-1}}^{(m-1)}$ such that
$$\begin{aligned} U_{m-1} = \mathrm{Span}(&B\boldsymbol{b}_1^{(m)},\ B\boldsymbol{b}_2^{(m)}, \ldots,\ B\boldsymbol{b}_{r_m}^{(m)};\\ &\boldsymbol{b}_1^{(m-1)}, \boldsymbol{b}_2^{(m-1)}, \ldots, \boldsymbol{b}_{r_{m-1}}^{(m-1)}) \end{aligned}$$
is a complement of V_{m-2} in V_{m-1}.
- Since $B\boldsymbol{b}_1^{(m)}, B\boldsymbol{b}_2^{(m)}, \ldots, B\boldsymbol{b}_{r_m}^{(m)}; \boldsymbol{b}_1^{(m-1)}, \boldsymbol{b}_2^{(m-1)}, \ldots, \boldsymbol{b}_{r_{m-1}}^{(m-1)} \in V_{m-1}$ are linearly independent modulo V_{m-2}, Proposition 6.7 implies that $B^2\boldsymbol{b}_1^{(m)}, B^2\boldsymbol{b}_2^{(m)}, \ldots, B^2\boldsymbol{b}_{r_m}^{(m)};\ B\boldsymbol{b}_1^{(m-1)}, B\boldsymbol{b}_2^{(m-1)}, \ldots, B\boldsymbol{b}_{r_{m-1}}^{(m-1)} \in V_{m-2}$ are linearly independent modulo V_{m-3}. By Proposition 6.6, we can choose vectors $\boldsymbol{b}_1^{(m-2)}, \boldsymbol{b}_2^{(m-2)}, \ldots, \boldsymbol{b}_{r_{m-2}}^{(m-2)}$ such that
$$\begin{aligned} U_{m-2} = \mathrm{Span}(&B^2\boldsymbol{b}_1^{(m)},\quad B^2\boldsymbol{b}_2^{(m)}, \ldots,\ B^2\boldsymbol{b}_{r_m}^{(m)};\\ &B\ \boldsymbol{b}_1^{(m-1)}, B\boldsymbol{b}_2^{(m-1)}, \ldots, B\boldsymbol{b}_{r_{m-1}}^{(m-1)};\\ &\boldsymbol{b}_1^{(m-2)},\quad \boldsymbol{b}_2^{(m-2)}, \ldots,\quad \boldsymbol{b}_{r_{m-2}}^{(m-2)}) \end{aligned}$$
is a complement of V_{m-3} in V_{m-2}.

- Continuing this process we can construct a complement U_1 of $V_0 = \{\mathbf{0}\}$ in V_1 of the following form

$$\begin{aligned} U_1 = \operatorname{Span}(&B^{m-1}\boldsymbol{b}_1^{(m)}, \quad B^{m-1}\boldsymbol{b}_2^{(m)}, \ldots, \quad B^{m-1}\boldsymbol{b}_{r_m}^{(m)}; \\ &B^{m-2}\boldsymbol{b}_1^{(m-1)}, B^{m-2}\boldsymbol{b}_2^{(m-1)}, \ldots, B^{m-2}\boldsymbol{b}_{r_{m-1}}^{(m-1)}; \\ &\cdots\cdots \\ &B\boldsymbol{b}_1^{(2)}, \quad B\boldsymbol{b}_2^{(2)}, \ldots, \quad B\boldsymbol{b}_{r_2}^{(2)}; \\ &\boldsymbol{b}_1^{(1)}, \quad \boldsymbol{b}_2^{(1)}, \ldots, \quad \boldsymbol{b}_{r_1}^{(1)}). \end{aligned}$$

By the construction above we have $V_m = U_m \oplus U_{m-1} \oplus \cdots \oplus U_1$, which implies that the collection of bases of respective subspaces $U_m, U_{m-1}, \ldots, U_1$ forms a basis of $V_m = \hat{E}_A(\lambda_i)$. The bases of $U_m, U_{m-1}, \ldots, U_1$ may be arranged in the following form.

U_m	$\boldsymbol{b}_1^{(m)}, \ldots, \boldsymbol{b}_{r_m}^{(m)}$			
U_{m-1}	$B\boldsymbol{b}_1^{(m)}, \ldots, B\boldsymbol{b}_{r_m}^{(m)}$	$\boldsymbol{b}_1^{(m-1)}, \ldots, \boldsymbol{b}_{r_{m-1}}^{(m-1)}$		
U_{m-2}	$B^2\boldsymbol{b}_1^{(m)}, \ldots, B^2\boldsymbol{b}_{r_m}^{(m)}$	$B\boldsymbol{b}_1^{(m-1)}, \ldots, B\boldsymbol{b}_{r_{m-1}}^{(m-1)}$		
$\vdots$	$\vdots$	$\vdots$	$\ddots$	
U_1	$B^{m-1}\boldsymbol{b}_1^{(m)}, \ldots, B^{m-1}\boldsymbol{b}_{r_m}^{(m)}$	$B^{m-2}\boldsymbol{b}_1^{(m-1)}, \ldots, B^{m-2}\boldsymbol{b}_{r_{m-1}}^{(m-1)}$	$\cdots$	$\boldsymbol{b}_1^{(1)}, \ldots, \boldsymbol{b}_{r_1}^{(1)}$

The basis $[\boldsymbol{x}_1^i, \boldsymbol{x}_2^i, \ldots, \boldsymbol{x}_{n_i}^i]$ of $\hat{E}_A(\lambda_i)$ in Theorem 6.38 can be obtained from the above vectors by ordering them from the left-bottom upward in the first column, then upward in the second column, *etc.*, as

$$\begin{aligned} &[B^{m-1}\boldsymbol{b}_1^{(m)}, \ldots, B^2\boldsymbol{b}_1^{(m)}, B\boldsymbol{b}_1^{(m)}, \boldsymbol{b}_1^{(m)}; \ldots; B^{m-1}\boldsymbol{b}_{r_m}^{(m)}, \ldots, B^2\boldsymbol{b}_{r_m}^{(m)}, B\boldsymbol{b}_{r_m}^{(m)}, \boldsymbol{b}_{r_m}^{(m)}; \\ &\ B^{m-2}\boldsymbol{b}_1^{(m-1)}, \ldots, B\boldsymbol{b}_1^{(m-1)}, \boldsymbol{b}_1^{(m-1)}; \ \ldots \ ; B^{m-2}\boldsymbol{b}_{r_{m-1}}^{(m-1)}, \ldots, B\boldsymbol{b}_{r_{m-1}}^{(m-1)}, \boldsymbol{b}_{r_{m-1}}^{(m-1)}; \\ &\ \cdots\cdots; \\ &\ \boldsymbol{b}_1^{(1)}; \ \ldots \ ; \boldsymbol{b}_{r_1}^{(1)}]. \end{aligned}$$

We can verify that the relation (6.76) is satisfied, as follows. Recall that $A = \lambda_i I + B$. First we consider vectors with superscript (m). For each $j = 1, 2, \ldots, r_m$, we have

$$\begin{aligned} &A[B^{m-1}\boldsymbol{b}_j^{(m)}, B^{m-2}\boldsymbol{b}_j^{(m)}, \ldots, \boldsymbol{b}_j^{(m)}] \\ &= \lambda_i[B^{m-1}\boldsymbol{b}_j^{(m)}, B^{m-2}\boldsymbol{b}_j^{(m)}, \ldots, \boldsymbol{b}_j^{(m)}] + B[B^{m-1}\boldsymbol{b}_j^{(m)}, B^{m-2}\boldsymbol{b}_j^{(m)}, \ldots, \boldsymbol{b}_j^{(m)}] \\ &= \lambda_i[B^{m-1}\boldsymbol{b}_j^{(m)}, B^{m-2}\boldsymbol{b}_j^{(m)}, \ldots, \boldsymbol{b}_j^{(m)}] + [\mathbf{0}, B^{m-1}\boldsymbol{b}_j^{(m)}, \ldots, B\boldsymbol{b}_j^{(m)}] \\ &= [B^{m-1}\boldsymbol{b}_j^{(m)}, B^{m-2}\boldsymbol{b}_j^{(m)}, \ldots, \boldsymbol{b}_j^{(m)}]J_m(\lambda_i), \end{aligned}$$

where $B^m \boldsymbol{b}_j^{(m)} = \mathbf{0}$ by $\boldsymbol{b}_j^{(m)} \in V_m$. Next we consider vectors with superscript $(m-1)$. For each $j = 1, 2, \ldots, r_{m-1}$, we have

$$\begin{aligned}
&A[B^{m-2}\boldsymbol{b}_j^{(m-1)}, B^{m-3}\boldsymbol{b}_j^{(m-1)}, \ldots, \boldsymbol{b}_j^{(m-1)}] \\
&= \lambda_i[B^{m-2}\boldsymbol{b}_j^{(m-1)}, B^{m-3}\boldsymbol{b}_j^{(m-1)}, \ldots, \boldsymbol{b}_j^{(m-1)}] \\
&\quad + B[B^{m-2}\boldsymbol{b}_j^{(m-1)}, B^{m-3}\boldsymbol{b}_j^{(m-1)}, \ldots, \boldsymbol{b}_j^{(m-1)}] \\
&= \lambda_i[B^{m-2}\boldsymbol{b}_j^{(m-1)}, B^{m-3}\boldsymbol{b}_j^{(m-1)}, \ldots, \boldsymbol{b}_j^{(m-1)}] \\
&\quad + [\mathbf{0}, B^{m-2}\boldsymbol{b}_j^{(m-1)}, \ldots, B\boldsymbol{b}_j^{(m-1)}] \\
&= [B^{m-2}\boldsymbol{b}_j^{(m-1)}, B^{m-3}\boldsymbol{b}_j^{(m-1)}, \ldots, \boldsymbol{b}_j^{(m-1)}] J_{m-1}(\lambda_i),
\end{aligned}$$

where $B^{m-1}\boldsymbol{b}_j^{(m-1)} = \mathbf{0}$ by $\boldsymbol{b}_j^{(m-1)} \in V_{m-1}$. Continuing this way, we arrive at vectors with superscript (1). For each $j = 1, 2, \ldots, r_1$, we have

$$A[\boldsymbol{b}_j^{(1)}] = \lambda_i[\boldsymbol{b}_j^{(1)}] + B[\boldsymbol{b}_j^{(1)}] = \lambda_i[\boldsymbol{b}_j^{(1)}] = [\boldsymbol{b}_j^{(1)}] J_1(\lambda_i),$$

where $B\boldsymbol{b}_j^{(1)} = \mathbf{0}$ by $\boldsymbol{b}_j^{(1)} \in V_1$. Thus the relation (6.76) is verified.

It remains to prove the expression (6.78) of $r_l = r_l^{(i)}$ for $l = 1, 2, \ldots, m$. By the construction of the bases, we have the following equations:

$$\begin{aligned}
r_m &= \dim U_m = \dim V_m - \dim V_{m-1}, \\
r_m + r_{m-1} &= \dim U_{m-1} = \dim V_{m-1} - \dim V_{m-2}, \\
&\cdots\cdots \\
r_m + r_{m-1} + \cdots + r_2 &= \dim U_2 = \dim V_2 - \dim V_1, \\
r_m + r_{m-1} + \cdots + r_2 + r_1 &= \dim U_1 = \dim V_1.
\end{aligned}$$

Therefore,

$$\begin{aligned}
r_m &= \dim V_m - \dim V_{m-1}, \\
r_{m-1} &= -\dim V_m + 2\dim V_{m-1} - \dim V_{m-2}, \\
&\cdots\cdots \\
r_2 &= -\dim V_3 + 2\dim V_2 - \dim V_1, \\
r_1 &= -\dim V_2 + 2\dim V_1.
\end{aligned} \tag{6.80}$$

By substituting $\dim V_l = n - \operatorname{rank}((A - \lambda_i I)^l)$ into these expressions we obtain (6.78).

This completes the proof of Theorem 6.38.

6.8.3 *Proof of Uniqueness*

In this section we prove the uniqueness of the Jordan normal form (the second statement in Theorem 6.36). Suppose that a Jordan matrix

$$J = \mathrm{diag}\,(\cdots, J_{n_{ij}}(\lambda_i), \cdots) \tag{6.81}$$

is obtained from A by a similarity transformation as in (6.69), where the orders of Jordan blocks are denoted by $n_{i1} \geqq n_{i2} \geqq \cdots \geqq n_{ig_i}$ with g_i denoting the number of Jordan blocks corresponding to eigenvalue λ_i. The numbers $n_{i1} \geqq n_{i2} \geqq \cdots \geqq n_{ig_i}$ can be expressed as

$$(n_{i1}, n_{i2}, \ldots\ldots, n_{ig_i}) = (\overbrace{\hat{m}_i, \ldots, \hat{m}_i}^{\hat{r}^{(i)}_{\hat{m}_i}} \; ; \; \overbrace{\hat{m}_i - 1, \ldots, \hat{m}_i - 1}^{\hat{r}^{(i)}_{\hat{m}_i - 1}} \; ; \; \cdots \; ; \; \overbrace{1, \ldots, 1}^{\hat{r}^{(i)}_1}) \tag{6.82}$$

with the use of notations

$$\hat{r}^{(i)}_l = |\{j \mid n_{ij} = l\}| \qquad (l = 1, 2, \ldots),$$
$$\hat{m}_i = \max\{l \mid \hat{r}^{(i)}_l \neq 0\}.$$

We establish the uniqueness of the Jordan normal form by proving that, for each $i = 1, 2, \ldots, s$, the numbers $\hat{r}^{(i)}_l$ $(l = 1, 2, \ldots)$ are determined uniquely from the given matrix A. Note that the eigenvalues λ_i $(i = 1, 2, \ldots, s)$ are determined uniquely from matrix A.

It follows from the block-diagonal form (6.81) of a Jordan matrix and the expression (6.82) that

$$\begin{aligned} \mathrm{rank}\,((J - \lambda_i I)^p) &= \sum_{k=1}^{s} \sum_{j=1}^{g_k} \mathrm{rank}\,((J_{n_{kj}}(\lambda_k) - \lambda_i I)^p) \\ &= \sum_{k=1}^{s} \sum_{l=1}^{\hat{m}_k} \hat{r}^{(k)}_l \cdot \mathrm{rank}\,((J_l(\lambda_k) - \lambda_i I)^p). \end{aligned}$$

Since

$$J_l(\lambda_k) - \lambda_i I = \begin{bmatrix} \lambda_k - \lambda_i & 1 & & \\ & \lambda_k - \lambda_i & \ddots & \\ & & \ddots & 1 \\ & & & \lambda_k - \lambda_i \end{bmatrix},$$

we have

$$\mathrm{rank}\,((J_l(\lambda_k) - \lambda_i I)^p) = \begin{cases} \max(l - p, 0) & (\text{if } k = i), \\ l & (\text{if } k \neq i), \end{cases}$$

and therefore

$$\operatorname{rank}\left((J-\lambda_i I)^p\right) = \sum_{l=1}^{\hat{m}_i} \hat{r}_l^{(i)} \cdot \max(l-p,0) + \sum_{k\neq i}\sum_{l=1}^{\hat{m}_k} \hat{r}_l^{(k)} l. \tag{6.83}$$

These equations for $p = 1, 2, \ldots$ can be solved for $\hat{r}_l^{(i)}$ as

$$\hat{r}_l^{(i)} = \operatorname{rank}\left((J-\lambda_i I)^{l+1}\right) - 2\operatorname{rank}\left((J-\lambda_i I)^{l}\right) + \operatorname{rank}\left((J-\lambda_i I)^{l-1}\right). \tag{6.84}$$

Indeed, by (6.83), the right-hand side of (6.84) reduces to $\hat{r}_l^{(i)}$.

Finally, we observe that (6.84) implies

$$\hat{r}_l^{(i)} = \operatorname{rank}\left((A-\lambda_i I)^{l+1}\right) - 2\operatorname{rank}\left((A-\lambda_i I)^{l}\right) + \operatorname{rank}\left((A-\lambda_i I)^{l-1}\right), \tag{6.85}$$

since

$$(J-\lambda_i I)^p = (X^{-1}AX - \lambda_i I)^p = X^{-1}(A-\lambda_i I)^p X$$

for each p and the rank of a matrix is invariant under a similarity transformation. Since the right-hand side of (6.85) refers to the given matrix A in place of a Jordan matrix J in (6.84), this expression (6.85) shows that the integers $\hat{r}_l^{(i)}$ $(l = 1, 2, \ldots)$ are determined uniquely from the given matrix A.[19]

Thus we have proved the uniqueness of the Jordan normal form.

6.8.4 *Examples*

The construction of the Jordan normal form in the proof of Theorem 6.36 in Sec. 6.8.2 is illustrated for two examples.

Example 6.10. The matrix

$$A = \begin{bmatrix} -3 & -5 & 3 & -2 \\ 9 & 11 & -5 & 4 \\ 3 & 3 & 0 & 1 \\ -3 & -3 & 2 & 1 \end{bmatrix}$$

in Example 6.8 in Sec. 6.8.1 has eigenvalues $\lambda_1 = 2$ (of multiplicity $n_1 = 3$) and $\lambda_2 = 3$ (of multiplicity $n_2 = 1$).

[19]Note that the expression (6.85) is consistent with (6.78).

First we consider the eigenvalue $\lambda_1 = 2$. The matrices $A - 2I$ and $(A-2I)^2$ and their reduced echelon forms (cf., Sec. 3.3) are given by

$$A - 2I = \begin{bmatrix} -5 & -5 & 3 & -2 \\ 9 & 9 & -5 & 4 \\ 3 & 3 & -2 & 1 \\ -3 & -3 & 2 & -1 \end{bmatrix} \longrightarrow \begin{bmatrix} 1 & 1 & 0 & 1 \\ 0 & 0 & 1 & 1 \\ 0 & 0 & 0 & 0 \\ 0 & 0 & 0 & 0 \end{bmatrix}, \quad (6.86)$$

$$(A - 2I)^2 = \begin{bmatrix} -5 & -5 & 0 & -5 \\ 9 & 9 & 0 & 9 \\ 3 & 3 & 0 & 3 \\ -3 & -3 & 0 & -3 \end{bmatrix} \longrightarrow \begin{bmatrix} 1 & 1 & 0 & 1 \\ 0 & 0 & 0 & 0 \\ 0 & 0 & 0 & 0 \\ 0 & 0 & 0 & 0 \end{bmatrix}. \quad (6.87)$$

Therefore,

$$\dim \mathrm{Ker}(A - 2I) = 2, \quad \dim \mathrm{Ker}((A - 2I)^2) = 3 = n_1 = \dim \hat{E}_A(\lambda_1)$$

and hence

$$\{\mathbf{0}\} \subsetneqq \mathrm{Ker}(A - 2I) \subsetneqq \mathrm{Ker}((A - 2I)^2) = \hat{E}_A(\lambda_1), \quad m_1 = 2.$$

The number $m_1 = 2$ is equal to the maximum order of a Jordan block, and furthermore, (6.80) gives

$$r_2^{(1)} = \dim \mathrm{Ker}((A - 2I)^2) - \dim \mathrm{Ker}(A - 2I) = 1,$$
$$r_1^{(1)} = -\dim \mathrm{Ker}((A - 2I)^2) + 2 \dim \mathrm{Ker}(A - 2I) = 1.$$

Therefore, there exist a single Jordan block of order two and a single Jordan block of order one.

We go on to find the basis $[\boldsymbol{x}_1^1, \boldsymbol{x}_2^1, \boldsymbol{x}_3^1]$ of the generalized eigenspace $\hat{E}_A(\lambda_1)$ satisfying (6.76). In the notation of the proof of Theorem 6.38 we have $B = A - 2I$ and $[\boldsymbol{x}_1^1, \boldsymbol{x}_2^1, \boldsymbol{x}_3^1] = [B\boldsymbol{b}_1^{(2)}, \boldsymbol{b}_1^{(2)}, \boldsymbol{b}_1^{(1)}]$.

- The vector $\boldsymbol{x}_2^1 = \boldsymbol{b}_1^{(2)}$ is the basis vector of the one-dimensional complementary subspace of $\mathrm{Ker}(A - 2I)$ in $\hat{E}_A(\lambda_1) = \mathrm{Ker}((A - 2I)^2)$. The reduced echelon form of $(A - 2I)^2$ shows

$$\mathrm{Ker}((A - 2I)^2) = \{\boldsymbol{x} = \xi_1(-1, 1, 0, 0)^\top + \xi_2(0, 0, 1, 0)^\top + \xi_3(-1, 0, 0, 1)^\top \mid \xi_1, \xi_2, \xi_3 \in \mathbb{C}\}$$

(cf., Sec. 5.3). As the basis vector $\boldsymbol{b}_1^{(2)}$ we may adopt any vector $\boldsymbol{x}$ of the above form that satisfies $(A - 2I)\boldsymbol{x} \neq \mathbf{0}$. Here we choose $\boldsymbol{b}_1^{(2)} = (-1, 0, 0, 1)^\top$, for which $\boldsymbol{x}_1^1 = B\boldsymbol{b}_1^{(2)} = (3, -5, -2, 2)^\top$.

- The vector $\boldsymbol{x}_3^1 = \boldsymbol{b}_1^{(1)}$ is chosen so that $\mathrm{Ker}(A - 2I) = \mathrm{Span}(B\boldsymbol{b}_1^{(2)}, \boldsymbol{b}_1^{(1)})$ holds. The reduced echelon form of $A - 2I$ shows
$$\mathrm{Ker}(A - 2I) = \{\boldsymbol{x} = \xi_1(-1,1,0,0)^\top + \xi_2(-1,0,-1,1)^\top \mid \xi_1, \xi_2 \in \mathbb{C}\}. \tag{6.88}$$
As the vector $\boldsymbol{b}_1^{(1)}$ we may adopt any vector $\boldsymbol{x}$ of the above form that is linearly independent of $B\boldsymbol{b}_1^{(2)} = (3,-5,-2,2)^\top$. Here we choose $\boldsymbol{b}_1^{(1)} = (-1,0,-1,1)^\top$.

Next we consider the other eigenvalue $\lambda_2 = 3$ (of multiplicity $n_2 = 1$). The matrix $A - 3I$ and its reduced echelon form are given by
$$A - 3I = \begin{bmatrix} -6 & -5 & 3 & -2 \\ 9 & 8 & -5 & 4 \\ 3 & 3 & -3 & 1 \\ -3 & -3 & 2 & -2 \end{bmatrix} \longrightarrow \begin{bmatrix} 1 & 0 & 0 & -5/3 \\ 0 & 1 & 0 & 3 \\ 0 & 0 & 1 & 1 \\ 0 & 0 & 0 & 0 \end{bmatrix}.$$
Hence we can choose
$$\boldsymbol{x}_1^2 = (5,-9,-3,3)^\top \tag{6.89}$$
as a basis of $\hat{E}_A(\lambda_2) = \mathrm{Ker}(A - 3I)$.

By collecting the basis vectors above we obtain the transformation matrix
$$X = [\boldsymbol{x}_1^1, \boldsymbol{x}_2^1, \boldsymbol{x}_3^1, \boldsymbol{x}_1^2] = \begin{bmatrix} 3 & -1 & -1 & 5 \\ -5 & 0 & 0 & -9 \\ -2 & 0 & -1 & -3 \\ 2 & 1 & 1 & 3 \end{bmatrix},$$
which brings A to its Jordan normal form as
$$X^{-1}AX = \mathrm{diag}\,(J_2(\lambda_1), J_1(\lambda_1), J_1(\lambda_2)) = \begin{bmatrix} 2 & 1 & & \\ 0 & 2 & & \\ & & 2 & \\ & & & 3 \end{bmatrix}. \tag{6.90}$$
This coincides with (6.73) in Example 6.8.

By choosing different basis vectors $\boldsymbol{b}_1^{(2)} = (-1,2,2,-1)^\top$ and $\boldsymbol{b}_1^{(1)} = (-1,1,0,0)^\top$, we obtain the transformation matrix $\tilde{X}$ in (6.74). ■

Example 6.11. We consider a matrix
$$A = \begin{bmatrix} 6 & -1 & -2 & -1 & 0 & -2 & -3 \\ 0 & 1 & -1 & 0 & -1 & 0 & -2 \\ -6 & 5 & 2 & 1 & -1 & 3 & 3 \\ 0 & 0 & 1 & 1 & 1 & 0 & 2 \\ 2 & -1 & -3 & -1 & 0 & -1 & -5 \\ 12 & -3 & -6 & -3 & 0 & -4 & -9 \\ 2 & -2 & 1 & 0 & 1 & -1 & 2 \end{bmatrix}.$$

The eigenvalues are given by $\lambda_1 = 1$ (of multiplicity $n_1 = 6$) and $\lambda_2 = 2$ (of multiplicity $n_2 = 1$).

First we consider the eigenvalue $\lambda_1 = 1$. The matrices $A - I$, $(A-I)^2$, and $(A-I)^3$ and their reduced echelon forms are given by

$$A - I = \begin{bmatrix} 5 & -1 & -2 & -1 & 0 & -2 & -3 \\ 0 & 0 & -1 & 0 & -1 & 0 & -2 \\ -6 & 5 & 1 & 1 & -1 & 3 & 3 \\ 0 & 0 & 1 & 0 & 1 & 0 & 2 \\ 2 & -1 & -3 & -1 & -1 & -1 & -5 \\ 12 & -3 & -6 & -3 & 0 & -5 & -9 \\ 2 & -2 & 1 & 0 & 1 & -1 & 1 \end{bmatrix} \to \begin{bmatrix} 1 & 0 & 0 & 0 & 0 & -1/3 & 0 \\ 0 & 1 & 0 & 0 & 0 & 1/6 & 1/2 \\ 0 & 0 & 1 & 0 & 1 & 0 & 2 \\ 0 & 0 & 0 & 1 & -2 & 1/6 & -3/2 \\ 0 & 0 & 0 & 0 & 0 & 0 & 0 \\ 0 & 0 & 0 & 0 & 0 & 0 & 0 \\ 0 & 0 & 0 & 0 & 0 & 0 & 0 \end{bmatrix},$$

$$(A - I)^2 = \begin{bmatrix} 7 & -3 & -3 & -1 & -1 & -3 & -6 \\ 0 & 0 & 0 & 0 & 0 & 0 & 0 \\ 4 & -3 & -3 & -1 & -1 & -2 & -6 \\ 0 & 0 & 0 & 0 & 0 & 0 & 0 \\ 4 & -3 & -3 & -1 & -1 & -2 & -6 \\ 18 & -9 & -9 & -3 & -3 & -8 & -18 \\ -4 & 3 & 3 & 1 & 1 & 2 & 6 \end{bmatrix} \to \begin{bmatrix} 1 & 0 & 0 & 0 & 0 & -1/3 & 0 \\ 0 & 1 & 1 & 1/3 & 1/3 & 2/9 & 2 \\ 0 & 0 & 0 & 0 & 0 & 0 & 0 \\ 0 & 0 & 0 & 0 & 0 & 0 & 0 \\ 0 & 0 & 0 & 0 & 0 & 0 & 0 \\ 0 & 0 & 0 & 0 & 0 & 0 & 0 \\ 0 & 0 & 0 & 0 & 0 & 0 & 0 \end{bmatrix},$$

$$(A - I)^3 = \begin{bmatrix} 3 & 0 & 0 & 0 & 0 & -1 & 0 \\ 0 & 0 & 0 & 0 & 0 & 0 & 0 \\ 0 & 0 & 0 & 0 & 0 & 0 & 0 \\ 0 & 0 & 0 & 0 & 0 & 0 & 0 \\ 0 & 0 & 0 & 0 & 0 & 0 & 0 \\ 6 & 0 & 0 & 0 & 0 & -2 & 0 \\ 0 & 0 & 0 & 0 & 0 & 0 & 0 \end{bmatrix} \to \begin{bmatrix} 1 & 0 & 0 & 0 & 0 & -1/3 & 0 \\ 0 & 0 & 0 & 0 & 0 & 0 & 0 \\ 0 & 0 & 0 & 0 & 0 & 0 & 0 \\ 0 & 0 & 0 & 0 & 0 & 0 & 0 \\ 0 & 0 & 0 & 0 & 0 & 0 & 0 \\ 0 & 0 & 0 & 0 & 0 & 0 & 0 \\ 0 & 0 & 0 & 0 & 0 & 0 & 0 \end{bmatrix}.$$

Therefore,

$$\begin{aligned} \dim \operatorname{Ker}(A - I) &= 3, \\ \dim \operatorname{Ker}((A - I)^2) &= 5, \\ \dim \operatorname{Ker}((A - I)^3) &= 6 = n_1 \end{aligned}$$

with $m_1 = 3$. Furthermore, (6.80) gives

$$\begin{aligned} r_3^{(1)} &= \dim \operatorname{Ker}((A - I)^3) - \dim \operatorname{Ker}((A - I)^2) = 1, \\ r_2^{(1)} &= -\dim \operatorname{Ker}((A - I)^3) + 2 \dim \operatorname{Ker}((A - I)^2) - \dim \operatorname{Ker}(A - I) = 1, \\ r_1^{(1)} &= -\dim \operatorname{Ker}((A - I)^2) + 2 \dim \operatorname{Ker}(A - I) = 1. \end{aligned}$$

Therefore there are three blocks of orders 3, 2, and 1, respectively.

Next we find the basis $[\boldsymbol{x}_1^1, \boldsymbol{x}_2^1, \boldsymbol{x}_3^1, \boldsymbol{x}_4^1, \boldsymbol{x}_5^1, \boldsymbol{x}_6^1] = [B^2\boldsymbol{b}_1^{(3)}, B\boldsymbol{b}_1^{(3)}, \boldsymbol{b}_1^{(3)}, B\boldsymbol{b}_1^{(2)}, \boldsymbol{b}_1^{(2)}, \boldsymbol{b}_1^{(1)}]$ of $\hat{E}_A(\lambda_1)$, where $B = A - I$.

- The vector $\boldsymbol{x}_3^1 = \boldsymbol{b}_1^{(3)}$ is the basis vector of the one-dimensional complementary subspace of $\mathrm{Ker}((A-I)^2)$ in $\hat{E}_A(\lambda_1) = \mathrm{Ker}((A-I)^3)$. The reduced echelon form of $(A-I)^3$ shows

$$\begin{aligned}\mathrm{Ker}((A-I)^3) = \{\boldsymbol{x} = \xi_1\boldsymbol{e}_2 + \xi_2\boldsymbol{e}_3 + \xi_3\boldsymbol{e}_4 + \xi_4\boldsymbol{e}_5 \\ + \xi_5(1,0,0,0,0,3,0)^\top + \xi_6\boldsymbol{e}_7\}\end{aligned}$$

(cf., Sec. 5.3; $\boldsymbol{e}_j$ denotes the jth unit vector). As the basis vector $\boldsymbol{b}_1^{(3)}$ we may adopt any vector $\boldsymbol{x}$ of the above form that satisfies $(A-I)^2\boldsymbol{x} \neq \boldsymbol{0}$. Here we choose $\boldsymbol{b}_1^{(3)} = \boldsymbol{e}_2 = (0,1,0,0,0,0,0)^\top$, for which

$$\begin{aligned}\boldsymbol{x}_1^1 &= B^2\boldsymbol{b}_1^{(3)} = (-3,0,-3,0,-3,-9,3)^\top, \\ \boldsymbol{x}_2^1 &= B\boldsymbol{b}_1^{(3)} = (-1,0,5,0,-1,-3,-2)^\top.\end{aligned}$$

- The vector $\boldsymbol{x}_5^1 = \boldsymbol{b}_1^{(2)}$ is chosen so that $\{B\boldsymbol{b}_1^{(3)}, \boldsymbol{b}_1^{(2)}\}$ is a basis of the two-dimensional complementary subspace of $\mathrm{Ker}(A-I)$ in $\mathrm{Ker}((A-I)^2)$. The reduced echelon form of $(A-I)^2$ shows

$$\begin{aligned}\mathrm{Ker}((A-I)^2) = \{\boldsymbol{x} = \xi_1(0,-1,1,0,0,0,0)^\top + \xi_2(0,-1,0,3,0,0,0)^\top \\ + \xi_3(0,-1,0,0,3,0,0)^\top + \xi_4(3,-2,0,0,0,9,0)^\top \\ + \xi_5(0,-2,0,0,0,0,1)^\top\}.\end{aligned}$$

As the vector $\boldsymbol{b}_1^{(2)}$ we may adopt any vector $\boldsymbol{x}$ of the above form with $(A-I)\boldsymbol{x} \neq \boldsymbol{0}$ that is linearly independent of $B\boldsymbol{b}_1^{(3)} = (-1,0,5,0,-1,-3,-2)^\top$. Here we choose $\boldsymbol{b}_1^{(2)} = (0,-1,1,0,0,0,0)^\top$, for which $\boldsymbol{x}_4^1 = B\boldsymbol{b}_1^{(2)} = (-1,-1,-4,1,-2,-3,3)^\top$.

- The vector $\boldsymbol{x}_6^1 = \boldsymbol{b}_1^{(1)}$ is chosen so that $\mathrm{Ker}(A-I) = \mathrm{Span}(B^2\boldsymbol{b}_1^{(3)}, B\boldsymbol{b}_1^{(2)}, \boldsymbol{b}_1^{(1)})$ holds. The reduced echelon form of $A-I$ shows

$$\begin{aligned}\mathrm{Ker}(A-I) = \{\boldsymbol{x} = \xi_1(0,0,-1,2,1,0,0)^\top + \xi_2(2,-1,0,-1,0,6,0)^\top \\ + \xi_3(0,-1,-4,3,0,0,2)^\top\}. \qquad (6.91)\end{aligned}$$

As the basis vector $\boldsymbol{b}_1^{(1)}$ we may adopt any vector $\boldsymbol{x}$ of the above form that is linearly independent of $B^2\boldsymbol{b}_1^{(3)} = (-3,0,-3,0,-3,-9,3)^\top$ and $B\boldsymbol{b}_1^{(2)} = (-1,-1,-4,1,-2,-3,3)^\top$. Here we choose $\boldsymbol{b}_1^{(1)} = (0,0,-1,2,1,0,0)^\top$.

Next we consider the other eigenvalue $\lambda_2 = 2$ (of multiplicity $n_2 = 1$). The matrix $A - 2I$ and its reduced echelon form are given by

$$A - 2I = \begin{bmatrix} 4 & -1 & -2 & -1 & 0 & -2 & -3 \\ 0 & -1 & -1 & 0 & -1 & 0 & -2 \\ -6 & 5 & 0 & 1 & -1 & 3 & 3 \\ 0 & 0 & 1 & -1 & 1 & 0 & 2 \\ 2 & -1 & -3 & -1 & -2 & -1 & -5 \\ 12 & -3 & -6 & -3 & 0 & -6 & -9 \\ 2 & -2 & 1 & 0 & 1 & -1 & 0 \end{bmatrix} \to \begin{bmatrix} 1 & 0 & 0 & 0 & 0 & -1/2 & 0 \\ 0 & 1 & 0 & 0 & 0 & 0 & 0 \\ 0 & 0 & 1 & 0 & 0 & 0 & 0 \\ 0 & 0 & 0 & 1 & 0 & 0 & 0 \\ 0 & 0 & 0 & 0 & 1 & 0 & 0 \\ 0 & 0 & 0 & 0 & 0 & 0 & 1 \\ 0 & 0 & 0 & 0 & 0 & 0 & 0 \end{bmatrix}.$$

Hence we can choose

$$\boldsymbol{x}_1^2 = (1, 0, 0, 0, 0, 2, 0)^\top \tag{6.92}$$

as a basis of $\hat{E}_A(\lambda_2) = \text{Ker}(A - 2I)$.

By collecting the basis vectors above we obtain the transformation matrix

$$X = [\boldsymbol{x}_1^1, \boldsymbol{x}_2^1, \boldsymbol{x}_3^1, \boldsymbol{x}_4^1, \boldsymbol{x}_5^1, \boldsymbol{x}_6^1, \boldsymbol{x}_1^2] = \begin{bmatrix} -3 & -1 & 0 & -1 & 0 & 0 & 1 \\ 0 & 0 & 1 & -1 & -1 & 0 & 0 \\ -3 & 5 & 0 & -4 & 1 & -1 & 0 \\ 0 & 0 & 0 & 1 & 0 & 2 & 0 \\ -3 & -1 & 0 & -2 & 0 & 1 & 0 \\ -9 & -3 & 0 & -3 & 0 & 0 & 2 \\ 3 & -2 & 0 & 3 & 0 & 0 & 0 \end{bmatrix},$$

which brings A to its Jordan normal form as

$$X^{-1}AX = \text{diag}\,(J_3(\lambda_1), J_2(\lambda_1), J_1(\lambda_1), J_1(\lambda_2))$$

$$= \left[\begin{array}{ccc|cc|c|c} 1 & 1 & 0 & & & & \\ 0 & 1 & 1 & & & & \\ 0 & 0 & 1 & & & & \\ \hline & & & 1 & 1 & & \\ & & & 0 & 1 & & \\ \hline & & & & & 1 & \\ \hline & & & & & & 2 \end{array}\right].$$

■

In the above we have shown how to construct the Jordan normal form by a systematic or mechanical procedure used in the proof of Theorem 6.36 in Sec. 6.8.2. In the following we show an alternative method that will be more natural and intuitive, although it may sometimes require ad hoc heuristic ideas. The basic idea of this alternative method is to aim at diagonalization of the given matrix and to call for additional considerations if the diagonalization is found to be impossible.

Suppose that we are given a matrix A. From the characteristic polynomial of A, we compute the eigenvalues of A, with their multiplicities, and the corresponding eigenvectors.

(1) If, for each eigenvalue, the number of linearly independent eigenvectors is equal to its multiplicity, the matrix A is diagonalizable. In this case, we construct the transformation matrix X as the collection of the eigenvectors for all eigenvalues. Then $X^{-1}AX$ is a diagonal matrix.
(2) Otherwise, we compute the generalized eigenvectors to construct a transformation matrix X for the Jordan normal form.

This alternative method is illustrated below for the matrices treated in Examples 6.10 and 6.11.

Example 6.12. We consider the matrix

$$A = \begin{bmatrix} -3 & -5 & 3 & -2 \\ 9 & 11 & -5 & 4 \\ 3 & 3 & 0 & 1 \\ -3 & -3 & 2 & 1 \end{bmatrix}$$

treated in Example 6.10. The eigenvalues are computed already: $\lambda_1 = 2$ of multiplicity $n_1 = 3$ and $\lambda_2 = 3$ of multiplicity $n_2 = 1$.

First we consider the eigenvalue $\lambda_1 = 2$. By (6.88) there are two linearly independent eigenvectors given by $(-1, 1, 0, 0)^\top$ and $(-1, 0, -1, 1)^\top$. This implies that the number g_1 of Jordan blocks corresponding to this eigenvalue is equal to 2, that is, $g_1 = 2$. It follows from $n_{11} + n_{12} = n_1$ $(= 3)$ in (6.70) that $n_{11} = 2$ and $n_{12} = 1$. That is, there are two Jordan blocks $J_2(\lambda_1)$ and $J_1(\lambda_1)$ for λ_1.

- The basis vectors corresponding to $J_2(\lambda_1)$ can be determined as follows. By definition there exist a pair of vectors $\boldsymbol{x}_1^1$ and $\boldsymbol{x}_2^1$ satisfying

$$A[\boldsymbol{x}_1^1, \boldsymbol{x}_2^1] = [\boldsymbol{x}_1^1, \boldsymbol{x}_2^1] J_2(\lambda_1) = [\boldsymbol{x}_1^1, \boldsymbol{x}_2^1] \begin{bmatrix} 2 & 1 \\ 0 & 2 \end{bmatrix}, \qquad (6.93)$$

or equivalently,

$$(A - 2I)\boldsymbol{x}_1^1 = \mathbf{0}, \qquad (6.94)$$

$$(A - 2I)\boldsymbol{x}_2^1 = \boldsymbol{x}_1^1. \qquad (6.95)$$

By (6.88) the solution to the first equation (6.94) is represented as

$$\boldsymbol{x}_1^1 = \xi_1(-1,1,0,0)^\top + \xi_2(-1,0,-1,1)^\top \qquad (\xi_1, \xi_2 \in \mathbb{C}). \tag{6.96}$$

This parametric representation is obtained from the reduced echelon form of $A - 2I$ given in (6.86). Let S denote the product of elementary matrices used to bring $A - 2I$ to its reduced echelon form.[20] By multiplying (6.95) with S from the left and noting $S(-1,1,0,0)^\top = (2,3,-2,0)^\top$ and $S(-1,0,-1,1)^\top = (5,8,-5,0)^\top$ for the basis vectors in (6.96) we can transform (6.95) to

$$\begin{bmatrix} 1 & 1 & 0 & 1 \\ 0 & 0 & 1 & 1 \\ 0 & 0 & 0 & 0 \\ 0 & 0 & 0 & 0 \end{bmatrix} \boldsymbol{x}_2^1 = \xi_1 \begin{bmatrix} 2 \\ 3 \\ -2 \\ 0 \end{bmatrix} + \xi_2 \begin{bmatrix} 5 \\ 8 \\ -5 \\ 0 \end{bmatrix}. \tag{6.97}$$

This equation has a solution if and only if $-2\xi_1 - 5\xi_2 = 0$. For the choice of $\xi_1 = -5$ and $\xi_2 = 2$, the equation reads

$$\begin{bmatrix} 1 & 1 & 0 & 1 \\ 0 & 0 & 1 & 1 \\ 0 & 0 & 0 & 0 \\ 0 & 0 & 0 & 0 \end{bmatrix} \boldsymbol{x}_2^1 = \begin{bmatrix} 0 \\ 1 \\ 0 \\ 0 \end{bmatrix}, \tag{6.98}$$

for which $\boldsymbol{x}_2^1 = (-1,0,0,1)^\top$ is a solution. The first vector $\boldsymbol{x}_1^1$ is given by (6.96) as $\boldsymbol{x}_1^1 = (3,-5,-2,2)^\top$.

- As the basis vector $\boldsymbol{x}_3^1$ for $J_1(\lambda_1)$, we may take $\boldsymbol{x}_3^1 = (-1,0,-1,1)^\top$ using one of the eigenvectors found above for $\lambda_1 = 2$, since it is linearly independent of $\boldsymbol{x}_1^1$ and satisfies

$$A\boldsymbol{x}_3^1 = 2\boldsymbol{x}_3^1 = \boldsymbol{x}_3^1 J_1(\lambda_1). \tag{6.99}$$

  We may also use the other eigenvector $(-1,1,0,0)^\top$ for $\boldsymbol{x}_3^1$.

Next we consider the other eigenvalue $\lambda_2 = 3$. The eigenvector is given by $\boldsymbol{x}_1^2 = (5,-9,-3,3)^\top$ (cf., (6.89)), for which

$$A\boldsymbol{x}_1^2 = 3\boldsymbol{x}_1^2 = \boldsymbol{x}_1^2 J_1(\lambda_2). \tag{6.100}$$

It follows from (6.93), (6.99), and (6.100) that

$$A[\boldsymbol{x}_1^1, \boldsymbol{x}_2^1, \boldsymbol{x}_3^1, \boldsymbol{x}_1^2] = [\boldsymbol{x}_1^1, \boldsymbol{x}_2^1, \boldsymbol{x}_3^1, \boldsymbol{x}_1^2] \cdot \mathrm{diag}\,(J_2(\lambda_1), J_1(\lambda_1), J_1(\lambda_2)).$$

[20]This matrix S is not uniquely determined. Here we have chosen $S = \begin{bmatrix} -2 & 0 & -3 & 0 \\ -3 & 0 & -5 & 0 \\ 3 & 1 & 2 & 0 \\ 0 & 0 & 1 & 1 \end{bmatrix}$.

This coincides with (6.90) in Example 6.10 and (6.73) in Example 6.8, including the transformation matrix $X = [\boldsymbol{x}_1^1, \boldsymbol{x}_2^1, \boldsymbol{x}_3^1, \boldsymbol{x}_1^2]$.

It is worth mentioning that the transformation matrix $\tilde{X}$ in (6.74) is obtained if we choose $\boldsymbol{x}_2^1 = (-1, 2, 2, -1)^\top$ and $\boldsymbol{x}_3^1 = (-1, 1, 0, 0)^\top$ to meet (6.98) and (6.99), respectively. ■

Example 6.13. We consider the matrix

$$A = \begin{bmatrix} 6 & -1 & -2 & -1 & 0 & -2 & -3 \\ 0 & 1 & -1 & 0 & -1 & 0 & -2 \\ -6 & 5 & 2 & 1 & -1 & 3 & 3 \\ 0 & 0 & 1 & 1 & 1 & 0 & 2 \\ 2 & -1 & -3 & -1 & 0 & -1 & -5 \\ 12 & -3 & -6 & -3 & 0 & -4 & -9 \\ 2 & -2 & 1 & 0 & 1 & -1 & 2 \end{bmatrix}$$

treated in Example 6.11. The eigenvalues are computed already: $\lambda_1 = 1$ of multiplicity $n_1 = 6$ and $\lambda_2 = 2$ of multiplicity $n_2 = 1$.

First we consider the eigenvalue $\lambda_1 = 1$. By (6.91) there are three linearly independent eigenvectors given by

$$(0, 0, -1, 2, 1, 0, 0)^\top, \quad (2, -1, 0, -1, 0, 6, 0)^\top, \quad (0, -1, -4, 3, 0, 0, 2)^\top.$$

This implies that the number g_1 of Jordan blocks corresponding to this eigenvalue is equal to 3, that is, $g_1 = 3$. For the orders of Jordan blocks, we have three possibilities $(n_{11}, n_{12}, n_{13}) = (4, 1, 1), (3, 2, 1)$, and $(2, 2, 2)$ compatible with the relation $n_{11} + n_{12} + n_{13} = n_1$ $(= 6)$ in (6.70). That is, the diagonal blocks corresponding to λ_1 are in one of the following forms:

$$\text{diag}\,(J_4(\lambda_1), J_1(\lambda_1), J_1(\lambda_1)), \tag{6.101}$$

$$\text{diag}\,(J_3(\lambda_1), J_2(\lambda_1), J_1(\lambda_1)), \tag{6.102}$$

$$\text{diag}\,(J_2(\lambda_1), J_2(\lambda_1), J_2(\lambda_1)). \tag{6.103}$$

These three cases can be distinguished by the number of Jordan blocks of order $\geqq 2$. Indeed, (6.101), (6.102), and (6.103) have one, two, and three such Jordan blocks, respectively. For each Jordan block of order $\geqq 2$ corresponding to λ_1, there exist a pair of vectors $\boldsymbol{x}$, $\boldsymbol{y}$ satisfying

$$A[\boldsymbol{x}, \boldsymbol{y}] = [\boldsymbol{x}, \boldsymbol{y}] \begin{bmatrix} \lambda_1 & 1 \\ 0 & \lambda_1 \end{bmatrix}, \tag{6.104}$$

and the vectors $\boldsymbol{x}$ corresponding to different Jordan blocks are linearly independent. Furthermore, it can be seen[21] that the number of linearly

[21] By similarity transformation we may assume that the matrix A in (6.104) is in the Jordan normal form. Then the statement is obviously true.

independent such vectors $\boldsymbol{x}$ is equal to the number of Jordan blocks of order $\geqq 2$ corresponding to λ_1.

We now focus on solving equation (6.104), which can be rewritten as

$$(A - I)\boldsymbol{x} = \boldsymbol{0}, \tag{6.105}$$

$$(A - I)\boldsymbol{y} = \boldsymbol{x}. \tag{6.106}$$

It follows from the reduced echelon form of $A-I$ computed in Example 6.11 that the solution of the first equation (6.105) is given in a parametric form (cf., (5.11) in Sec. 5.3) as

$$\begin{aligned}\boldsymbol{x} = \xi_1(0,0,-1,2,1,0,0)^\top &+ \xi_2(2,-1,0,-1,0,6,0)^\top \\ &+ \xi_3(0,-1,-4,3,0,0,2)^\top.\end{aligned} \tag{6.107}$$

Let S denote the product of elementary matrices used to bring $A-I$ to its reduced echelon form.[22] By multiplying (6.106) with S from the left and noting

$$\begin{aligned}S(0,0,-1,2,1,0,0)^\top &= (0,-1/2,0,-3/2,0,2,-1)^\top, \\ S(2,-1,0,-1,0,6,0)^\top &= (0,1,1,-3,0,-2,1)^\top, \\ S(0,-1,-4,3,0,0,2)^\top &= (0,-1,1,-3,0,2,-1)^\top\end{aligned}$$

for the basis vectors in (6.107) we can transform (6.106) to

$$\begin{bmatrix} 1 & 0 & 0 & 0 & 0 & -1/3 & 0 \\ 0 & 1 & 0 & 0 & 0 & 1/6 & 1/2 \\ 0 & 0 & 1 & 0 & 1 & 0 & 2 \\ 0 & 0 & 0 & 1 & -2 & 1/6 & -3/2 \\ 0 & 0 & 0 & 0 & 0 & 0 & 0 \\ 0 & 0 & 0 & 0 & 0 & 0 & 0 \\ 0 & 0 & 0 & 0 & 0 & 0 & 0 \end{bmatrix} \boldsymbol{y} = \xi_1 \begin{bmatrix} 0 \\ -1/2 \\ 0 \\ -3/2 \\ 0 \\ 2 \\ -1 \end{bmatrix} + \xi_2 \begin{bmatrix} 0 \\ 1 \\ 1 \\ -3 \\ 0 \\ -2 \\ 1 \end{bmatrix} + \xi_3 \begin{bmatrix} 0 \\ -1 \\ 1 \\ -3 \\ 0 \\ 2 \\ -1 \end{bmatrix}. \tag{6.108}$$

This equation has a solution if and only if $\xi_1 - \xi_2 + \xi_3 = 0$. For the choices of $(\xi_1, \xi_2, \xi_3) = (2, 1, -1), (0, 1/2, 1/2)$, the right-hand side of (6.108) is equal, respectively, to

$$(0,1,0,-3,0,0,0)^\top, \quad (0,0,1,-3,0,0,0)^\top$$

[22]This matrix S is not uniquely determined. Here we have chosen $S = \begin{bmatrix} 1 & 0 & 0 & 0 & 0 & -1/3 & 0 \\ 0 & 0 & 1/2 & 0 & 0 & 1/6 & 1/2 \\ 0 & -1 & 0 & 0 & 0 & 0 & 0 \\ 0 & 4 & 3/2 & 0 & 0 & 1/6 & 7/2 \\ 0 & 0 & 1 & 0 & 1 & 0 & 2 \\ 0 & 1 & 0 & 1 & 0 & 0 & 0 \\ -2 & 1 & 1 & 0 & 0 & 1 & 2 \end{bmatrix}$.

(these vectors are linearly independent), and we obtain respective solutions

$$\boldsymbol{y}_1^1 = (0,1,0,-3,0,0,0)^\top, \qquad \boldsymbol{y}_2^1 = (0,0,1,-3,0,0,0)^\top.$$

The corresponding vectors $\boldsymbol{x}$ are obtained from (6.107) with $(\xi_1, \xi_2, \xi_3) = (2,1,-1), (0,1/2,1/2)$ as

$$\boldsymbol{x}_1^1 = (2,0,2,0,2,6,-2)^\top, \qquad \boldsymbol{x}_2^1 = (1,-1,-2,1,0,3,1)^\top.$$

This shows that the equation (6.104) has two linearly independent solutions $\boldsymbol{x}$. Therefore, we have (6.102) with $(n_{11}, n_{12}, n_{13}) = (3,2,1)$ for the eigenvalue λ_1.

The basis vectors corresponding to $J_3(\lambda_1)$ can be determined as follows. By definition there exist vectors $\boldsymbol{x}$, $\boldsymbol{y}$, $\boldsymbol{z}$ satisfying

$$A[\boldsymbol{x}, \boldsymbol{y}, \boldsymbol{z}] = [\boldsymbol{x}, \boldsymbol{y}, \boldsymbol{z}] \begin{bmatrix} 1 & 1 & 0 \\ 0 & 1 & 1 \\ 0 & 0 & 1 \end{bmatrix},$$

or equivalently,

$$(A - I)\boldsymbol{x} = \mathbf{0}, \tag{6.109}$$

$$(A - I)\boldsymbol{y} = \boldsymbol{x}, \tag{6.110}$$

$$(A - I)\boldsymbol{z} = \boldsymbol{y}. \tag{6.111}$$

We have already investigated the first two equations (6.109) and (6.110) in (6.105) and (6.106). Our analysis implies that $\boldsymbol{y}$ is represented parametrically as

$$\boldsymbol{y} = \eta_1 \begin{bmatrix} 0 \\ 0 \\ -1 \\ 2 \\ \hline 1 \\ 0 \\ 0 \end{bmatrix} + \eta_2 \begin{bmatrix} 2 \\ -1 \\ 0 \\ -1 \\ \hline 0 \\ 6 \\ 0 \end{bmatrix} + \eta_3 \begin{bmatrix} 0 \\ -1 \\ -4 \\ 3 \\ \hline 0 \\ 0 \\ 2 \end{bmatrix} + \eta_4 \begin{bmatrix} 0 \\ 1 \\ 0 \\ -3 \\ \hline 0 \\ 0 \\ 0 \end{bmatrix} + \eta_5 \begin{bmatrix} 0 \\ 0 \\ 1 \\ -3 \\ \hline 0 \\ 0 \\ 0 \end{bmatrix}. \tag{6.112}$$

Note that the last two vectors, with coefficients η_4 and η_5, are equal to $\boldsymbol{y}_1^1$ and $\boldsymbol{y}_2^1$, respectively. They are particular solutions of (6.110) for $\boldsymbol{x}_1^1$ and $\boldsymbol{x}_2^1$, respectively, which form a basis of the solutions of (6.109) for which (6.110) admits a solution. Accordingly, a particular solution of (6.110) for $\boldsymbol{x} = \eta_4 \boldsymbol{x}_1^1 + \eta_5 \boldsymbol{x}_2^1$ is given by the sum of the last two vectors in (6.112). On the other hand, the first three vectors in (6.112), with coefficients η_1, η_2, and η_3, are the basis vectors of $\operatorname{Ker}(A - I)$ given in (6.107). Therefore,

(6.112) provides an expression of $\boldsymbol{y}$ in the general solution $(\boldsymbol{x}, \boldsymbol{y})$ of (6.109) and (6.110).

To solve the third equation (6.111), we multiply (6.111) with the matrix S from the left. Then the coefficient matrix $A - I$ is transformed to its reduced echelon form, and the right-hand side vector $\boldsymbol{y}$ is changed to $S\boldsymbol{y}$, where

$$S(0,1,0,-3,0,0,0)^\top = (0,0,-1,4,0,-2,1)^\top,$$
$$S(0,0,1,-3,0,0,0)^\top = (0,1/2,0,3/2,1,-3,1)^\top$$

for the fourth and fifth basis vectors on the right-hand side of (6.112), while we already have the transformed vectors for the first three basis vectors. We have thus arrived at the following equivalent form of equation (6.111):

$$\begin{bmatrix} 1 & 0 & 0 & 0 & 0 & -1/3 & 0 \\ 0 & 1 & 0 & 0 & 0 & 1/6 & 1/2 \\ 0 & 0 & 1 & 0 & 1 & 0 & 2 \\ 0 & 0 & 0 & 1 & -2 & 1/6 & -3/2 \\ 0 & 0 & 0 & 0 & 0 & 0 & 0 \\ 0 & 0 & 0 & 0 & 0 & 0 & 0 \\ 0 & 0 & 0 & 0 & 0 & 0 & 0 \end{bmatrix} \boldsymbol{z}$$
$$= \eta_1 \begin{bmatrix} 0 \\ -1/2 \\ 0 \\ -3/2 \\ 0 \\ 2 \\ -1 \end{bmatrix} + \eta_2 \begin{bmatrix} 0 \\ 1 \\ 1 \\ -3 \\ 0 \\ -2 \\ 1 \end{bmatrix} + \eta_3 \begin{bmatrix} 0 \\ -1 \\ 1 \\ -3 \\ 0 \\ 2 \\ -1 \end{bmatrix} + \eta_4 \begin{bmatrix} 0 \\ 0 \\ -1 \\ 4 \\ 0 \\ -2 \\ 1 \end{bmatrix} + \eta_5 \begin{bmatrix} 0 \\ 1/2 \\ 0 \\ 3/2 \\ 1 \\ -3 \\ 1 \end{bmatrix}.$$

This equation has a solution if and only if $\eta_5 = 0$ and $\eta_1 - \eta_2 + \eta_3 - \eta_4 = 0$. For the choice of $(\eta_1, \eta_2, \eta_3, \eta_4, \eta_5) = (0,0,1,1,0)$, the right-hand side vector becomes $(0,-1,0,1,0,0,0)^\top$ and the solution $\boldsymbol{z}$ is given by

$$\boldsymbol{z}_1^1 = (0,-1,0,1,0,0,0)^\top.$$

The corresponding $\boldsymbol{y}$ is obtained from (6.112) with $(\eta_1, \eta_2, \eta_3, \eta_4, \eta_5) = (0,0,1,1,0)$ as

$$\tilde{\boldsymbol{y}}_1^1 = (0,0,-4,0,0,0,2)^\top,$$

for which $\boldsymbol{x} = (A - I)\tilde{\boldsymbol{y}}_1^1$ coincides with $\boldsymbol{x}_1^1 = (2,0,2,0,2,6,-2)^\top$.

From the above argument, we obtain the following generalized eigenvectors corresponding to eigenvalue λ_1.

- For $J_3(\lambda_1)$, the generalized eigenvectors are given by $\boldsymbol{x}_1^1, \tilde{\boldsymbol{y}}_1^1, \boldsymbol{z}_1^1$.
- For $J_2(\lambda_1)$, the generalized eigenvectors are given by $\boldsymbol{x}_2^1, \boldsymbol{y}_2^1$.
- For $J_1(\lambda_1)$, we may take any one of the three eigenvectors chosen at the beginning, since each of them is linearly independent of $\boldsymbol{x}_1^1$ and $\boldsymbol{x}_2^1$. We choose $\boldsymbol{x}_3^1 = (0, 0, -1, 2, 1, 0, 0)^\top$ here.

For the other eigenvalue $\lambda_2 = 2$, the eigenvector is given by

$$\boldsymbol{x}_1^2 = (1, 0, 0, 0, 0, 2, 0)^\top$$

(cf., (6.92)).

By collecting all generalized eigenvectors found above, we finally obtain

$$\begin{aligned} &A[\boldsymbol{x}_1^1, \tilde{\boldsymbol{y}}_1^1, \boldsymbol{z}_1^1, \boldsymbol{x}_2^1, \boldsymbol{y}_2^1, \boldsymbol{x}_3^1, \boldsymbol{x}_1^2] \\ &= [\boldsymbol{x}_1^1, \tilde{\boldsymbol{y}}_1^1, \boldsymbol{z}_1^1, \boldsymbol{x}_2^1, \boldsymbol{y}_2^1, \boldsymbol{x}_3^1, \boldsymbol{x}_1^2] \cdot \mathrm{diag}\,(J_3(\lambda_1), J_2(\lambda_1), J_1(\lambda_1), J_1(\lambda_2)). \end{aligned}$$

That is, with the matrix

$$X = [\boldsymbol{x}_1^1, \tilde{\boldsymbol{y}}_1^1, \boldsymbol{z}_1^1, \boldsymbol{x}_2^1, \boldsymbol{y}_2^1, \boldsymbol{x}_3^1, \boldsymbol{x}_1^2] = \begin{bmatrix} 2 & 0 & 0 & 1 & 0 & 0 & 1 \\ 0 & 0 & -1 & -1 & 0 & 0 & 0 \\ 2 & -4 & 0 & -2 & 1 & -1 & 0 \\ 0 & 0 & 1 & 1 & -3 & 2 & 0 \\ 2 & 0 & 0 & 0 & 0 & 1 & 0 \\ 6 & 0 & 0 & 3 & 0 & 0 & 2 \\ -2 & 2 & 0 & 1 & 0 & 0 & 0 \end{bmatrix},$$

we can transform A to its Jordan normal form

$$X^{-1}AX = \mathrm{diag}\,(J_3(\lambda_1), J_2(\lambda_1), J_1(\lambda_1), J_1(\lambda_2)).$$

This Jordan normal form coincides with the one obtained in Example 6.11, but the transformation matrix X is not the same. ■

6.8.5 *Powers of Matrices*

The Jordan normal form offers an effective method to analyze the behavior of powers A^p of a matrix A. The following theorem shows a basic fact.

Theorem 6.39. *For a nonnegative integer $p \geqq 0$, the pth power of a Jordan* *block* $J_k(\lambda)$ *of order* k *is given by*

$$J_k(\lambda)^p = \begin{bmatrix} \lambda^p & \binom{p}{1}\lambda^{p-1} & \binom{p}{2}\lambda^{p-2} & \cdots & \binom{p}{k-1}\lambda^{p-k+1} \\ & \lambda^p & \binom{p}{1}\lambda^{p-1} & \ddots & \vdots \\ & & \lambda^p & \ddots & \binom{p}{2}\lambda^{p-2} \\ & O & & \ddots & \binom{p}{1}\lambda^{p-1} \\ & & & & \lambda^p \end{bmatrix}, \tag{6.113}$$

where, by convention, an entry with a negative exponent to λ *is understood to vanish, that is, the* (i, j) *entry of the above matrix is equal to zero if* $j - i > p$.

Proof. Decompose $J_k(\lambda)$ as $J_k(\lambda) = \lambda I_k + N_k$ with

$$N_k = \begin{bmatrix} 0 & 1 & & \\ & 0 & \ddots & \\ & & \ddots & 1 \\ & & & 0 \end{bmatrix}.$$

Since λI_k and N_k commute, we have

$$J_k(\lambda)^p = (\lambda I_k + N_k)^p = (\lambda I_k)^p + \binom{p}{1}\lambda^{p-1}N_k + \binom{p}{2}\lambda^{p-2}{N_k}^2 + \cdots + \binom{p}{p}{N_k}^p.$$

The (i, j) entry of ${N_k}^q$ is equal to 1 if $j - i = q$, and 0 otherwise. In particular, ${N_k}^q = O$ if $q \geqq k$. Hence follows (6.113). $\square$

To simplify notation, let

$$X^{-1}AX = \mathrm{diag}\,(\cdots, J_k(\lambda), \cdots)$$

mean the expression (6.69) for the Jordan normal form of matrix A. Since $(X^{-1}AX)^p = X^{-1}A^pX$, we have

$$A^p = X \cdot \mathrm{diag}\,(\cdots, J_k(\lambda)^p, \cdots) \cdot X^{-1}, \tag{6.114}$$

in which each diagonal block $J_k(\lambda)^p$ is given by (6.113).

Example 6.14. We consider the 4×4 matrix A treated in Example 6.8 in Sec. 6.8.1. The power A^p for $p \geqq 1$ is given as

$$A^p = X \cdot \left[\begin{array}{cc|c|c} 2^p & p\,2^{p-1} & & \\ 0 & 2^p & & \\ \hline & & 2^p & \\ \hline & & & 3^p \end{array}\right] \cdot X^{-1}$$

$$= \begin{bmatrix} 3 & -1 & -1 & 5 \\ -5 & 0 & 0 & -9 \\ -2 & 0 & -1 & -3 \\ 2 & 1 & 1 & 3 \end{bmatrix} \left[\begin{array}{cc|c|c} 2^p & p\,2^{p-1} & & \\ 0 & 2^p & & \\ \hline & & 2^p & \\ \hline & & & 3^p \end{array}\right] \begin{bmatrix} 9/5 & 8/5 & 0 & 9/5 \\ 0 & 0 & 1 & 1 \\ -3/5 & -1/5 & -1 & -3/5 \\ -1 & -1 & 0 & -1 \end{bmatrix}$$

$$= \left[\begin{array}{c|c|c|c} 6 \cdot 2^p - 5 \cdot 3^p & 5 \cdot 2^p - 5 \cdot 3^p & 3 \cdot p\,2^{p-1} & 5 \cdot 2^p + 3 \cdot p\,2^{p-1} - 5 \cdot 3^p \\ -9 \cdot 2^p + 9 \cdot 3^p & -8 \cdot 2^p + 9 \cdot 3^p & -5 \cdot p\,2^{p-1} & -9 \cdot 2^p - 5 \cdot p\,2^{p-1} + 9 \cdot 3^p \\ -3 \cdot 2^p + 3 \cdot 3^p & -3 \cdot 2^p + 3 \cdot 3^p & 2^p - 2 \cdot p\,2^{p-1} & -3 \cdot 2^p - 2 \cdot p\,2^{p-1} + 3 \cdot 3^p \\ 3 \cdot 2^p - 3 \cdot 3^p & 3 \cdot 2^p - 3 \cdot 3^p & 2 \cdot p\,2^{p-1} & 4 \cdot 2^p + 2 \cdot p\,2^{p-1} - 3 \cdot 3^p \end{array}\right].$$

■

Limit of powers

The existence of $\lim_{p\to\infty} A^p$ is often of interest in applications (see, *e.g.*, Sec. 2.3 of Volume II [9]). The following theorem shows a most fundamental fact.

Theorem 6.40. *For a square matrix A, the limit $B = \lim_{p\to\infty} A^p$ exists in the following two cases*[23] *(and not in other cases).*

(a) *For each eigenvalue λ of A, we have $|\lambda| < 1$. Then we have $B = O$.*
(b) *For each eigenvalue λ of A, we have $|\lambda| < 1$ or $\lambda = 1$, and moreover, every Jordan block corresponding to $\lambda = 1$ is of order 1. Then $B = X \cdot \mathrm{diag}\,(I_{n_1}, O_{n-n_1}) \cdot X^{-1}$ with the understanding that $\lambda_1 = 1$ in (6.69).*

Proof. By Theorem 6.39, $\lim_{p\to\infty} J_k(\lambda)^p$ exists if and only if (i) $|\lambda| < 1$ (and k is arbitrary) or (ii) $\lambda = 1$ and $k = 1$. In case (i) we have $\lim_{p\to\infty} J_k(\lambda)^p = O$, whereas $\lim_{p\to\infty} J_k(\lambda)^p = 1$ in case (ii). The theorem follows from this and (6.114). □

Remark 6.4. If a square matrix A is known to satisfy the condition (b) of Theorem 6.40, we can easily compute the value of the limit $\lim_{p\to\infty} A^p =$

[23]Case (a) may be regarded as a special case of (b), but it is singled out here because of its fundamental nature. Case (a) is the case where the spectral radius $\rho(A) < 1$.

$X \cdot \mathrm{diag}\,(I_{n_1}, O_{n-n_1}) \cdot X^{-1}$ without solving the eigenvalue problem. First, solve the equation $(A - I)\boldsymbol{x} = \mathbf{0}$ to obtain a matrix $X_1 = [\boldsymbol{x}_1, \boldsymbol{x}_2, \ldots, \boldsymbol{x}_{n_1}]$ representing a basis of $\mathrm{Ker}\,(A - I)$. Next, solve the equation $(A^\top - I)\boldsymbol{y} = \mathbf{0}$ to obtain a matrix $Y_1 = [\boldsymbol{y}_1, \boldsymbol{y}_2, \ldots, \boldsymbol{y}_{n_1}]$ representing a basis of $\mathrm{Ker}\,(A^\top - I)$. Then the limit value is given by[24]

$$\lim_{p \to \infty} A^p = X_1(Y_1^\top X_1)^{-1} Y_1^\top. \tag{6.115}$$

This formula can be derived as follows. We may assume that $X = [X_1 \mid X_2]$ with an appropriate $n \times (n - n_1)$ matrix X_2. We partition X^{-1} accordingly as $X^{-1} = \begin{bmatrix} Z_1^\top \\ Z_2^\top \end{bmatrix}$, where Z_1 is an $n \times n_1$ matrix. By (6.69) for the Jordan normal form, we have

$$X^{-1}A = \mathrm{diag}\,(I_{n_1}, \hat{J}) \cdot X^{-1},$$

where $\hat{J}$ is a matrix consisting of Jordan blocks for other eigenvalues than 1. This implies $Z_1^\top A = Z_1^\top$, that is, $(A^\top - I)Z_1 = O$. Since the column vectors of Y_1 are basis vectors of $\mathrm{Ker}\,(A^\top - I)$, we have $Z_1 = Y_1 S$ for some matrix S. It follows from $X^{-1}X = I$ that $Z_1^\top X_1 = I$, whereas $Z_1^\top X_1 = (Y_1 S)^\top X_1 = S^\top (Y_1^\top X_1)$. Therefore, $S^\top = (Y_1^\top X_1)^{-1}$ and hence $B = X \cdot \mathrm{diag}\,(I_{n_1}, O_{n-n_1}) \cdot X^{-1} = X_1 Z_1^\top = X_1 S^\top Y_1^\top = X_1(Y_1^\top X_1)^{-1} Y_1^\top$.

■

Application to simultaneous recurrence relations

Powers of a matrix are applied successfully to the analysis of simultaneous recurrence relations. Suppose that n sequences $\{x_1(k)\}_{k=0}^{\infty}, \{x_2(k)\}_{k=0}^{\infty}, \ldots, \{x_n(k)\}_{k=0}^{\infty}$ are subject to simultaneous recurrence relations of the form

$$\begin{aligned} x_1(k+1) &= a_{11}x_1(k) + a_{12}x_2(k) + \cdots + a_{1n}x_n(k), \\ x_2(k+1) &= a_{21}x_1(k) + a_{22}x_2(k) + \cdots + a_{2n}x_n(k), \\ &\vdots \\ x_n(k+1) &= a_{n1}x_1(k) + a_{n2}x_2(k) + \cdots + a_{nn}x_n(k) \end{aligned}$$

for $k = 0, 1, 2, \ldots$, where the initial values $x_1(0), x_2(0), \ldots, x_n(0)$ are assumed to be given.

By introducing a vector $\boldsymbol{x}(k) = (x_1(k), x_2(k), \ldots, x_n(k))^\top$ and a matrix $A = (a_{ij})$, we can express the above simultaneous recurrence relations compactly as

$$\boldsymbol{x}(k+1) = A\boldsymbol{x}(k) \qquad (k = 0, 1, 2, \ldots).$$

---

[24]The formula (6.115) will be used in Theorem 2.5 in Sec. 2.3 of Volume II [9].

This expression immediately shows that

$$\boldsymbol{x}(k) = A^k \boldsymbol{x}(0) \qquad (k = 0, 1, 2, \ldots).$$

If the Jordan normal form of matrix A is available together with the transformation matrix X in (6.69), a concrete expression of $A^k \boldsymbol{x}(0)$ can be obtained from (6.113) and (6.114).

Thus the Jordan normal form offers an effective means for transparent analysis of simultaneous recurrence relations.

6.8.6 *Exponential Functions of Matrices*

Exponential function e^A of a square matrix A is defined by a (convergent) series as

$$\mathrm{e}^A = \sum_{p=0}^{\infty} \frac{1}{p!} A^p. \tag{6.116}$$

In applications we often encounter a variant of the form e^{tA} with a parameter $t \in \mathbb{R}$ representing time. The Jordan normal form offers an effective method to analyze the behavior of e^{tA} as a function in t with a fixed matrix A. The following theorem shows a basic fact.

Theorem 6.41. *The exponential function of a Jordan block $J_k(\lambda)$ of order k is given by*

$$\mathrm{e}^{tJ_k(\lambda)} = \begin{bmatrix} \mathrm{e}^{t\lambda} & t\mathrm{e}^{t\lambda} & \dfrac{t^2}{2!}\mathrm{e}^{t\lambda} & \cdots & \dfrac{t^{k-1}}{(k-1)!}\mathrm{e}^{t\lambda} \\ & \mathrm{e}^{t\lambda} & t\mathrm{e}^{t\lambda} & \ddots & \vdots \\ & & \mathrm{e}^{t\lambda} & \ddots & \dfrac{t^2}{2!}\mathrm{e}^{t\lambda} \\ & O & & \ddots & t\mathrm{e}^{t\lambda} \\ & & & & \mathrm{e}^{t\lambda} \end{bmatrix}. \tag{6.117}$$

Proof. The substitution of expression (6.113) of $J_k(\lambda)^p$ into $\mathrm{e}^{tJ_k(\lambda)} = \sum_{p=0}^{\infty} \frac{t^p}{p!} J_k(\lambda)^p$ gives (6.117) above. □

Using the simplified expression $X^{-1}AX = \mathrm{diag}\,(\cdots, J_k(\lambda), \cdots)$ for (6.69), we have

$$\mathrm{e}^{tA} = X \cdot \mathrm{diag}\,(\cdots, \mathrm{e}^{tJ_k(\lambda)}, \cdots) \cdot X^{-1}, \tag{6.118}$$

in which each diagonal block $\mathrm{e}^{tJ_k(\lambda)}$ is given by (6.117).

Example 6.15. We consider the 4×4 matrix A treated in Example 6.8 in Sec. 6.8.1. The exponential function e^{tA} is given as

$$\mathrm{e}^{tA} = X \cdot \begin{bmatrix} \mathrm{e}^{2t} & t\mathrm{e}^{2t} & & \\ 0 & \mathrm{e}^{2t} & & \\ & & \mathrm{e}^{2t} & \\ & & & \mathrm{e}^{3t} \end{bmatrix} \cdot X^{-1}$$

$$= \begin{bmatrix} 3 & -1 & -1 & 5 \\ -5 & 0 & 0 & -9 \\ -2 & 0 & -1 & -3 \\ 2 & 1 & 1 & 3 \end{bmatrix} \begin{bmatrix} \mathrm{e}^{2t} & t\mathrm{e}^{2t} & & \\ 0 & \mathrm{e}^{2t} & & \\ & & \mathrm{e}^{2t} & \\ & & & \mathrm{e}^{3t} \end{bmatrix} \begin{bmatrix} 9/5 & 8/5 & 0 & 9/5 \\ 0 & 0 & 1 & 1 \\ -3/5 & -1/5 & -1 & -3/5 \\ -1 & -1 & 0 & -1 \end{bmatrix}$$

$$= \begin{bmatrix} 6\mathrm{e}^{2t} - 5\mathrm{e}^{3t} & 5\mathrm{e}^{2t} - 5\mathrm{e}^{3t} & 3t\mathrm{e}^{2t} & (5+3t)\mathrm{e}^{2t} - 5\mathrm{e}^{3t} \\ -9\mathrm{e}^{2t} + 9\mathrm{e}^{3t} & -8\mathrm{e}^{2t} + 9\mathrm{e}^{3t} & -5t\mathrm{e}^{2t} & (-9-5t)\mathrm{e}^{2t} + 9\mathrm{e}^{3t} \\ -3\mathrm{e}^{2t} + 3\mathrm{e}^{3t} & -3\mathrm{e}^{2t} + 3\mathrm{e}^{3t} & (1-2t)\mathrm{e}^{2t} & (-3-2t)\mathrm{e}^{2t} + 3\mathrm{e}^{3t} \\ 3\mathrm{e}^{2t} - 3\mathrm{e}^{3t} & 3\mathrm{e}^{2t} - 3\mathrm{e}^{3t} & 2t\mathrm{e}^{2t} & (4+2t)\mathrm{e}^{2t} - 3\mathrm{e}^{3t} \end{bmatrix}.$$

■

Limit of exponential functions

The existence of $\lim_{t\to+\infty} \mathrm{e}^{tA}$ is often of interest in applications (cf., Sec. 5.7). The following theorem shows a most fundamental fact. We denote by $\mathrm{Re}(\cdot)$ the real part of a complex number.

Theorem 6.42. *For a square matrix A, the limit $B = \lim_{t\to+\infty} \mathrm{e}^{tA}$ exists in the following two cases (and not in other cases).*

(a) *For each eigenvalue λ of A, we have $\mathrm{Re}(\lambda) < 0$. Then we have $B = O$.*
(b) *For each eigenvalue λ of A, we have $\mathrm{Re}(\lambda) < 0$ or $\lambda = 0$, and moreover, every Jordan block corresponding to $\lambda = 0$ is of order 1. Then $B = X \cdot \mathrm{diag}\,(I_{n_1}, O_{n-n_1}) \cdot X^{-1}$ with the understanding that $\lambda_1 = 0$ in (6.69).*

Proof. By Theorem 6.41, $\lim_{t\to+\infty} \mathrm{e}^{tJ_k(\lambda)}$ exists if and only if (i) $\mathrm{Re}(\lambda) < 0$ (and k is arbitrary) or (ii) $\lambda = 0$ and $k = 1$. In case (i) we have $\lim_{t\to+\infty} \mathrm{e}^{tJ_k(\lambda)} = O$, whereas $\lim_{t\to+\infty} \mathrm{e}^{tJ_k(\lambda)} = 1$ in case (ii). The theorem follows from this and (6.118). □

Remark 6.5. If a square matrix A is known to satisfy the condition (b) of Theorem 6.42, we can easily compute the value of the limit $\lim_{t\to+\infty} \mathrm{e}^{tA} =$

$X \cdot \mathrm{diag}\,(I_{n_1}, O_{n-n_1}) \cdot X^{-1}$ without solving the eigenvalue problem (cf., Remark 6.4). Let X_0 and Y_0 be matrices consisting of basis vectors of $\mathrm{Ker}\,A$ and $\mathrm{Ker}\,A^\top$, respectively. Then the limit value is given by

$$\lim_{t\to+\infty} \mathrm{e}^{tA} = X_0(Y_0^\top X_0)^{-1}Y_0^\top. \tag{6.119}$$

■

Application to ordinary differential equations

Exponential functions of a matrix are applied successfully to the initial value problem of linear ordinary differential equations. Suppose that n functions $y_1(t), y_2(t), \ldots, y_n(t)$ are subject to a system of linear ordinary differential equations of the form

$$\begin{aligned}
\frac{\mathrm{d}y_1}{\mathrm{d}t}(t) &= a_{11}y_1(t) + a_{12}y_2(t) + \cdots + a_{1n}y_n(t),\\
\frac{\mathrm{d}y_2}{\mathrm{d}t}(t) &= a_{21}y_1(t) + a_{22}y_2(t) + \cdots + a_{2n}y_n(t),\\
&\vdots\\
\frac{\mathrm{d}y_n}{\mathrm{d}t}(t) &= a_{n1}y_1(t) + a_{n2}y_2(t) + \cdots + a_{nn}y_n(t),
\end{aligned}$$

where $t > 0$ and the initial values $y_1(0), y_2(0), \ldots, y_n(0)$ are assumed to be given.

By introducing a vector $\boldsymbol{y}(t) = (y_1(t), y_2(t), \ldots, y_n(t))^\top$ and a matrix $A = (a_{ij})$, we can express the above system compactly as

$$\frac{\mathrm{d}\boldsymbol{y}}{\mathrm{d}t}(t) = A\boldsymbol{y}(t) \qquad (t > 0). \tag{6.120}$$

From this expression the solution $\boldsymbol{y}(t)$ is given immediately as

$$\boldsymbol{y}(t) = \mathrm{e}^{tA}\boldsymbol{y}(0) \qquad (t \geqq 0). \tag{6.121}$$

Indeed, it follows from (6.116) that

$$\begin{aligned}
\frac{\mathrm{d}}{\mathrm{d}t}\left(\mathrm{e}^{tA}\boldsymbol{y}(0)\right) &= \frac{\mathrm{d}}{\mathrm{d}t}\sum_{p=0}^{\infty}\frac{t^p}{p!}A^p\boldsymbol{y}(0) = \sum_{p=0}^{\infty}\frac{pt^{p-1}}{p!}A^p\boldsymbol{y}(0)\\
&= A\sum_{p=1}^{\infty}\frac{t^{p-1}}{(p-1)!}A^{p-1}\boldsymbol{y}(0) = A\left(\mathrm{e}^{tA}\boldsymbol{y}(0)\right).
\end{aligned}$$

If the Jordan normal form of matrix A is available together with the transformation matrix X in (6.69), a concrete expression of $\mathrm{e}^{tA}\boldsymbol{y}(0)$ can be obtained from (6.117) and (6.118).

Thus the Jordan normal form offers a transparent structural understanding of linear ordinary differential equations. On the other hand, it is not necessary to find the Jordan normal form itself to obtain a solution. It is more convenient to use the classical method of undetermined coefficients [74,75] to obtain a concrete form of a solution.

6.8.7 *Relation to Elementary Divisors*

There is a close relation between the Jordan normal form of A and the elementary divisors of $\lambda I - A$. The structural indices of the Jordan normal form of A can be recovered from the elementary divisors of $\lambda I - A$. In this way, the Jordan normal form can be identified without making explicit reference to eigenvectors.

While elementary divisors of a polynomial matrix is discussed in detail in Chapter 5 of Volume II [9], minimum background needed for our argument is presented here. First note that a minor (subdeterminant) of matrix $\lambda I - A$ is a polynomial in variable λ. For $k = 1, 2, \ldots, n$, the greatest common divisor of all minors of order k of $\lambda I - A$ is called the kth *determinantal divisor* of $\lambda I - A$, which we denote by $d_k(\lambda)$. It is assumed that $d_k(\lambda)$ is a monic polynomial with the leading coefficient equal to one. It is known that $d_{k-1}(\lambda)$ divides $d_k(\lambda)$ for $k = 2, 3, \ldots, n$, and the polynomials $e_1(\lambda), e_2(\lambda), \ldots, e_n(\lambda)$ defined by

$$e_k(\lambda) = \frac{d_k(\lambda)}{d_{k-1}(\lambda)} \qquad (k = 1, 2, \ldots, n) \tag{6.122}$$

with $d_0(\lambda) = 1$ are called the *elementary divisors* of $\lambda I - A$. Furthermore, it is known that $e_{k-1}(\lambda)$ divides $e_k(\lambda)$ for $k = 2, 3, \ldots, n$.

Our argument here starts with the following fundamental theorem.[25]

Theorem 6.43. *Two square matrices A and B are similar*[26] *if and only if $\lambda I - A$ and $\lambda I - B$ have the same elementary divisors for all orders.*

This theorem immediately implies the following statement that connects a given matrix A to its Jordan normal form J in terms of the elementary divisors.

[25] The reader is referred to Chapter 6 of [28] for the proof of Theorem 6.43.

[26] Recall that A and B are said to be similar if $B = S^{-1}AS$ for some nonsingular matrix S (cf., Sec. 6.1.1).

Theorem 6.44. *Let J denote the Jordan normal form of A. Then $\lambda I - A$ and $\lambda I - J$ have the same elementary divisors for all orders.*

Our next step is to derive, from the elementary divisors of $\lambda I - J$, the structural information about the Jordan normal form such as the number and order of Jordan blocks for each eigenvalue. Let J be a Jordan matrix in (6.69), characterized by distinct eigenvalues λ_i, geometric multiplicities g_i, and orders n_{ij} of Jordan blocks. Denote the elementary divisors of $\lambda I - J$ by $e_1(\lambda), e_2(\lambda), \ldots, e_n(\lambda)$, and define an $n \times n$ diagonal matrix

$$D_J(\lambda) = \mathrm{diag}\,(e_1(\lambda), e_2(\lambda), \ldots, e_n(\lambda)),$$

where each diagonal entry is a polynomial in λ. For each of the distinct eigenvalues $\lambda_1, \lambda_2, \ldots, \lambda_s$ we define an $n \times n$ diagonal (polynomial) matrix by

$$\begin{aligned} D_J^i(\lambda) = \mathrm{diag}\,(\overbrace{1, 1, \ldots, 1}^{(n-g_i)}, (\lambda - \lambda_i)^{n_{ig_i}}, (\lambda - \lambda_i)^{n_{i,g_i-1}}, \\ \ldots, (\lambda - \lambda_i)^{n_{i2}}, (\lambda - \lambda_i)^{n_{i1}}), \end{aligned}$$

where $1 \leqq i \leqq s$ and $n_{ig_i} \leqq n_{i,g_i-1} \leqq \cdots \leqq n_{i2} \leqq n_{i1}$. This matrix may be regarded as an encoding of the structural information about the submatrix $\mathrm{diag}\,(J_{n_{i1}}(\lambda_i), J_{n_{i2}}(\lambda_i), \ldots, J_{n_{ig_i}}(\lambda_i))$ in (6.69) corresponding to the eigenvalue λ_i. Furthermore, it is known[27] that the product of these matrices over all distinct eigenvalues is equal to the matrix $D_J(\lambda)$ defined from the elementary divisors, *i.e.*,

$$D_J(\lambda) = D_J^1(\lambda)\, D_J^2(\lambda) \cdots D_J^s(\lambda). \tag{6.123}$$

This relation is illustrated in the following.

Example 6.16. For the Jordan matrix $J = \mathrm{diag}\,(J_2(\lambda_1), J_1(\lambda_1), J_1(\lambda_2))$ in Example 6.8, where $\lambda_1 = 2$ and $\lambda_2 = 3$, we have

$$\begin{aligned} D_J(\lambda) &= \mathrm{diag}\,(1, 1,\ (\lambda - 2),\ (\lambda - 2)^2) \cdot \mathrm{diag}\,(1, 1, 1,\ (\lambda - 3)) \\ &= \mathrm{diag}\,(1, 1,\ (\lambda - 2),\ (\lambda - 2)^2(\lambda - 3)). \end{aligned}$$

For the Jordan matrix $J = \mathrm{diag}\,(J_3(\lambda_1), J_2(\lambda_1), J_1(\lambda_1), J_1(\lambda_2))$ in Example 6.11, where $\lambda_1 = 1$ and $\lambda_2 = 2$, we have

$$\begin{aligned} D_J(\lambda) &= \mathrm{diag}\,(1, 1, 1, 1,\ (\lambda - 1),\ (\lambda - 1)^2,\ (\lambda - 1)^3) \\ &\quad \cdot \mathrm{diag}\,(1, 1, 1, 1, 1, 1,\ (\lambda - 2)) \\ &= \mathrm{diag}\,(1, 1, 1, 1,\ (\lambda - 1),\ (\lambda - 1)^2,\ (\lambda - 1)^3(\lambda - 2)). \end{aligned}$$

■

[27] The reader is referred to Chapter 6 of [28] for the proof of (6.123).

The equation (6.123) is a key relation that enables us to determine the Jordan normal form of a given matrix A from the elementary divisors of $\lambda I - A$. Since the elementary divisors of $\lambda I - A$ are the same as those of $\lambda I - J$ (Theorem 6.44), the matrix $D_J(\lambda)$ can be defined from the elementary divisors of $\lambda I - A$ as well. The representation (6.123) of $D_J(\lambda)$ as the product of matrices $D_J^1(\lambda), D_J^2(\lambda), \ldots, D_J^s(\lambda)$ is unique, whereas these matrices encode the structural information such as g_i and n_{ij}. The correspondence between the elementary divisors and the Jordan normal form may be described as follows, where (1) is rather obvious and (2) is the main part of our interest.

(1) Suppose that the Jordan normal form J of A is given. From the Jordan blocks in J, define matrices $D_J^1(\lambda), D_J^2(\lambda), \cdots, D_J^s(\lambda)$. The diagonal entries of their product (6.123) give the elementary divisors of $\lambda I - A$.

(2) Suppose that the elementary divisors $e_1(\lambda), e_2(\lambda), \ldots, e_n(\lambda)$ of $\lambda I - A$ are given. The distinct eigenvalues $\lambda_1, \lambda_2, \ldots, \lambda_s$ of A are determined as the distinct roots of $e_n(\lambda) = 0$. For each $i = 1, 2, \ldots, s$, consider a factor of the form $(\lambda - \lambda_i)^p$ in $e_1(\lambda), e_2(\lambda), \ldots, e_n(\lambda)$, and let $p_{i1}, p_{i2}, \ldots, p_{in}$ denote the respective exponents, where we have $0 \leqq p_{i1} \leqq p_{i2} \leqq \cdots \leqq p_{in}$ since $e_{k-1}(\lambda)$ divides $e_k(\lambda)$ for $k = 2, 3, \ldots, n$. By considering factors of the form $(\lambda - \lambda_i)^p$ on both sides of (6.123), we obtain

$$D_J^i(\lambda) = \mathrm{diag}\left((\lambda - \lambda_i)^{p_{i1}}, (\lambda - \lambda_i)^{p_{i2}}, \ldots, (\lambda - \lambda_i)^{p_{in}}\right),$$

and hence the structural indices g_i and n_{ij} are determined from these exponents $p_{i1}, p_{i2}, \ldots, p_{in}$ by the relation

$$(p_{i1}, p_{i2}, \ldots, p_{in}) = (\overbrace{0, 0, \ldots, 0}^{n-g_i}, n_{ig_i}, n_{i,g_i-1}, \ldots, n_{i2}, n_{i1}),$$

where $1 \leqq n_{ig_i} \leqq n_{i,g_i-1} \leqq \cdots \leqq n_{i2} \leqq n_{i1}$. In this way the Jordan normal form of A is determined from the elementary divisors of $\lambda I - A$.

Chapter 7

Quadratic Forms

In this chapter, we deal with quadratic forms defined by symmetric (or Hermitian) matrices, and present fundamental facts about positive definiteness, completion of squares, and Sylvester's law of inertia. Examples of positive definite matrices in several engineering disciplines are shown. Physically, a quadratic form defined by a positive definite matrix corresponds to energy.

7.1 Definition of Quadratic Forms

Definition 7.1. Let A be a (real) symmetric matrix and $\boldsymbol{x} = (x_i)$ denote an n-dimensional real vector. A polynomial in $(x_1, x_2, \ldots, x_n)$ consisting of terms of degree two:

$$\boldsymbol{x}^\top A \boldsymbol{x} = \sum_{i=1}^{n} \sum_{j=1}^{n} a_{ij} x_i x_j \tag{7.1}$$

is called the *quadratic form* defined by A. ■

Definition 7.2. Let A be a (complex) Hermitian matrix and $\boldsymbol{x} = (x_i)$ denote an n-dimensional complex vector. A polynomial in $(x_1, x_2, \ldots, x_n)$ consisting of terms of degree two:[1]

$$\boldsymbol{x}^* A \boldsymbol{x} = \sum_{i=1}^{n} \sum_{j=1}^{n} a_{ij} \overline{x_i} x_j \tag{7.2}$$

is called the *quadratic form* defined by A. ■

[1] In (7.2), each of a_{ij}, x_i, x_j is complex, but $\boldsymbol{x}^* A \boldsymbol{x}$ is real.

7.2 Positive Definiteness of Symmetric Matrices

7.2.1 *Definition and Equivalent Conditions*

Definition 7.3. A symmetric matrix A is called *positive definite* if it satisfies the condition

$$\boldsymbol{x}^\top A\boldsymbol{x} > 0 \quad \text{for any vector } \boldsymbol{x} \neq \mathbf{0}. \tag{7.3}$$

When $-A$ is positive definite, A is called *negative definite.* ■

The following theorem gives necessary and sufficient conditions for positive definiteness.

Theorem 7.1. *The following six conditions* (a)–(f) *are equivalent for a symmetric matrix A.*

(a) *A is positive definite, that is, the condition* (7.3) *holds.*
(b) *Every eigenvalue of A is positive.*[2]
(c) *We can represent A as $A = QDQ^\top$ with an orthogonal matrix Q and a diagonal matrix D with positive diagonal entries.*
(d) *We can represent A as $A = SS^\top$ with a nonsingular matrix S.*
(e) *Every principal minor of A is positive.*
(f) *Every leading principal minor of A is positive.*

Proof. [(a) ⇔ (b) ⇔ (c)]: Let A be a symmetric matrix of order n. By Theorem 6.9 in Sec. 6.3.1, there exist an orthogonal matrix Q and a diagonal matrix $D = \text{diag}\,(\lambda_1, \lambda_2, \ldots, \lambda_n)$ with which A is decomposed as

$$A = QDQ^\top = Q\,\text{diag}\,(\lambda_1, \lambda_2, \ldots, \lambda_n)Q^\top, \tag{7.4}$$

where λ_i $(i = 1, 2, \ldots, n)$ are eigenvalues of A. Let $\boldsymbol{y} = Q^\top\boldsymbol{x}$. Then we have

$$\boldsymbol{x}^\top A\boldsymbol{x} = \boldsymbol{x}^\top QDQ^\top\boldsymbol{x} = \boldsymbol{y}^\top D\boldsymbol{y} = \sum_{i=1}^{n} \lambda_i {y_i}^2, \tag{7.5}$$

and $\boldsymbol{x}$ runs over all nonzero vectors if and only if $\boldsymbol{y}$ runs over all nonzero vectors. Therefore, A is positive definite if and only if $\lambda_i > 0$ for $i = 1, 2, \ldots, n$.

[(c) ⇒ (d)]: In the decomposition (7.4) we have $\lambda_i > 0$ $(i = 1, 2, \ldots, n)$ by (c). Therefore, we can take

$$S = Q\,\text{diag}\,(\sqrt{\lambda_1}, \sqrt{\lambda_2}, \ldots, \sqrt{\lambda_n})$$

to obtain $A = SS^\top$ in (d).

[2]Note that every eigenvalue of a symmetric matrix is a real number.

[(d) $\Rightarrow$ (a)]: By nonsingularity of S in (d), $\boldsymbol{x} \neq \mathbf{0}$ implies $S^\top \boldsymbol{x} \neq \mathbf{0}$. Therefore,

$$\boldsymbol{x}^\top A\boldsymbol{x} = \boldsymbol{x}^\top SS^\top \boldsymbol{x} = (S^\top \boldsymbol{x})^\top (S^\top \boldsymbol{x}) > 0,$$

which shows (a).

[(a) $\Rightarrow$ (e)]: We first consider the determinant itself (the principal minor of order n). We have $\det A = \lambda_1\lambda_2 \cdots \lambda_n$ where λ_i $(i = 1, 2, \ldots, n)$ are the eigenvalues of A. Since [(a) $\Rightarrow$ (b)] is already proved, we have $\lambda_i > 0$ $(i = 1, 2, \ldots, n)$, and hence $\det A > 0$.

A similar argument works also for any principal submatrix B of order $< n$, since we can show that B is positive definite as follows. For notational simplicity, assume that B is the principal submatrix corresponding to row and column indices $i = 1, 2, \ldots, k$. For any k-dimensional vector $\boldsymbol{y} = (y_1, y_2, \ldots, y_k)^\top \in \mathbb{R}^k$, we consider an n-dimensional vector defined by $\boldsymbol{x} = (y_1, y_2, \ldots, y_k, 0, \ldots, 0)^\top \in \mathbb{R}^n$. If $\boldsymbol{y} \neq \mathbf{0}$, then $\boldsymbol{x} \neq \mathbf{0}$, and moreover, $\boldsymbol{y}^\top B\boldsymbol{y} = \boldsymbol{x}^\top A\boldsymbol{x} > 0$, where the last inequality is due to (a).

[(e) $\Rightarrow$ (f)]: This is obvious, since a leading principal minor is a special case of a principal minor.

[(f) $\Rightarrow$ (a)]: We prove by induction on the order n of a symmetric matrix A. The statement is obviously true for $n = 1$. Assume $n \geqq 2$, and partition A as

$$A = \left[\begin{array}{c|c} B & \boldsymbol{c} \\ \hline \boldsymbol{c}^\top & d \end{array}\right],$$

where B is a symmetric matrix of order $n-1$, $\boldsymbol{c}$ is an $(n-1)$-dimensional vector, and d is a real number. Noting that B is nonsingular by (f), define a (nonsingular) matrix

$$S = \left[\begin{array}{c|c} I_{n-1} & -B^{-1}\boldsymbol{c} \\ \hline \mathbf{0}^\top & 1 \end{array}\right]$$

and transform A to

$$\begin{aligned} S^\top AS &= \left[\begin{array}{c|c} I_{n-1} & \mathbf{0} \\ \hline -\boldsymbol{c}^\top B^{-1} & 1 \end{array}\right] \left[\begin{array}{c|c} B & \boldsymbol{c} \\ \hline \boldsymbol{c}^\top & d \end{array}\right] \left[\begin{array}{c|c} I_{n-1} & -B^{-1}\boldsymbol{c} \\ \hline \mathbf{0}^\top & 1 \end{array}\right] \\ &= \left[\begin{array}{c|c} B & \mathbf{0} \\ \hline \mathbf{0}^\top & d - \boldsymbol{c}^\top B^{-1}\boldsymbol{c} \end{array}\right]. \end{aligned} \tag{7.6}$$

For any n-dimensional vector $\boldsymbol{x}$, partition $S^{-1}\boldsymbol{x}$ as $S^{-1}\boldsymbol{x} = \left[\begin{array}{c} \boldsymbol{y} \\ \hline z \end{array}\right]$ with an $(n-1)$-dimensional vector $\boldsymbol{y}$ and a real number z. Then we have

$$\boldsymbol{x}^\top A\boldsymbol{x} = \boldsymbol{y}^\top B\boldsymbol{y} + (d - \boldsymbol{c}^\top B^{-1}\boldsymbol{c})z^2. \tag{7.7}$$

Since the matrix B satisfies the condition (f), it is positive definite by the induction hypothesis. That is, $\boldsymbol{y}^\top B\boldsymbol{y} > 0$ for any $\boldsymbol{y} \neq \mathbf{0}$. On the other hand, it follows from (7.6) that

$$\det S \cdot \det A \cdot \det S = \det B \cdot (d - \mathbf{c}^\top B^{-1}\mathbf{c}).$$

Since $\det S = 1$, $\det A > 0$, and $\det B > 0$, we obtain $d - \mathbf{c}^\top B^{-1}\mathbf{c} > 0$. Therefore, $(d - \mathbf{c}^\top B^{-1}\mathbf{c})z^2 > 0$ for any $z \neq 0$. For any nonzero vector $\boldsymbol{x}$, we have $\boldsymbol{y} \neq \mathbf{0}$ and/or $z \neq 0$, and hence (7.7) is positive. □

Definition 7.4. A symmetric matrix A is called *positive semidefinite* if it satisfies the condition

$$\boldsymbol{x}^\top A\boldsymbol{x} \geqq 0 \quad \text{for any vector } \boldsymbol{x} \neq \mathbf{0}. \tag{7.8}$$

The condition (7.8) is a weakening of the condition (7.3) for positive definiteness in that the strict inequality $>$ in (7.3) is replaced by $\geqq$ in (7.8). In (7.8) the requirement that $\boldsymbol{x}$ be a nonzero vector ($\boldsymbol{x} \neq \mathbf{0}$) is not essential, and hence (7.8) is equivalent to the following condition:

$$\boldsymbol{x}^\top A\boldsymbol{x} \geqq 0 \quad \text{for any vector } \boldsymbol{x}. \tag{7.9}$$

A positive semidefinite matrix is also called a *nonnegative definite* matrix. When $-A$ is positive semidefinite, A is called *negative semidefinite*. ■

The following theorem gives necessary and sufficient conditions for positive semidefiniteness.

Theorem 7.2. *The following five conditions* (a)–(e) *are equivalent for a symmetric matrix A.*

(a) *A is positive semidefinite, that is, the condition* (7.9) *holds.*
(b) *Every eigenvalue of A is nonnegative.*
(c) *We can represent A as $A = QDQ^\top$ with an orthogonal matrix Q and a diagonal matrix D with nonnegative diagonal entries.*
(d) *We can represent A as $A = SS^\top$ with a square matrix S.*
(e) *Every principal minor of A is nonnegative.*

Proof. The proofs are mostly similar to those for Theorem 7.1, except that the proof of [(e) ⇒ (a)] uses Theorem 7.1.

[(a) ⇔ (b) ⇔ (c)]: In (7.5), $\boldsymbol{x}$ runs over all vectors if and only if $\boldsymbol{y}$ runs over all vectors. Therefore, A is positive semidefinite if and only if $\lambda_i \geqq 0$ for $i = 1, 2, \ldots, n$.

[(c) ⇒ (d)]: In the decomposition $A = Q\,\mathrm{diag}\,(\lambda_1, \lambda_2, \ldots, \lambda_n)Q^\top$ in (7.4), we have $\lambda_i \geqq 0$ $(i = 1, 2, \ldots, n)$. Hence we can take $S = Q\,\mathrm{diag}\,(\sqrt{\lambda_1}, \sqrt{\lambda_2}, \ldots, \sqrt{\lambda_n})$.

[(d) ⇒ (a)]: This follows from $\boldsymbol{x}^\top A\boldsymbol{x} = \boldsymbol{x}^\top SS^\top\boldsymbol{x} = (S^\top\boldsymbol{x})^\top(S^\top\boldsymbol{x}) \geqq 0$.

[(a) ⇒ (e)]: This is proved by replacing strict inequality "$>$" with "$\geqq$" in the proof of [(a) ⇒ (e)] in Theorem 7.1.

[(e) ⇒ (a)]: Define $A_\varepsilon = A + \varepsilon I$ for a positive number $\varepsilon > 0$, where I is the unit matrix. Consider a principal submatrix B_ε of order k of A_ε. Its determinant $\det B_\varepsilon$ is a polynomial of degree k in ε of the form

$$\det B_\varepsilon = \varepsilon^k + c_{k-1}\varepsilon^{k-1} + \cdots + c_1\varepsilon + c_0.$$

Here we have $c_i \geqq 0$ $(i = 0, 1, \ldots, k-1)$, since each c_i is a sum of principal minors of order $k - i$ of A and (e) is assumed. Therefore, $\det B_\varepsilon > 0$, which implies, by [(e) ⇒ (a)] of Theorem 7.1, that $A_\varepsilon = A + \varepsilon I$ is positive definite, that is,

$$\boldsymbol{x}^\top A_\varepsilon \boldsymbol{x} = \boldsymbol{x}^\top A\boldsymbol{x} + \varepsilon(\boldsymbol{x}^\top\boldsymbol{x}) > 0$$

for any $\boldsymbol{x} \neq \boldsymbol{0}$. Letting $\varepsilon \to 0$, we obtain (7.8). □

Remark 7.1. In Theorem 7.2(d), we do not have to restrict the matrix S to a square matrix. If A is an $n \times n$ positive semidefinite matrix of rank r, we can choose S to be an $n \times r$ matrix of rank r. Indeed, in the proof of [(c) ⇒ (d)] of Theorem 7.2, we may assume that the eigenvalues are numbered so that $\lambda_i > 0$ $(1 \leqq i \leqq r)$ and $\lambda_i = 0$ $(r < i \leqq n)$. Consider the partition $Q = [Q_1 \mid Q_2]$ with an $n \times r$ matrix Q_1 and an $n \times (n-r)$ matrix Q_2. Then we can take $S = Q_1\,\mathrm{diag}\,(\sqrt{\lambda_1}, \sqrt{\lambda_2}, \ldots, \sqrt{\lambda_r})$. ■

Remark 7.2. By Theorem 7.1, all principal minors are positive if and only if all leading principal minors are positive. However, we cannot replace "positive" by "nonnegative" here. That is, some principal minors can be negative even if all leading principal minors are nonnegative. This is shown by a simple example

$$A = \begin{bmatrix} 0 & 0 \\ 0 & -1 \end{bmatrix}.$$

Thus we see the following:

Every leading principal minor of A is nonnegative

$\not\Rightarrow$ A is positive semidefinite.

It is also noted that there are $2^n - 1$ principal minors while there are only n leading principal minors. ■

As an application of Theorem 7.2, we can show that a diagonally dominant matrix is positive semidefinite.[3]

Theorem 7.3.

(1) *A diagonally dominant symmetric matrix with nonnegative diagonal entries is positive semidefinite.*
(2) *A strictly diagonally dominant symmetric matrix with positive diagonal entries is positive definite.*

Proof. (1) Let $A = (a_{ij})$ be a diagonally dominant symmetric matrix with $a_{ii} \geqq 0$ for all i. Let λ be an eigenvalue of A. It follows from the Gershgorin theorem (Theorem 6.5 in Sec. 6.1.3) and the diagonal dominance that there exists i such that

$$|\lambda - a_{ii}| \leqq \sum_{j \neq i} |a_{ij}| \leqq a_{ii},$$

which implies $\lambda \geqq 0$. Since every eigenvalue is nonnegative, A is positive semidefinite by Theorem 7.2.

(2) By Theorem 1.15 in Sec. 1.7.1, a strictly diagonally dominant matrix is nonsingular, and hence every eigenvalue is distinct from zero, whereas $\lambda \geqq 0$ as shown above. Hence $\lambda > 0$. Since every eigenvalue is positive, A is positive definite by Theorem 7.1. □

Remark 7.3. We can test for positive (semi)definiteness with $O(n^3)$ arithmetic operations by means of a method similar to the Gaussian elimination. See Remark 7.7 in Sec. 7.5.1. ■

7.2.2 *Properties*

Theorem 7.4.

(1) *Every principal submatrix of a positive definite matrix is positive definite.*
(2) *Every principal submatrix of a positive semidefinite matrix is positive semidefinite.*

Proof. (1) and (2) follow from [(a) ⇔ (e)] in Theorems 7.1 and 7.2, respectively, in Sec. 7.2.1. □

---

[3]Theorem 7.3 is used in Examples 7.7 and 7.8 in Sec. 7.4.4.

Theorem 7.5. *Let S be a nonsingular matrix.*

(1) *If A is positive definite, then $S^\top AS$ is positive definite.*
(2) *If A is positive semidefinite, then $S^\top AS$ is positive semidefinite.*

That is, positive definiteness and positive semidefiniteness of a symmetric matrix are invariant under congruence transformations.[4]

Proof. Let $B = S^\top AS$. For any $\boldsymbol{y} \neq \boldsymbol{0}$, the vector $\boldsymbol{x} = S\boldsymbol{y}$ satisfies $\boldsymbol{x} \neq \boldsymbol{0}$ and $\boldsymbol{y}^\top B\boldsymbol{y} = \boldsymbol{x}^\top A\boldsymbol{x}$.

(1) Since A is positive definite, we have $\boldsymbol{y}^\top B\boldsymbol{y} = \boldsymbol{x}^\top A\boldsymbol{x} > 0$. Therefore, B is positive definite.

(2) Since A is positive semidefinite, we have $\boldsymbol{y}^\top B\boldsymbol{y} = \boldsymbol{x}^\top A\boldsymbol{x} \geqq 0$. Therefore, B is positive semidefinite. □

Theorem 7.6.

(1) *If A and B are positive definite and $\alpha, \beta > 0$, then $\alpha A + \beta B$ is positive definite.*
(2) *If A and B are positive semidefinite and $\alpha, \beta \geqq 0$, then $\alpha A + \beta B$ is positive semidefinite.*

Proof. (1) For any vector $\boldsymbol{x} \neq \boldsymbol{0}$, we have

$$\boldsymbol{x}^\top(\alpha A + \beta B)\boldsymbol{x} = \alpha(\boldsymbol{x}^\top A\boldsymbol{x}) + \beta(\boldsymbol{x}^\top B\boldsymbol{x}) > 0.$$

(2) We have "$\geqq 0$" in place of "> 0" in the above. □

Theorem 7.6(2) may be rephrased to a (geometric) statement that, in the linear space consisting of all symmetric matrices of order n, positive semidefinite matrices form a convex cone. See Remark 7.4 below as well as Example 9.10 in Sec. 9.2.2 and Example 9.16 in Sec. 9.4.2.

Remark 7.4. A subset S of a linear space is called a *convex set* if, for any pair of points in S, the line segment connecting the two points is contained in S, that is, if the following implication holds:

$$\boldsymbol{x}, \boldsymbol{y} \in S, \quad 0 \leqq \alpha \leqq 1 \quad \Longrightarrow \quad \alpha\boldsymbol{x} + (1-\alpha)\boldsymbol{y} \in S. \tag{7.10}$$

A set S is called a *cone* if, for any point $\boldsymbol{x} \in S$ and any nonnegative real number $\alpha \geqq 0$, the point $\alpha\boldsymbol{x}$ belongs to S. A cone that is a convex set is called a *convex cone*. A set S is a convex cone if and only if the following implication holds:

$$\boldsymbol{x}, \boldsymbol{y} \in S, \quad \alpha, \beta \geqq 0 \quad \Longrightarrow \quad \alpha\boldsymbol{x} + \beta\boldsymbol{y} \in S. \tag{7.11}$$

■

[4]See Remark 3.3 in Sec. 3.2 for the definition of a congruence transformation.

Remark 7.5. Positive semidefinite matrices are used as an important tool for modeling in optimization and control. Convexity property in Theorem 7.6 is exploited successfully in optimization algorithms. The modeling methodology based on positive semidefinite matrices is called *positive semidefinite programming* in optimization [46, 59, 80] and *linear matrix inequalities* in control theory [48, 53, 55]. ■

For symmetric matrices A and B, we write $A \preceq B$ (or $B \succeq A$) to mean that $B - A$ is positive semidefinite. With this notation we can write $O \preceq B$ (or $B \succeq O$) when B is positive semidefinite.

Theorem 7.7. *Let A and B be symmetric matrices of order n.*

(1) *Let S be a nonsingular matrix. Then $A \preceq B$ if and only if $S^\top AS \preceq S^\top BS$.*
(2) *If $A \preceq B$, then*[5] $\lambda_k(A) \leqq \lambda_k(B)$ *for $k = 1, 2, \ldots, n$.*

Proof. (1) This follows from Theorem 7.5(2) since $S^\top BS - S^\top AS = S^\top(B - A)S$.

(2) By the assumption $A \preceq B$, we have $\boldsymbol{x}^\top A\boldsymbol{x} \leqq \boldsymbol{x}^\top B\boldsymbol{x}$ for any $\boldsymbol{x}$. Then the claim follows from the Courant–Fischer theorem (Theorem 6.12 in Sec. 6.3.2). □

Theorem 7.8. *Let A and B be symmetric matrices. If $0 \leqq \alpha \leqq 1$, then*

$$(\alpha A + (1-\alpha)B)^2 \preceq \alpha A^2 + (1-\alpha)B^2.$$

Proof. (Right-hand side) – (Left-hand side) $= \alpha(1-\alpha)(A-B)^\top(A-B) \succeq O$. □

Theorem 7.9. *Let A and B be positive definite matrices. If $0 \leqq \alpha \leqq 1$, then*

$$(\alpha A + (1-\alpha)B)^{-1} \preceq \alpha A^{-1} + (1-\alpha)B^{-1}. \tag{7.12}$$

In particular,

$$\left(\frac{A+B}{2}\right)^{-1} \preceq \frac{A^{-1}+B^{-1}}{2}.$$

Proof. By Theorem 7.1(d) we can represent $A = SS^\top$ with a nonsingular matrix S. By multiplying (7.12) with $S^\top$ from the left and with S from the right, we obtain

$$(\alpha I + (1-\alpha)\tilde{B})^{-1} \preceq \alpha I + (1-\alpha)\tilde{B}^{-1},$$

---

[5] $\lambda_k(A)$ denotes the kth largest eigenvalue of A.

where $\tilde{B} = S^{-1}B(S^{-1})^\top$ and Theorem 7.7(1) is used. By Theorem 7.1(c) we can represent $\tilde{B} = QDQ^\top$ with an orthogonal matrix Q and a diagonal matrix $D = \operatorname{diag}(\lambda_1, \lambda_2, \ldots, \lambda_n)$ with $\lambda_i > 0$ $(i = 1, 2, \ldots, n)$. With the use of $\tilde{B} = QDQ^\top$ we can transform the above to

$$(\alpha I + (1-\alpha)D)^{-1} \preceq \alpha I + (1-\alpha)D^{-1}.$$

This is equivalent to a set of scalar inequalities

$$(\alpha + (1-\alpha)\lambda_i)^{-1} \leqq \alpha + (1-\alpha)\lambda_i^{-1} \qquad (i = 1, 2, \ldots, n),$$

which are easily seen to be true. □

Theorem 7.10. *Let A and B be positive definite matrices. If $0 \leqq \alpha \leqq 1$, then*

$$\det(\alpha A + (1-\alpha)B) \geqq (\det A)^\alpha \cdot (\det B)^{1-\alpha}. \tag{7.13}$$

In particular,

$$\det\left(\frac{A+B}{2}\right) \geqq \sqrt{(\det A) \cdot (\det B)}.$$

Proof. Similarly to the proof of Theorem 7.9, the proof of (7.13) can be reduced to the special case where $A = I$ and $B = \operatorname{diag}(\lambda_1, \lambda_2, \ldots, \lambda_n)$. In this special case the inequality (7.13) is reduced to scalar inequalities

$$\log(\alpha + (1-\alpha)\lambda_i) \geqq (1-\alpha)\log\lambda_i \qquad (i = 1, 2, \ldots, n),$$

which are easily seen to be true. □

Theorem 7.11.

(1) *The Hadamard product of two positive definite matrices is positive definite.*
(2) *The Hadamard product of two positive semidefinite matrices is positive semidefinite.*

Proof. Let $A = (a_{ij})$ and $B = (b_{ij})$ be symmetric matrices of order n, and $C = A \odot B$ be their Hadamard product. Since $C = (a_{ij}b_{ij})$, C is also a symmetric matrix. By Theorem 6.9 in Sec. 6.3.1, we can represent $A = Q\operatorname{diag}(\lambda_1, \lambda_2, \ldots, \lambda_n)Q^\top$ with an orthogonal matrix $Q = (q_{ij})$, that is,

$$a_{ij} = \sum_k \lambda_k q_{ik} q_{jk}.$$

Given any vector $\boldsymbol{x} = (x_1, x_2, \ldots, x_n)^\top$, we define $y_{ik} = q_{ik}x_i$ to obtain

$$\begin{aligned}\boldsymbol{x}^\top C\boldsymbol{x} &= \sum_i \sum_j a_{ij}b_{ij}x_ix_j \\ &= \sum_i \sum_j \left(\sum_k \lambda_k q_{ik}q_{jk} \right) b_{ij}x_ix_j \\ &= \sum_k \lambda_k \left(\sum_i \sum_j y_{ik}y_{jk}b_{ij} \right) \\ &= \sum_k \lambda_k \cdot (y_{1k}, y_{2k}, \ldots, y_{nk})B(y_{1k}, y_{2k}, \ldots, y_{nk})^\top.\end{aligned}$$

In the following we prove (2) first, and then (1).

(2) Suppose that A and B are positive semidefinite. For each $k = 1, 2, \ldots, n$, we have

$$\lambda_k \geqq 0, \qquad (y_{1k}, y_{2k}, \ldots, y_{nk})B(y_{1k}, y_{2k}, \ldots, y_{nk})^\top \geqq 0.$$

Therefore $\boldsymbol{x}^\top C\boldsymbol{x} \geqq 0$, showing that C is positive semidefinite.

(1) Suppose that A and B are positive definite. We have $\lambda_i > 0$ for $i = 1, 2, \ldots, n$. If $\boldsymbol{x} \neq \boldsymbol{0}$, then $x_i \neq 0$ for some i. Fix such i. Since Q is an orthogonal matrix, we have $q_{ik} \neq 0$ for some k. For such (i, k) we have $y_{ik} = q_{ik}x_i \neq 0$, and hence $(y_{1k}, y_{2k}, \ldots, y_{nk})B(y_{1k}, y_{2k}, \ldots, y_{nk})^\top > 0$ since B is positive definite. Therefore $\boldsymbol{x}^\top C\boldsymbol{x} > 0$, showing that C is positive definite. □

7.3 Positive Definiteness of Hermitian Matrices

The concepts of positive definiteness and positive semidefiniteness can be defined for Hermitian complex matrices. The characterizations of positive (semi)definiteness for symmetric matrices can naturally be adapted to Hermitian matrices.

Definition 7.5. A Hermitian matrix A is called *positive definite* if it satisfies the condition

$$\boldsymbol{x}^* A\boldsymbol{x} > 0 \quad \text{for any complex vector } \boldsymbol{x} \neq \boldsymbol{0}. \tag{7.14}$$

When $-A$ is positive definite, A is called *negative definite*. ■

The following theorem gives necessary and sufficient conditions for positive definiteness of a Hermitian matrix.

Theorem 7.12. *The following six conditions* (a)–(f) *are equivalent for a Hermitian matrix A.*

(a) *A is positive definite, that is, the condition* (7.14) *holds.*
(b) *Every eigenvalue of A is positive.*[6]
(c) *We can represent A as $A = UDU^*$ with a unitary matrix U and a diagonal matrix D with positive diagonal entries.*
(d) *We can represent A as $A = SS^*$ with a nonsingular matrix S.*
(e) *Every principal minor of A is positive.*
(f) *Every leading principal minor of A is positive.*

Definition 7.6. A Hermitian matrix A is called *positive semidefinite* if it satisfies the condition

$$\boldsymbol{x}^* A \boldsymbol{x} \geqq 0 \quad \text{for any complex vector } \boldsymbol{x} \neq \mathbf{0}. \tag{7.15}$$

The condition (7.15) is a weakening of the condition (7.14) for positive definiteness in that the strict inequality $>$ in (7.14) is replaced by $\geqq$ in (7.15). In (7.15) the requirement that $\boldsymbol{x}$ be a nonzero vector ($\boldsymbol{x} \neq \mathbf{0}$) is not essential, and hence (7.15) is equivalent to the following condition:

$$\boldsymbol{x}^* A \boldsymbol{x} \geqq 0 \quad \text{for any complex vector } \boldsymbol{x}. \tag{7.16}$$

A positive semidefinite matrix is also called a *nonnegative definite* matrix. When $-A$ is positive semidefinite, A is called *negative semidefinite.* ■

The following theorem gives necessary and sufficient conditions for positive semidefiniteness of a Hermitian matrix.

Theorem 7.13. *The following five conditions* (a)–(e) *are equivalent for a Hermitian matrix A.*

(a) *A is positive semidefinite, that is, the condition* (7.16) *holds.*
(b) *Every eigenvalue of A is nonnegative.*
(c) *We can represent A as $A = UDU^*$ with a unitary matrix U and a diagonal matrix D with nonnegative diagonal entries.*
(d) *We can represent A as $A = SS^*$ with a square matrix S.*
(e) *Every principal minor of A is nonnegative.*

Theorems 7.3–7.11 and Remarks 7.1–7.3, with appropriate adaptations, hold for Hermitian matrices.

[6] Note that every eigenvalue of a Hermitian matrix is a real number (Theorem 6.17 in Sec. 6.4.1).

7.4 Positive Definite Matrices in Engineering

Positive (semi)definite matrices arise naturally in many areas of engineering and the concept of positive (semi)definiteness captures essential features of respective engineering systems. We give some examples below.

7.4.1 *Random Variables*

We start with three examples related to random variables [47,49,51,58,66, 79].

Example 7.1. Let $X_1, X_2, \ldots, X_n$ be mutually correlated random variables, and assume that their

$$\begin{aligned} \text{Expectation:} \quad & m_i = \mathrm{E}[X_i], \\ \text{Variance:} \quad & V_i = \mathrm{E}[(X_i - m_i)^2] \end{aligned}$$

exist (as finite values).[7] It then follows that

$$\text{Covariance:} \quad \mathrm{E}[(X_i - m_i)(X_j - m_j)] \qquad (i \neq j)$$

also exists, and we can consider a (symmetric) matrix $C = (c_{ij})$ with the (i, j) entry defined by

$$c_{ij} = \mathrm{E}[(X_i - m_i)(X_j - m_j)] \qquad (i, j = 1, 2, \ldots, n).$$

This matrix is called the *covariance matrix* (or *variance–covariance matrix*). The covariance matrix C is positive semidefinite, as follows. By defining a vector Y of random variables by

$$Y = [X_1 - m_1, X_2 - m_2, \ldots, X_n - m_n]^\top,$$

we can express C as $C = \mathrm{E}[YY^\top]$, and moreover, for any vector $\boldsymbol{u} = (u_1, u_2, \ldots, u_n)^\top$, we have[8]

$$\boldsymbol{u}^\top C\, \boldsymbol{u} = \boldsymbol{u}^\top \mathrm{E}[YY^\top]\, \boldsymbol{u} = \mathrm{E}[\boldsymbol{u}^\top YY^\top \boldsymbol{u}] = \mathrm{E}[(\boldsymbol{u}^\top Y)^2] \geqq 0.$$

The correlation coefficients r_{ij} are defined by

$$r_{ij} = \frac{c_{ij}}{\sqrt{V_i}\sqrt{V_j}} \qquad (i, j = 1, 2, \ldots, n),$$

and the *correlation coefficient matrix* $R = (r_{ij})$ is also a positive semidefinite symmetric matrix. ■

[7] Notation $\mathrm{E}[\cdot]$ means the expected value of a random variable.

[8] The quadratic form $\boldsymbol{u}^\top C\, \boldsymbol{u}$ coincides with the variance of a random variable $u_1 X_1 + u_2 X_2 + \cdots + u_n X_n$.

Example 7.2. Let X be a random variable. The *moment matrix* of X is defined to be a matrix $M = (m_{ij})$ with

$$m_{ij} = \mathrm{E}[X^{i+j-2}] \qquad (i, j = 1, 2, \ldots, n).$$

That is, the moment matrix is a symmetric matrix whose (i, j) entry is the $(i + j - 2)$th moment.[9] The moment matrix M is positive semidefinite. Indeed, by defining a vector Y of random variables by

$$Y = [1, X, X^2, \ldots, X^{n-1}]^\top$$

we can express M as $M = \mathrm{E}[YY^\top]$, and moreover, for any vector $\boldsymbol{u} = (u_1, u_2, \ldots, u_n)^\top$, we have

$$\boldsymbol{u}^\top M\, \boldsymbol{u} = \boldsymbol{u}^\top \mathrm{E}[YY^\top]\, \boldsymbol{u} = \mathrm{E}[\boldsymbol{u}^\top YY^\top \boldsymbol{u}] = \mathrm{E}[(\boldsymbol{u}^\top Y)^2] \geqq 0.$$

The moment matrix M is a Hankel matrix (cf., Definition 1.50 in Sec. 1.7.2). ■

Example 7.3. The *characteristic function* of a random variable X is a complex-valued function $\varphi(t)$ in a real variable t defined by

$$\varphi(t) = \mathrm{E}[\exp(\mathrm{i}\, tX)],$$

where i denotes the imaginary unit. The function $\varphi(t)$ is the Fourier transform of the probability distribution and it satisfies

$$\varphi(0) = 1, \qquad \varphi(-t) = \overline{\varphi(t)}. \tag{7.17}$$

For any choice of real numbers $t_1, t_2, \ldots, t_n$, we consider a complex matrix $\varPhi$ whose (i, j) entry is given by $\varphi(t_i - t_j)$, that is,

$$\varPhi = (\varphi(t_i - t_j) \mid i, j = 1, 2, \ldots, n).$$

By (7.17), this matrix $\varPhi$ is a Hermitian matrix with diagonal entries all equal to one.[10] Moreover, matrix $\varPhi$ is positive semidefinite. Indeed, for any complex vector $\boldsymbol{u} = (u_1, u_2, \ldots, u_n)^\top$, we have

$$\begin{aligned}
\boldsymbol{u}^* \varPhi\, \boldsymbol{u} &= \sum_{i,j} \overline{u_i}\, u_j \varphi(t_i - t_j) \\
&= \mathrm{E}\Big[\sum_{i,j} \overline{u_i}\, u_j \exp(\mathrm{i}\,(t_i - t_j)X)\Big] \\
&= \mathrm{E}\Big[\Big(\sum_{i} \overline{u_i} \exp(\mathrm{i}\, t_i X)\Big)\Big(\sum_{j} u_j \exp(-\mathrm{i}\, t_j X)\Big)\Big] \\
&= \mathrm{E}\Big[\Big|\sum_{j} u_j \exp(-\mathrm{i}\, t_j X)\Big|^2\Big] \geqq 0.
\end{aligned}$$

---

[9]The moment matrix M of order n can be defined when X has a (finite) kth moment for all $k \leqq 2n - 2$.

[10]If $t_1 < t_2 < \cdots < t_n$ are chosen with constant differences, $\varPhi$ is a Toeplitz matrix (cf., Sec. 1.7.2).

In this way a characteristic function $\varphi(t)$ of a random variable has the property that, for any n and any real numbers $t_1, t_2, \ldots, t_n$, the matrix $\Phi = (\varphi(t_i - t_j))$ is positive semidefinite. This property captures the essential nature of a characteristic function, and it is known as *Bochner's theorem* that a function $\varphi : \mathbb{R} \to \mathbb{C}$ endowed with this property[11] is necessarily a characteristic function of some random variable. Nonnegativity of a probability density function is reflected in the characteristic function as positive semidefiniteness of the matrix Φ. ■

7.4.2 *Structural Mechanics*

In this section we present two examples from structural mechanics.

Example 7.4. Positive semidefinite matrices arise naturally in *truss* structures. Consider a planar truss, shown in Fig. 7.1, which consists of nine *members* and four *free nodes* (nodes 1 to 4) and two *supports* (fixed nodes 5, 6). We denote the number of free nodes by n, *i.e.*, $n = 4$.

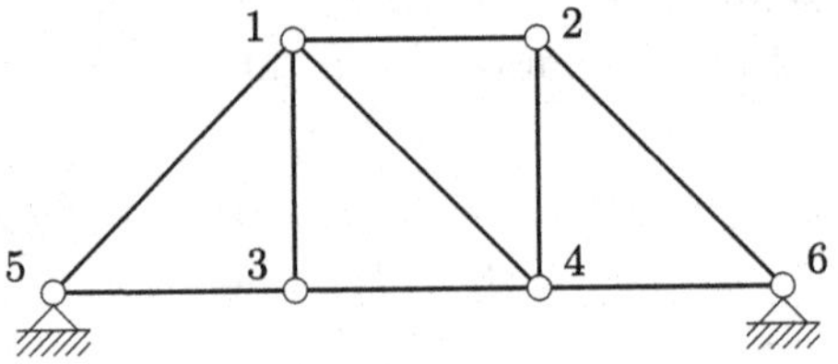

Fig. 7.1 Truss structure.

First we concentrate on a single member and derive a quadratic form representing the stored elastic energy. Let i and j denote the end nodes of the member, $E = E_{ij}$ the modulus of elasticity, $L = L_{ij}$ the length, and $A = A_{ij}$ the cross-sectional area of the member. Denote by (x_i^0, y_i^0) and (x_j^0, y_j^0) the locations of nodes i and j before deformation, respectively. Then we have

$$L = \sqrt{(x_i^0 - x_j^0)^2 + (y_i^0 - y_j^0)^2}.$$

Denote the coordinates of nodes i and j after deformation by

$$(x_i, y_i) = (x_i^0 + u_i, y_i^0 + v_i), \qquad (x_j, y_j) = (x_j^0 + u_j, y_j^0 + v_j),$$

[11]To be precise, (7.17) and the continuity of φ at $t = 0$ are assumed here. See [47, 49, 51, 58, 78] for details.

respectively. Then the elongation ΔL of the member is given as

$$\Delta L = \sqrt{((x_i^0 + u_i) - (x_j^0 + u_j))^2 + ((y_i^0 + v_i) - (y_j^0 + v_j))^2} - L.$$

Under the assumption of small deformation, we approximate ΔL, via Taylor expansion, with a linear function in (u_i, v_i) and (u_j, v_j) as

$$\Delta L \approx \left[\begin{array}{cc|cc} -\alpha & -\beta & \alpha & \beta \end{array}\right] \left[\begin{array}{c} u_i \\ v_i \\ \hline u_j \\ v_j \end{array}\right] = \boldsymbol{b}^\top \boldsymbol{u}.$$

Here

$$\alpha = \frac{x_j^0 - x_i^0}{L}, \qquad \beta = \frac{y_j^0 - y_i^0}{L} \tag{7.18}$$

represent the *direction cosine*[12] of the vector from node i to node j, and

$$\boldsymbol{b} = (-\alpha, -\beta, \alpha, \beta)^\top, \qquad \boldsymbol{u} = (u_i, v_i, u_j, v_j)^\top.$$

The *elastic energy* stored in this member is given by

$$\frac{1}{2} \cdot \frac{EA}{L} \cdot (\Delta L)^2 = \frac{1}{2} \cdot \boldsymbol{u}^\top \left(\frac{EA}{L} \cdot \boldsymbol{b}\boldsymbol{b}^\top \right) \boldsymbol{u} = \frac{1}{2} \boldsymbol{u}^\top K \boldsymbol{u},$$

where the matrix K defining this quadratic form is given as[13]

$$K = \frac{EA}{L} \cdot \boldsymbol{b}\boldsymbol{b}^\top = \frac{EA}{L} \left[\begin{array}{c} -\alpha \\ -\beta \\ \hline \alpha \\ \beta \end{array}\right] \left[\begin{array}{cc|cc} -\alpha & -\beta & \alpha & \beta \end{array}\right]. \tag{7.19}$$

This is a positive semidefinite symmetric matrix of rank one that represents the stiffness of a single member.[14] The matrix K can also be expressed as

$$K = \frac{EA}{L} \left[\begin{array}{cc|cc} \alpha^2 & \alpha\beta & -\alpha^2 & -\alpha\beta \\ \alpha\beta & \beta^2 & -\alpha\beta & -\beta^2 \\ \hline -\alpha^2 & -\alpha\beta & \alpha^2 & \alpha\beta \\ -\alpha\beta & -\beta^2 & \alpha\beta & \beta^2 \end{array}\right]. \tag{7.20}$$

---

[12] Note that $\alpha^2 + \beta^2 = 1$.

[13] In the finite element method, the matrix K in (7.19) (or (7.20)) is called the *element stiffness matrix*.

[14] In general terms, stiffness means the physical extent to resist the deformation due to an applied force. In a (linear) spring, for example, the spring constant, often denoted by k, represents the stiffness of the spring. The stiffness matrix K is a matrix version thereof, and relates deformations and forces at the free nodes. However, we do not discuss forces in this example (and in Example 7.5), because we are mainly interested in positive semidefinite matrices and the associated quadratic forms here. Positive semidefiniteness of K may be interpreted as a generalization of the fact that a spring constant is a nonnegative real number.

Yet another expression

$$K = \frac{EA}{L} \left[\begin{array}{cc|cc} \alpha & -\beta & & \\ \beta & \alpha & & \\ \hline & & \alpha & -\beta \\ & & \beta & \alpha \end{array} \right] \left[\begin{array}{cc|cc} 1 & 0 & -1 & 0 \\ 0 & 0 & 0 & 0 \\ \hline -1 & 0 & 1 & 0 \\ 0 & 0 & 0 & 0 \end{array} \right] \left[\begin{array}{cc|cc} \alpha & \beta & & \\ -\beta & \alpha & & \\ \hline & & \alpha & \beta \\ & & -\beta & \alpha \end{array} \right]$$

reveals that K is composed of the stiffness matrix in the local coordinate system along the member axial direction

$$\frac{EA}{L} \begin{bmatrix} 1 & -1 \\ -1 & 1 \end{bmatrix}$$

and a matrix

$$\begin{bmatrix} \alpha & -\beta \\ \beta & \alpha \end{bmatrix}$$

representing the coordinate transformation.

The stiffness matrix $\tilde{K}$ of the whole truss structure can be obtained by assembling the stiffness matrices K for all members. We illustrate this procedure for the structure in Fig. 7.1. As this planar truss has $n = 4$ free nodes, the stiffness matrix $\tilde{K}$ of the whole structure is a symmetric matrix of order $2n = 8$, in which the matrices of the form of (7.20) are embedded. For example, for the member connecting node $i = 1$ and node $j = 3$, we consider an 8×8 matrix

$$K^{(1,3)} = \frac{EA}{L} \times \begin{array}{c} u_1 \\ v_1 \\ u_2 \\ v_2 \\ u_3 \\ v_3 \\ u_4 \\ v_4 \end{array} \overset{\begin{array}{cccccccc} u_1 & v_1 & u_2 & v_2 & u_3 & v_3 & u_4 & v_4 \end{array}}{\left[\begin{array}{cc|cc|cc|cc} \alpha^2 & \alpha\beta & & & -\alpha^2 & -\alpha\beta & & \\ \alpha\beta & \beta^2 & & & -\alpha\beta & -\beta^2 & & \\ \hline & & & & & & & \\ & & & & & & & \\ \hline -\alpha^2 & -\alpha\beta & & & \alpha^2 & \alpha\beta & & \\ -\alpha\beta & -\beta^2 & & & \alpha\beta & \beta^2 & & \\ \hline & & & & & & & \\ & & & & & & & \end{array} \right]},$$

where α and β are defined by (7.18) with $(i, j) = (1, 3)$, and $E = E_{13}$, $A = A_{13}$, and $L = L_{13}$. In contrast, for the member connecting a free node

$i = 2$ and a fixed node $j = 6$, we consider the following matrix

$$K^{(2,6)} = \frac{\hat{E}\hat{A}}{\hat{L}} \times \begin{array}{c} \\ u_1 \\ v_1 \\ u_2 \\ v_2 \\ u_3 \\ v_3 \\ u_4 \\ v_4 \end{array} \begin{array}{cccccccc} u_1 & v_1 & u_2 & v_2 & u_3 & v_3 & u_4 & v_4 \end{array} \hspace{-20em} \left[\begin{array}{cc|cc|cc|cc} & & & & & & & \\ & & & & & & & \\ \hline & & \hat{\alpha}^2 & \hat{\alpha}\hat{\beta} & & & & \\ & & \hat{\alpha}\hat{\beta} & \hat{\beta}^2 & & & & \\ \hline & & & & & & & \\ & & & & & & & \\ \hline & & & & & & & \\ & & & & & & & \end{array} \right],$$

where $\hat{\alpha}$ and $\hat{\beta}$ are defined by (7.18) with $(i, j) = (2, 6)$, and $\hat{E} = E_{26}$, $\hat{A} = A_{26}$, and $\hat{L} = L_{26}$. The stiffness matrix $\tilde{K}$ of the whole truss structure[15] is given as the sum of all such matrices $K^{(i,j)}$ associated with respective members, that is,

$$\tilde{K} = \sum_{(i,j)} K^{(i,j)}.$$

The quadratic form

$$\frac{1}{2}\tilde{\boldsymbol{u}}^\top \tilde{K} \tilde{\boldsymbol{u}},$$

defined by the stiffness matrix $\tilde{K}$ for the vector $\tilde{\boldsymbol{u}} = (u_1, v_1, u_2, v_2, \dots, u_n, v_n)^\top$ of displacements of free nodes, represents the elastic energy stored in the whole truss. Since each $K^{(i,j)}$ is a positive semidefinite matrix and the sum of positive semidefinite matrices is positive semidefinite, the stiffness matrix $\tilde{K}$ for the whole structure is also positive semidefinite. With the knowledge of the stiffness matrix we can carry out structural equilibrium analysis (static analysis) of a truss.

For dynamic analysis, including analysis of natural frequencies of a truss structure, we need to employ mass matrices in addition to stiffness matrices. There are variants of mass matrices, depending on the modeling. For a single member of a planar truss, we can use a *lamped mass matrix*

$$M_{\mathrm{L}} = \frac{\rho AL}{2} \begin{bmatrix} 1 & 0 & 0 & 0 \\ 0 & 1 & 0 & 0 \\ 0 & 0 & 1 & 0 \\ 0 & 0 & 0 & 1 \end{bmatrix} \tag{7.21}$$

[15]In the finite element method, the matrix $\tilde{K}$ is called the *global stiffness matrix*.

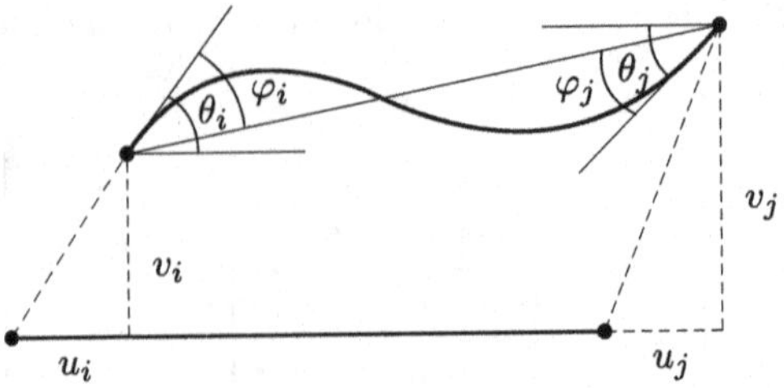

Fig. 7.2 Beam element.

or a *consistent mass matrix*

$$M_C = \frac{\rho AL}{6} \begin{bmatrix} 2 & 0 & 1 & 0 \\ 0 & 2 & 0 & 1 \\ 1 & 0 & 2 & 0 \\ 0 & 1 & 0 & 2 \end{bmatrix}, \tag{7.22}$$

where ρ denotes the density of a truss member. Either variant of a mass matrix is a positive definite symmetric matrix.[16] ■

Example 7.5. The stiffness matrix of a *beam element* for analysis of frames is also a positive semidefinite symmetric matrix. Here we consider a planar beam element in Fig. 7.2, where the deformation is described by a 6-dimensional vector $\boldsymbol{u} = (u_i, v_i, \theta_i, u_j, v_j, \theta_j)^\top$ representing the displacements (u_i, v_i, u_j, v_j) and the rotations (*i.e.*, the slopes of the beam axis) (θ_i, θ_j) at both ends; we have $\theta_i > 0$ and $\theta_j > 0$ in Fig. 7.2. It is known[17] that the stiffness matrix is given by

[16] Positive definiteness of a mass matrix may be interpreted as a generalization of the fact that the mass of a particle is a positive real number. The natural angular frequency ω of a single truss member can be determined by solving the generalized eigenvalue problem $K\boldsymbol{u} = \omega^2 M\boldsymbol{u}$ for $M = M_L$ or M_C above and K in (7.20) with $(\alpha, \beta) = (1, 0)$.

[17] The expression (7.23) is derived under some assumptions, including small deformation, for a beam (Euler–Bernoulli beam element) with uniform cross section. For details, see textbooks on structural mechanics such as [54, 65, 69–71, 81].

$$K = \left[\begin{array}{ccc|ccc} \dfrac{EA}{L} & 0 & 0 & -\dfrac{EA}{L} & 0 & 0 \\ 0 & \dfrac{12EI}{L^3} & \dfrac{6EI}{L^2} & 0 & -\dfrac{12EI}{L^3} & \dfrac{6EI}{L^2} \\ 0 & \dfrac{6EI}{L^2} & \dfrac{4EI}{L} & 0 & -\dfrac{6EI}{L^2} & \dfrac{2EI}{L} \\ \hline -\dfrac{EA}{L} & 0 & 0 & \dfrac{EA}{L} & 0 & 0 \\ 0 & -\dfrac{12EI}{L^3} & -\dfrac{6EI}{L^2} & 0 & \dfrac{12EI}{L^3} & -\dfrac{6EI}{L^2} \\ 0 & \dfrac{6EI}{L^2} & \dfrac{2EI}{L} & 0 & -\dfrac{6EI}{L^2} & \dfrac{4EI}{L} \end{array}\right], \qquad (7.23)$$

where E denotes the modulus of elasticity, L the length, A the cross-sectional area, and I the moment of inertia of the beam element. This matrix K (with suitable permutations of rows and columns) is a direct sum of two matrices,

$$K_{\mathrm{a}} = \frac{EA}{L} \begin{bmatrix} 1 & -1 \\ -1 & 1 \end{bmatrix}$$

for (u_i, u_j) and

$$K_{\mathrm{b}} = \frac{EI}{L^3} \begin{bmatrix} 12 & 6L & -12 & 6L \\ 6L & 4L^2 & -6L & 2L^2 \\ -12 & -6L & 12 & -6L \\ 6L & 2L^2 & -6L & 4L^2 \end{bmatrix}$$

for $(v_i, \theta_i, v_j, \theta_j)$. The matrices K_{a} and K_{b} are positive semidefinite symmetric matrices of rank one and two, respectively. Hence K is a positive semidefinite symmetric matrix of rank three.

It is noteworthy that the matrix K in (7.23) can be decomposed as

$$K = \left[\begin{array}{c|cc} -1 & 0 & 0 \\ 0 & \frac{1}{L} & \frac{1}{L} \\ 0 & 1 & 0 \\ \hline 1 & 0 & 0 \\ 0 & -\frac{1}{L} & -\frac{1}{L} \\ 0 & 0 & 1 \end{array}\right] \left[\begin{array}{c|cc} \frac{EA}{L} & 0 & 0 \\ \hline 0 & \frac{4EI}{L} & \frac{2EI}{L} \\ 0 & \frac{2EI}{L} & \frac{4EI}{L} \end{array}\right] \left[\begin{array}{ccc|ccc} -1 & 0 & 0 & 1 & 0 & 0 \\ \hline 0 & \frac{1}{L} & 1 & 0 & -\frac{1}{L} & 0 \\ 0 & \frac{1}{L} & 0 & 0 & -\frac{1}{L} & 1 \end{array}\right] \qquad (7.24)$$

with a 3×3 positive definite matrix in the middle. This decomposition motivates us to change variables using the 3×6 matrix in the decomposition.

Namely, in view of

$$\begin{bmatrix} -1 & 0 & 0 & 1 & 0 & 0 \\ 0 & \frac{1}{L} & 1 & 0 & -\frac{1}{L} & 0 \\ 0 & \frac{1}{L} & 0 & 0 & -\frac{1}{L} & 1 \end{bmatrix} \begin{bmatrix} u_i \\ v_i \\ \theta_i \\ \hline u_j \\ v_j \\ \theta_j \end{bmatrix} = \begin{bmatrix} u_j - u_i \\ \theta_i - \dfrac{v_j - v_i}{L} \\ \theta_j - \dfrac{v_j - v_i}{L} \end{bmatrix}$$

we introduce new variables

$$\varphi_i = \theta_i - \frac{v_j - v_i}{L}, \qquad \varphi_j = \theta_j - \frac{v_j - v_i}{L}$$

to denote the second and third components. When the deformation is small, these variables φ_i and φ_j are approximately equal to the rotations from the line connecting the two nodes (see Fig. 7.2, where $\varphi_i > 0$, $\varphi_j > 0$). Thus the decomposition (7.24) admits a natural physical interpretation as well.

It follows from the decomposition (7.24) that the *elastic energy* stored in the beam, which is given by $\frac{1}{2}\boldsymbol{u}^\top K\boldsymbol{u}$, consists of two parts as

$$\frac{1}{2}\boldsymbol{u}^\top K\boldsymbol{u} = \frac{1}{2} \cdot \frac{EA}{L}(u_i - u_j)^2 + \frac{1}{2} \cdot \frac{4EI}{L}\left(\varphi_i^2 + \varphi_i\varphi_j + \varphi_j^2\right).$$

The first term represents the energy due to the axial deformation and the second is the energy due to the bending deformation. ■

7.4.3 *Differential Equations*

We next discuss positive semidefiniteness related to differential equations.

Example 7.6. In Sec. 5.5 we have dealt with a finite difference approximation of the Dirichlet problem (the Laplace equation with a Dirichlet boundary condition). With the use of a natural finite difference approximation (5.15) of the second order partial derivatives, we have obtained a system of equations (5.16) whose coefficient matrix A is given by (5.18) when $N = 5$, for instance. For any N ($\geqq 2$) the eigenvalues of matrix A are all positive, as shown in Example 6.2 in Sec. 6.1.3, and therefore A is a positive definite matrix.[18]

As is well known in *calculus of variation* [59], the Dirichlet problem can also be derived from a *variational problem* to minimize an integral (a

[18] Positive definiteness of A follows also from a combination of two facts that A is positive semidefinite as it is a diagonally dominant symmetric matrix with positive diagonal entries (Theorem 7.3) and that A is nonsingular (Sec. 5.5).

functional[19])

$$I[u] = \int_{\Omega} \left[\left(\frac{\partial u}{\partial x} \right)^2 + \left(\frac{\partial u}{\partial y} \right)^2 \right] \mathrm{d}x\mathrm{d}y$$

under the boundary condition (5.14). The quadratic form $\boldsymbol{u}^\top A \boldsymbol{u}$ defined by the matrix A above corresponds to a discretization of this integral.[20]

If the so-called *Neumann boundary condition* (specifying the normal derivative of a solution on the boundary) is imposed, the coefficient matrix appearing in the finite difference approximation is no longer positive definite but positive semidefinite. ■

It is often the case that discretization of an elliptic second order partial differential equation results in a positive definite matrix.

7.4.4 *Electric Circuits and Graphs*

Finally, we show examples from electric circuits and graph structures.

Example 7.7. Positive definite matrices arise also from electric circuits (consisting of positive linear resistors). Consider, for example, the electric circuit in Fig. 7.3 consisting of five branches (linear resistors) connected at four nodes. For each branch, numbered by $j = 1, 2, \ldots, 5$, let $g_j > 0$ denote the *conductance* (reciprocal of resistance) of branch j, i_j the current in branch j, and v_j the voltage across branch j. For each node, numbered by $k = 0, 1, 2, 3$, let p_k denote the potential of node k, where node 0 is grounded and $p_0 = 0$. For each $k = 1, 2, 3$, let x_k denote the current flowing into the circuit from the outside. Then the current flowing out of the circuit at node 0 is equal to $x_1 + x_2 + x_3$.

The voltage vector $\boldsymbol{v} = (v_1, v_2, \ldots, v_5)^\top$ and the node potential vector $\boldsymbol{p} = (p_1, p_2, p_3)^\top$ are related as $\boldsymbol{v} = N^\top \boldsymbol{p}$ with

$$N = \begin{bmatrix} -1 & 1 & 0 & 0 & -1 \\ 1 & 0 & 0 & -1 & 0 \\ 0 & -1 & 1 & 0 & 0 \end{bmatrix}.$$

---

[19]The integral $I[u]$ is a function of a function u, meaning that it takes a function u as an input and returns a real number as an output. In general, a function of a function is called a *functional*.

[20]To be more precise, the objective function in the discretization of the variational problem is a quadratic function of the form $\boldsymbol{u}^\top A \boldsymbol{u} - 2\boldsymbol{g}^\top \boldsymbol{u} +$ (a quadratic term in $\boldsymbol{g}$ arising from the boundary condition), reflecting the boundary condition (5.14). The condition for $\boldsymbol{u}$ to be a minimizer of this function coincides with the equation $A\boldsymbol{u} = \boldsymbol{g}$ in (5.17).

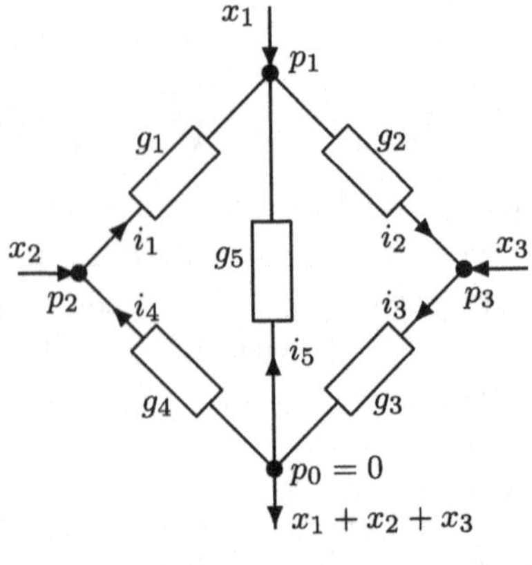

Fig. 7.3 A circuit consisting of resistors.

By Ohm's law, the voltage vector $\boldsymbol{v}$ and the current vector $\boldsymbol{i} = (i_1, i_2, \ldots, i_5)^\top$ are related as $\boldsymbol{i} = G\boldsymbol{v}$, where G is a diagonal matrix

$$G = \operatorname{diag}(g_1, g_2, \ldots, g_5)$$

consisting of conductances g_j. By Kirchhoff's current law, the current vector $\boldsymbol{i}$ and the current supply vector $\boldsymbol{x} = (x_1, x_2, x_3)^\top$ are related as $N\boldsymbol{i} = \boldsymbol{x}$ with the same matrix N as above. It follows from

$$N\boldsymbol{i} = \boldsymbol{x}, \qquad \boldsymbol{i} = G\boldsymbol{v}, \qquad \boldsymbol{v} = N^\top \boldsymbol{p}$$

that $NGN^\top \boldsymbol{p} = \boldsymbol{x}$, which shows the relation between the node potential vector $\boldsymbol{p}$ and the current supply vector $\boldsymbol{x}$. The coefficient matrix

$$A = NGN^\top = \begin{bmatrix} g_1 + g_2 + g_5 & -g_1 & -g_2 \\ -g_1 & g_1 + g_4 & 0 \\ -g_2 & 0 & g_2 + g_3 \end{bmatrix} \tag{7.25}$$

of this relation is called the *node conductance matrix*. The matrix A in (7.25) is a strictly diagonally dominant symmetric matrix with positive diagonal entries, and hence positive definite by Theorem 7.3 in Sec. 7.2.1. The positive definiteness of A follows also from the decomposition $A = SS^\top$ with $S = N \operatorname{diag}(\sqrt{g_1}, \sqrt{g_2}, \ldots, \sqrt{g_5})$, where $\operatorname{rank} S = \operatorname{rank} N = 3$ (cf., Remark 7.1 in Sec. 7.2.1).

In general, the node conductance matrix A of an arbitrary electric circuit consisting of positive linear resistors is positive definite. This mathematical property corresponds to a physical property called *passivity*, which says that the power (or energy) consumed in such an electric circuit, represented by $\boldsymbol{p}^\top \boldsymbol{x} = \boldsymbol{p}^\top A \boldsymbol{p}$, must be nonnegative for all $\boldsymbol{p}$. ■

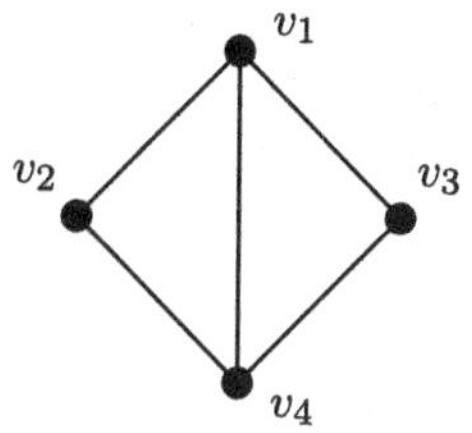

Fig. 7.4 Graph.

Example 7.8. A *graph* is a discrete structure composed of *vertices* and *edges*,[21] such as the one shown in Fig. 7.4. Graphs are easier to understand when they are illustrated as a drawing, but the way they are drawn is not important, because the relevant information is which edges connect which vertices. The set V of the vertices of a graph G is referred to as the *vertex set* of G, and the set E of the edges as the *edge set* of G. A graph with vertex set V and edge set E is often denoted as $G = (V, E)$. We consider here an *undirected graph*, where each edge has distinct end vertices and there are no multiple edges connecting two vertices.

The structure of a graph can be described by a symmetric matrix. For a graph G with vertex set $V = \{v_1, v_2, \ldots, v_n\}$, we define a (symmetric) matrix $A = (a_{ij})$ of order n as

$$a_{ii} = \text{number of edges incident to vertex } v_i \quad (i = 1, 2, \ldots, n),$$

$$a_{ij} = \begin{cases} -1 & (v_i \text{ and } v_j \text{ are connected by an edge}) \\ \ \ 0 & (v_i \text{ and } v_j \text{ are not connected by an edge}) \end{cases} \quad (i \neq j).$$

This matrix $A = (a_{ij})$ is sometimes called the *Laplacian matrix* of graph G. For the graph in Fig. 7.4, for example, we have

$$A = \begin{bmatrix} 3 & -1 & -1 & -1 \\ -1 & 2 & 0 & -1 \\ -1 & 0 & 2 & -1 \\ -1 & -1 & -1 & 3 \end{bmatrix}.$$

In general, the Laplacian matrix of a graph is positive semidefinite, as it is a diagonally dominant symmetric matrix with nonnegative diagonal entries (cf., Theorem 7.3 in Sec. 7.2.1).

A variant of the Laplacian matrix has appeared in the Dirichlet problem in Sec. 5.5 and Example 7.6. The grid for the discretized spatial domain,

[21] Vertices are also called *points* or *nodes*, and edges are also called *arcs* or *branches*. For more information about the interaction between graphs and matrices, see Chapter 1 of Volume II [9].

depicted in Fig. 5.1, may be regarded as a graph with vertex set $V = \{u_{ij} \mid i,j = 0,1,\ldots,N\}$, where $N = 5$. The Laplacian matrix of this graph is a symmetric matrix of order $(N+1)^2 = 36$. The matrix A in (5.18) coincides with the principal submatrix of this Laplacian matrix with rows and columns corresponding to the $(N-1)^2 = 16$ vertices $\{u_{ij} \mid i,j = 1,\ldots,N-1\}$ in the middle of the grid in Fig. 5.1.

Suppose that a certain (positive) weight is associated with each edge, and let $c_{ij} > 0$ denote the weight for the edge connecting v_i and v_j. In such a case it is often convenient to introduce a weighted generalization of the Laplacian matrix. For the graph in Fig. 7.4, for example, the weighted version is given by

$$B = \begin{bmatrix} c_{12}+c_{13}+c_{14} & -c_{12} & -c_{13} & -c_{14} \\ -c_{12} & c_{12}+c_{24} & 0 & -c_{24} \\ -c_{13} & 0 & c_{13}+c_{34} & -c_{34} \\ -c_{14} & -c_{24} & -c_{34} & c_{14}+c_{24}+c_{34} \end{bmatrix}.$$

If a graph represents an electric circuit and weight c_{ij} is the conductance, the weighted Laplacian matrix B is essentially the same as the node conductance matrix introduced in Example 7.7. For example, the 3×3 submatrix of B above with the fourth row and column deleted coincides with the node conductance matrix in (7.25) if $c_{12} = g_1$, $c_{13} = g_2$, *etc.* ■

7.5 Sylvester's Law of Inertia

7.5.1 *Symmetric Matrices*

Let A be a symmetric matrix of order n. We denote by $\nu_+(A)$ the number of positive eigenvalues of A, counted with multiplicity. Similarly, we denote by $\nu_0(A)$ and $\nu_-(A)$, respectively, the numbers of zero and negative eigenvalues of A, counted with multiplicity. We have

$$\nu_+(A) + \nu_0(A) + \nu_-(A) = n.$$

The triple $(\nu_+(A), \nu_0(A), \nu_-(A))$ is called the *signature* of a symmetric matrix A. For a positive definite matrix A, we have $(\nu_+(A), \nu_0(A), \nu_-(A)) = (n, 0, 0)$.

The following theorem is known as *Sylvester's law of inertia.*

Theorem 7.14. *Let A be a symmetric matrix, S a nonsingular matrix, and $B = S^\top AS$. Then*

$$\nu_+(A) = \nu_+(B), \qquad \nu_0(A) = \nu_0(B), \qquad \nu_-(A) = \nu_-(B).$$

That is, the signature of a symmetric matrix is invariant under congruence transformations.

Proof. Since every nonsingular matrix can be represented as a product of elementary matrices (Theorem 3.2(2)), we may assume that S is equal to one of the elementary matrices $E_1(p,q)$, $E_2(p;b)$, and $E_3(p,q;c)$ introduced in Definition 3.2 in Sec. 3.1.

First, if $S = E_1(p,q)$, S is an orthogonal matrix and therefore not only the signs of the eigenvalues but their numerical values are invariant in $B = S^\top AS$.

Next, suppose that $S = E_2(p;b)$. We first consider the case of $b > 0$. If the parameter b is changed continuously as $1 \to b$, the eigenvalues of

$$B(b) = E_2(p;b)^\top A\, E_2(p;b)$$

change continuously in b. Since $E_2(p;1) = I$ (unit matrix), we have $B(1) = A$. In the course of parameter change $1 \to b$, the matrix $E_2(p;b)$ is always nonsingular and hence $\operatorname{rank} B(b) = \operatorname{rank} A$. In particular, every nonzero eigenvalue remains to be nonzero, and hence, by continuity, the numbers $\nu_+(B(b))$ and $\nu_-(B(b))$ of positive and negative eigenvalues of $B(b)$ are invariant during this change $1 \to b$. Therefore we have $\nu_+(A) = \nu_+(B(1)) = \nu_+(B(b)) = \nu_+(B)$ and $\nu_-(A) = \nu_-(B(1)) = \nu_-(B(b)) = \nu_-(B)$. In the other case of $b < 0$, we change the parameter b continuously as $-1 \to b$. Since $E_2(p;-1)$ is an orthogonal matrix, $B(-1)$ and A have the same eigenvalues. The rest is similar to the case of $b > 0$.

Finally, suppose that $S = E_3(p,q;c)$. In a similar manner as for $S = E_2(p;b)$, we change the parameter c continuously as $0 \to c$. Note that $E_3(p,q;c)$ is nonsingular for all c. □

The significance of Sylvester's law of inertia may be summarized as follows. If S is an orthogonal matrix, the transformed matrix $B = S^\top AS$ has the same eigenvalues as the matrix A. If S is a general nonsingular matrix, the eigenvalues of A and B are no longer the same in value. Nevertheless, we can retrieve information about the sign of eigenvalues of A from the transformed matrix $B = S^\top AS$. Sylvester's law of inertia forms a basis of a numerical algorithm for computing eigenvalues; see Remark 7.7.

In view of Sylvester's law of inertia, it is natural to aim at reducing a given symmetric matrix A to a simplest possible form by means of a congruence transformation $S^\top AS$ with a nonsingular matrix S. The following theorem answers this, and the matrix of the form (7.26) below is called the *Sylvester normal form* of A.

Theorem 7.15. *For any symmetric matrix A, there exists a nonsingular matrix S such that*

$$S^\top AS = \mathrm{diag}\,(+1,\ldots,+1;0,\ldots,0;-1,\ldots,-1), \tag{7.26}$$

where the numbers of $+1$, 0, and -1 are equal to $\nu_+(A)$, $\nu_0(A)$, and $\nu_-(A)$, respectively.

Proof. By Theorem 6.9 in Sec. 6.3.1, there exists an orthogonal matrix Q such that

$$Q^\top AQ = \mathrm{diag}\,(\alpha_1,\alpha_2,\ldots,\alpha_p;0,\ldots,0;-\beta_1,-\beta_2,\ldots,-\beta_q),$$

where $\alpha_i > 0$ for $i = 1,2,\ldots,p$ and $\beta_j > 0$ for $j = 1,2,\ldots,q$. With the choice of

$$\begin{aligned} S = Q\cdot\mathrm{diag}\,(&1/\sqrt{\alpha_1},1/\sqrt{\alpha_2},\ldots,1/\sqrt{\alpha_p};\\ &1,\ldots,1;\ 1/\sqrt{\beta_1},1/\sqrt{\beta_2},\ldots,1/\sqrt{\beta_q}) \end{aligned}$$

we obtain (7.26) where the number of "+1" is equal to p and that of "−1" is q. Moreover, it follows from Sylvester's law of inertia (Theorem 7.14)[22] that

$$p = \nu_+(Q^\top AQ) = \nu_+(A), \qquad q = \nu_-(Q^\top AQ) = \nu_-(A).$$

□

Remark 7.6. A *completion of squares* means expressing a given quadratic form $\boldsymbol{x}^\top A\boldsymbol{x}$ as a weighted sum of squares as

$$\boldsymbol{x}^\top A\boldsymbol{x} = \sum_k c_k(\text{linear form in } x_1,x_2,\ldots,x_n)^2,$$

where the coefficients c_k can be positive or negative. Such an expression is not unique for a given quadratic form, but Sylvester's law of inertia implies that the numbers of positive and negative coefficients c_k are uniquely determined, respectively, under a certain natural condition. For example, the quadratic form ${x_1}^2 - 4x_1x_2 + 2{x_2}^2$ defined by $A = \begin{bmatrix} 1 & -2 \\ -2 & 2 \end{bmatrix}$ admits two different completions of squares:

$$\begin{aligned} {x_1}^2 - 4x_1x_2 + 2{x_2}^2 &= ({x_1}^2 - 4x_1x_2 + 4{x_2}^2) - 2{x_2}^2 = (x_1 - 2x_2)^2 - 2{x_2}^2, \\ {x_1}^2 - 4x_1x_2 + 2{x_2}^2 &= 2({x_1}^2 - 2x_1x_2 + {x_2}^2) - {x_1}^2 = 2(x_1 - x_2)^2 - {x_1}^2. \end{aligned}$$

[22] Alternatively, we may use Theorem 6.8 in Sec. 6.3.1, showing that $Q^\top AQ$ has the same eigenvalues as A.

In either of these expressions we have one positive coefficient and one negative coefficient in accordance with $\nu_+(A) = 1$ and $\nu_-(A) = 1$. It is noted, however, that in completing the squares, we impose a natural condition that the linear forms should be linearly independent. For instance, the quadratic form

$$2{x_1}^2 + 2{x_2}^2 + 2{x_3}^2 + 4x_1x_2 + 2x_2x_3 + 2x_3x_1$$

can be expressed as a sum of three squares with positive coefficients as

$$(x_1 + x_2 + x_3)^2 + (x_1 + x_2)^2 + {x_3}^2,$$

but this is not an admissible expression since the three linear forms $x_1 + x_2 + x_3$, $x_1 + x_2$, and x_3 are defined by linearly dependent vectors $(1, 1, 1)$, $(1, 1, 0)$, and $(0, 0, 1)$. A completion of squares meeting the required linear independence is given, for example, by

$$2(x_1 + x_2 + \frac{1}{2}x_3)^2 + \frac{3}{2}{x_3}^2$$

consisting of two terms with positive coefficients. Note that the given quadratic form is defined by

$$A = \begin{bmatrix} 2 & 2 & 1 \\ 2 & 2 & 1 \\ 1 & 1 & 2 \end{bmatrix},$$

for which we have $\operatorname{rank} A = 2$ and $(\nu_+(A), \nu_0(A), \nu_-(A)) = (2, 1, 0)$. ■

Remark 7.7. Sylvester's law of inertia is used effectively in numerical computation of eigenvalues of a symmetric matrix. Suppose that a procedure is available that determines, for any real number σ, the numbers of eigenvalues that are greater than σ, equal to σ, and smaller than σ, respectively. By repeatedly applying this procedure with varying σ, we can identify intervals that contain an eigenvalue with arbitrarily high precision.

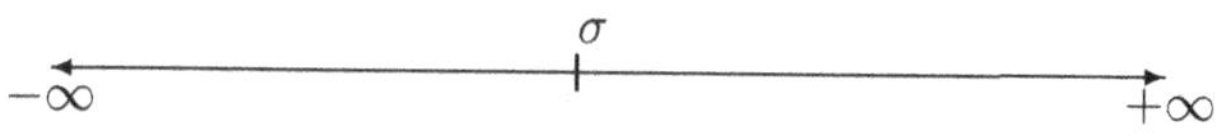

To be more specific, this method goes as follows. Let A be a symmetric matrix whose eigenvalues are sought. For any real number σ, we decompose the matrix $A - \sigma I$ as $A - \sigma I = LDL^\top$ using a lower triangular matrix L and a diagonal matrix D.[23] By Sylvester's law of inertia, the matrices

[23]We can carry out this decomposition with $O(n^3)$ arithmetic operations in a similar manner to the Gaussian elimination. To be precise, the matrix D may possibly contain 2×2 diagonal blocks. The diagonal entries of L are usually chosen to be 1. Details can be found in textbooks on numerical computations such as [37, 43].

$A - \sigma I$ and D have the same signature. The signature $(\nu_+(A - \sigma I), \nu_0(A - \sigma I), \nu_-(A - \sigma I))$ of $A - \sigma I$ gives the numbers of eigenvalues of A that are greater than σ, equal to σ, and smaller than σ, respectively, whereas the signature $(\nu_+(D), \nu_0(D), \nu_-(D))$ of the diagonal matrix D is readily known from the numbers of its diagonal entries that are positive, zero, and negative, respectively.

When we have an interval (σ_1, σ_2) at hand that contains at least one eigenvalue of A, we take the middle point $(\sigma_1 + \sigma_2)/2$ of this interval as the next σ to try with. In so doing we can divide the interval into two smaller intervals, at least one of which contains at least one eigenvalue of A. Such method to numerically approximate eigenvalues of a symmetric matrix is called the *bisection method*.

This method with the choice of $\sigma = 0$ can be used to test for positive definiteness, since a symmetric matrix is positive definite if and only if all of its eigenvalues are positive. ■

7.5.2 *Hermitian Matrices*

We have similar theorems for a Hermitian (complex) matrix A. Recalling that a Hermitian matrix has real eigenvalues, we denote by $\nu_+(A)$ (resp., $\nu_0(A)$ and $\nu_-(A)$) the number of positive (resp., zero and negative) eigenvalues of A, counted with multiplicity. The triple $(\nu_+(A), \nu_0(A), \nu_-(A))$ is called the *signature* of A.

Theorem 7.16. *Let A be a Hermitian matrix, S a nonsingular complex matrix, and $B = S^* AS$. Then*

$$\nu_+(A) = \nu_+(B), \qquad \nu_0(A) = \nu_0(B), \qquad \nu_-(A) = \nu_-(B).$$

That is, the signature of a Hermitian matrix is invariant under congruence transformations.

Theorem 7.17. *For any Hermitian matrix A, there exists a nonsingular complex matrix S such that*

$$S^* AS = \mathrm{diag}\,(+1, \ldots, +1; 0, \ldots, 0; -1, \ldots, -1), \tag{7.27}$$

where the numbers of $+1$, 0, and -1 are equal to $\nu_+(A)$, $\nu_0(A)$, and $\nu_-(A)$, respectively.

The matrix of the form of (7.27) in Theorem 7.17 is called the *Sylvester normal form* of a Hermitian matrix A.

Chapter 8

Singular Values and the Method of Least Squares

In this chapter we consider singular values in comparison with eigenvalues. The theoretical basis of the method of least squares is presented as an application of this concept.

8.1 Definition of Singular Values

8.1.1 *Real Matrices*

The following decomposition is possible for any (not necessarily square) real matrices.

Theorem 8.1. *For any $m \times n$ real matrix A of rank r, there exist an orthogonal matrix U of order m and an orthogonal matrix V of order n such that*

$$A = U\ \Sigma\ V^\top, \tag{8.1}$$

where Σ is an $m \times n$ matrix of the form

$$\Sigma = \begin{bmatrix} \mathrm{diag}\,(\sigma_1, \sigma_2, \ldots, \sigma_r) & O_{r,n-r} \\ O_{m-r,r} & O_{m-r,n-r} \end{bmatrix} \quad (\sigma_1 \geqq \sigma_2 \geqq \cdots \geqq \sigma_r > 0). \tag{8.2}$$

Proof. Consider the matrix $A^\top A$. By Theorem 4.16 in Sec. 4.3.1, we have $\mathrm{rank}\,(A^\top A) = \mathrm{rank}\,A = r$. Since $A^\top A$ is a positive semidefinite symmetric matrix, all of its eigenvalues are nonnegative and r is equal to the number of positive eigenvalues. Let $\lambda_1 \geqq \lambda_2 \geqq \cdots \geqq \lambda_r$ denote the positive eigenvalues of $A^\top A$, while the other $n - r$ eigenvalues are 0. By Theorem 6.9 in Sec. 6.3.1, there exists an orthogonal matrix $V = [\boldsymbol{v}_1, \boldsymbol{v}_2, \ldots, \boldsymbol{v}_n]$ consisting of eigenvectors of $A^\top A$ such that

$$A^\top A V = V \cdot \mathrm{diag}\,(\lambda_1, \lambda_2, \ldots, \lambda_r, \overbrace{0, \ldots, 0}^{n-r}).$$

Let

$$\sigma_i = \sqrt{\lambda_i}, \quad \boldsymbol{u}_i = A\boldsymbol{v}_i/\sigma_i \qquad (i = 1, 2, \ldots, r).$$

Then $\boldsymbol{u}_1, \boldsymbol{u}_2, \ldots, \boldsymbol{u}_r$ form an orthonormal system in $\mathbb{R}^m$. With appropriate vectors $\boldsymbol{u}_{r+1}, \boldsymbol{u}_{r+2}, \ldots, \boldsymbol{u}_m$, we can form an orthogonal matrix $U = [\boldsymbol{u}_1, \boldsymbol{u}_2, \ldots, \boldsymbol{u}_m]$.

In the following we show $U^\top AV = \Sigma$ by calculating $\boldsymbol{u}_i^\top A\boldsymbol{v}_j$, which is the (i, j) entry of the matrix $U^\top AV$.

- If $1 \leqq i \leqq r$, $1 \leqq j \leqq r$, we have

$$\boldsymbol{u}_i^\top A\boldsymbol{v}_j = \frac{1}{\sigma_i}\boldsymbol{v}_i^\top A^\top A\boldsymbol{v}_j = \frac{1}{\sigma_i}\boldsymbol{v}_i^\top \cdot {\sigma_j}^2\boldsymbol{v}_j = \begin{cases} \sigma_i & (i = j), \\ 0 & (i \neq j). \end{cases}$$

- If $r < i \leqq m$, $1 \leqq j \leqq r$, we have $\boldsymbol{u}_i^\top A\boldsymbol{v}_j = \boldsymbol{u}_i^\top \cdot \sigma_j \boldsymbol{u}_j = \sigma_j \boldsymbol{u}_i^\top \boldsymbol{u}_j = 0$.
- If $1 \leqq i \leqq m$, $r < j \leqq n$, we have $A^\top A\boldsymbol{v}_j = \boldsymbol{0}$, which implies $A\boldsymbol{v}_j = \boldsymbol{0}$ since $\mathrm{Ker}(A^\top A) = \mathrm{Ker}\, A$ by Theorem 9.71(1) in Sec. 9.8.1. Therefore, $\boldsymbol{u}_i^\top A\boldsymbol{v}_j = 0$.

Thus we have shown $U^\top AV = \Sigma$, which is equivalent to $A = U\,\Sigma\,V^\top$ in (8.1). □

We refer to (8.1) as the *singular value decomposition* of A, and to $\sigma_i = \sigma_i(A)$ $(i = 1, 2, \ldots, r)$ as the *singular values*[1] of A. The matrices U and V in the decomposition (8.1) are not unique, but the singular values are uniquely determined, as is seen from Theorem 8.2 below.

Singular values are closely related to eigenvalues.

Theorem 8.2.

(1) *The singular values of a matrix A coincide with the positive square roots of the nonzero eigenvalues of $A^\top A$, and the column vectors of the matrix $V = [\boldsymbol{v}_1, \boldsymbol{v}_2, \ldots, \boldsymbol{v}_n]$ in (8.1) are the eigenvectors of $A^\top A$. That is,*

$$(A^\top A)\boldsymbol{v}_j = \begin{cases} {\sigma_j}^2\boldsymbol{v}_j & (1 \leqq j \leqq r), \\ \boldsymbol{0} & (r < j \leqq n). \end{cases} \tag{8.3}$$

[1]In this book singular values are defined to be positive diagonal elements of Σ, which implies that the number of singular values varies with the rank of the matrix. In some books, however, a different definition is adopted so that the number of singular values is determined by the type (size) of the matrix. The alternative definition introduces $\sigma_i = 0$ for i satisfying $r < i \leqq \min(m, n)$ and call σ_i $(i = 1, 2, \ldots, \min(m, n))$ the singular values of A.

(2) *The singular values of a matrix A coincide with the positive square roots of the nonzero eigenvalues of $AA^\top$ and the column vectors of the matrix $U = [\boldsymbol{u}_1, \boldsymbol{u}_2, \ldots, \boldsymbol{u}_m]$ in (8.1) are the eigenvectors of $AA^\top$. That is,*

$$(AA^\top)\boldsymbol{u}_i = \begin{cases} {\sigma_i}^2 \boldsymbol{u}_i & (1 \leqq i \leqq r), \\ \mathbf{0} & (r < i \leqq m). \end{cases} \tag{8.4}$$

Proof. (1) It follows from (8.1) that

$$A^\top A = (U\Sigma V^\top)^\top (U\Sigma V^\top) = V\Sigma^\top \Sigma V^\top.$$

This implies $(A^\top A)V = V(\Sigma^\top \Sigma)$, which is equivalent to (8.3).

(2) Similarly we obtain

$$AA^\top = (U\Sigma V^\top)(U\Sigma V^\top)^\top = U\Sigma \Sigma^\top U^\top.$$

This implies $(AA^\top)U = U(\Sigma \Sigma^\top)$, which is equivalent to (8.4). □

Theorem 8.3. *The singular values of a positive semidefinite symmetric matrix A coincide with its positive eigenvalues.*

Proof. By Theorem 6.9 in Sec. 6.3.1 we have a decomposition $A = Q \cdot \mathrm{diag}\,(\lambda_1, \lambda_2, \ldots, \lambda_n) \cdot Q^\top$ with an orthogonal matrix Q, where λ_i are all nonnegative by the assumed semidefiniteness of A. Hence this decomposition is qualified as the decomposition in (8.1). □

The singular value decomposition has a natural geometric meaning. An $m \times n$ real matrix A can be regarded as a representation of a linear mapping T from $\mathbb{R}^n$ to $\mathbb{R}^m$ (cf., Sec. 9.5.3). Suppose that the spaces $\mathbb{R}^n$ and $\mathbb{R}^m$ are equipped, respectively, with standard inner products (cf., Example 9.39 in Sec. 9.6.2). Then it is natural to consider an orthogonal transformation for basis changes in $\mathbb{R}^n$ and $\mathbb{R}^m$ (cf., Theorem 9.39 in Sec. 9.6.3). If V and U are orthogonal matrices representing basis changes in $\mathbb{R}^n$ and $\mathbb{R}^m$, respectively, the matrix representing linear mapping T changes from A to $\tilde{A} = U^\top AV$. On recalling (8.1) in an equivalent form

$$U^\top AV = \Sigma, \tag{8.5}$$

we see that the singular value decomposition aims at desirable basis changes in the sense that the transformed matrix representation $\tilde{A} = U^\top AV$ takes on a simple form given by Σ. The basis vector $\boldsymbol{v}_j$ is mapped by T to σ_j times the basis vector $\boldsymbol{u}_j$ if $1 \leqq j \leqq r$, and to $\mathbf{0}$ if $r < j \leqq n$.

8.1.2 *Complex Matrices*

Singular values can also be defined for complex matrices.

Theorem 8.4. *For any $m \times n$ complex matrix A of rank r, there exist a unitary matrix U of order m and a unitary matrix V of order n such that*

$$A = U\ \Sigma\ V^*, \tag{8.6}$$

where Σ is an $m \times n$ matrix of the form of (8.2).

We refer to (8.6) as the *singular value decomposition* of a complex matrix A, and to $\sigma_i = \sigma_i(A)$ $(i = 1, 2, \ldots, r)$ as the *singular values* of A.

Singular values are closely related to eigenvalues as follows.

Theorem 8.5.

(1) *The singular values of a complex matrix A coincide with the positive square roots of the nonzero eigenvalues of A^*A, and the column vectors of the matrix $V = [\boldsymbol{v}_1, \boldsymbol{v}_2, \ldots, \boldsymbol{v}_n]$ in* (8.6) *are the eigenvectors of A^*A. That is,*

$$(A^*A)\boldsymbol{v}_j = \begin{cases} {\sigma_j}^2 \boldsymbol{v}_j & (1 \leqq j \leqq r), \\ \boldsymbol{0} & (r < j \leqq n). \end{cases} \tag{8.7}$$

(2) *The singular values of a complex matrix A coincide with the positive square roots of the nonzero eigenvalues of AA^* and the column vectors of the matrix $U = [\boldsymbol{u}_1, \boldsymbol{u}_2, \ldots, \boldsymbol{u}_m]$ in* (8.6) *are the eigenvectors of AA^*. That is,*

$$(AA^*)\boldsymbol{u}_i = \begin{cases} {\sigma_i}^2 \boldsymbol{u}_i & (1 \leqq i \leqq r), \\ \boldsymbol{0} & (r < i \leqq m). \end{cases} \tag{8.8}$$

Theorem 8.6.

(1) *The singular values of a normal matrix coincide with the absolute values of its nonzero eigenvalues.*
(2) *The singular values of a positive semidefinite Hermitian matrix coincide with its positive eigenvalues.*

Proof. (1) A normal matrix A of order n is diagonalized by a unitary matrix U as

$$U^*AU = \mathrm{diag}\,(\lambda_1, \lambda_2, \ldots, \lambda_n) \qquad (|\lambda_1| \geqq |\lambda_2| \geqq \cdots \geqq |\lambda_n|)$$

(cf., Theorem 6.29 in Sec. 6.6). This implies

$$U^*(AA^*)U = \mathrm{diag}\,(|\lambda_1|^2, |\lambda_2|^2, \ldots, |\lambda_n|^2),$$

which shows, by Theorem 8.5(2), that ${\sigma_i}^2 = |\lambda_i|^2$ for $i = 1, 2, \ldots, r$.

(2) A Hermitian matrix is a normal matrix (cf., Example 6.4 in Sec. 6.6). For a positive semidefinite Hermitian matrix, we have $\lambda_i \geqq 0$ and hence $\sigma_i = \lambda_i$ for $i = 1, 2, \ldots, r$. □

8.2 Properties of Singular Values

In this section we present fundamental properties of singular values of an $m \times n$ complex matrix $A = (a_{ij})$. First we discuss the relations between singular values and norms. Recall that $\|A\|_2$ denotes the 2-norm of A and $\|A\|_{\mathrm{F}}$ the Frobenius norm of A, defined respectively by (1.24) and (1.25) in Sec. 1.6.

Theorem 8.7. *Let $\sigma_1 \geqq \sigma_2 \geqq \cdots \geqq \sigma_r > 0$ be the singular values of a complex matrix A. Then*

$$\|A\|_2 = \sigma_1, \tag{8.9}$$

$$\|A\|_{\mathrm{F}} = \left(\sum_{i=1}^{r} {\sigma_i}^2 \right)^{1/2}. \tag{8.10}$$

Proof. We use the singular value decomposition $A = U\Sigma V^*$ in (8.6) with unitary matrices U and V. For any vector $\boldsymbol{x}$ with $\|\boldsymbol{x}\|_2 = 1$, let $\boldsymbol{y} = V^*\boldsymbol{x}$. We have $\|\boldsymbol{y}\|_2 = 1$ and

$$\|A\boldsymbol{x}\|_2{}^2 = \boldsymbol{x}^* A^* A \boldsymbol{x} = \boldsymbol{y}^* \Sigma^* \Sigma \boldsymbol{y} = \sum_{j=1}^{r} {\sigma_j}^2 |y_j|^2 \leqq {\sigma_1}^2 \sum_{j=1}^{r} |y_j|^2 \leqq {\sigma_1}^2,$$

where the inequality holds with equality if $\boldsymbol{y} = (1, 0, \ldots, 0)^\top$, which means that $\boldsymbol{x}$ is the first column vector of V. Therefore the maximum of $\|A\boldsymbol{x}\|_2$ over all $\boldsymbol{x}$ with $\|\boldsymbol{x}\|_2 = 1$ is equal to σ_1, whereas the former is equal to $\|A\|_2$ by the definition (1.24) in Sec. 1.6. Thus (8.9) is proved.

The second identity (8.10) is shown as follows:

$$\|A\|_{\mathrm{F}}{}^2 = \mathrm{Tr}\,(A^* A) = \mathrm{Tr}\,(V \Sigma^* \Sigma V^*) = \mathrm{Tr}\,(\Sigma^* \Sigma) = \sum_{i=1}^{r} {\sigma_i}^2,$$

where the first equality is due to Theorem 1.13(2) in Sec. 1.6 and the third equality is to Theorem 1.9(5) in Sec. 1.5. □

The following is a minimax theorem for singular values, corresponding to the Courant–Fischer theorem for eigenvalues given as Theorem 6.22 in Sec. 6.4.2. The special cases of (8.11) and (8.12) with $k = 1$ reduce to (8.9).

Theorem 8.8. *For the singular values $\sigma_1 \geqq \sigma_2 \geqq \cdots \geqq \sigma_r > 0$ of an $m \times n$ complex matrix A, we have the following formulas:*

$$\sigma_k = \max_{S:\, \dim S = k} \ \min_{\boldsymbol{x} \in S \setminus \{\boldsymbol{0}\}} \frac{\|A\boldsymbol{x}\|_2}{\|\boldsymbol{x}\|_2} \qquad (k = 1, 2, \ldots, r), \tag{8.11}$$

$$\sigma_k = \min_{S:\, \dim S = n-k+1} \ \max_{\boldsymbol{x} \in S \setminus \{\boldsymbol{0}\}} \frac{\|A\boldsymbol{x}\|_2}{\|\boldsymbol{x}\|_2} \qquad (k = 1, 2, \ldots, r), \tag{8.12}$$

where S ($\subseteq \mathbb{C}^n$) runs over all subspaces of the specified dimension.

Proof. We have

$$\frac{\|A\boldsymbol{x}\|_2}{\|\boldsymbol{x}\|_2} = \left(\frac{\boldsymbol{x}^* A^* A \boldsymbol{x}}{\boldsymbol{x}^* \boldsymbol{x}} \right)^{1/2}.$$

By Theorem 8.5(1), σ_k is the positive square root of the kth largest eigenvalue of A^*A. The expressions (6.46) and (6.47) in Theorem 6.22 (Sec. 6.4.2), applied to A^*A, imply (8.11) and (8.12), respectively. □

The following is a perturbation theorem for singular values, corresponding to Theorem 6.25 in Sec. 6.4.3 for eigenvalues.

Theorem 8.9. *For $m \times n$ complex matrices A and B, we have*

$$\sigma_k(A + B) \leqq \min_{1 \leqq j \leqq k} (\sigma_j(A) + \sigma_{k-j+1}(B)) \quad (k = 1, 2, \ldots, \operatorname{rank}(A + B)), \tag{8.13}$$

where we adopt the convention that $\sigma_i(A) = 0$ if $i > \operatorname{rank} A$ and $\sigma_i(B) = 0$ if $i > \operatorname{rank} B$.

Proof. Let $C = A + B$ and assume $1 \leqq j \leqq k \leqq \operatorname{rank}(A + B)$. Using notations in Proposition 6.2 in Sec. 6.3.2, let U_k^C be the subspace spanned by k eigenvectors of C^*C corresponding to the k largest eigenvalues. Similarly, let W_j^A and W_{k-j+1}^B be the subspaces defined from A^*A and B^*B, respectively, using their eigenvectors corresponding to the smaller eigenvalues $\lambda_i(A^*A)$ ($j \leqq i \leqq n$) and $\lambda_i(B^*B)$ ($k - j + 1 \leqq i \leqq n$). We have

$$\dim U_k^C = k, \quad \dim W_j^A = n - j + 1, \quad \dim W_{k-j+1}^B = n - k + j.$$

By the inequality in Remark 9.11 in Sec. 9.4.3 we have

$$\dim(U_k^C \cap W_j^A \cap W_{k-j+1}^B) \geqq k + (n - j + 1) + (n - k + j) - 2n = 1,$$

which implies that $U_k^C \cap W_j^A \cap W_{k-j+1}^B$ contains a unit-length vector, say, $\boldsymbol{x}$. Then it follows from Proposition 6.2 that

$$\begin{aligned}
\sigma_k(C) &= \sqrt{\lambda_k(C^*C)} \leqq \sqrt{\boldsymbol{x}^* C^* C \boldsymbol{x}} = \|C\boldsymbol{x}\|_2 = \|A\boldsymbol{x} + B\boldsymbol{x}\|_2 \\
&\leqq \|A\boldsymbol{x}\|_2 + \|B\boldsymbol{x}\|_2 = \sqrt{\boldsymbol{x}^* A^* A \boldsymbol{x}} + \sqrt{\boldsymbol{x}^* B^* B \boldsymbol{x}} \\
&\leqq \sqrt{\lambda_j(A^*A)} + \sqrt{\lambda_{k-j+1}(B^*B)} = \sigma_j(A) + \sigma_{k-j+1}(B),
\end{aligned}$$

which implies (8.13). □

The following two theorems indicate that singular values are larger than eigenvalues in some sense.

Theorem 8.10. *The absolute value of any eigenvalue of a square complex matrix is bounded by its largest singular value. In other words,*

$$\rho(A) \leqq \sigma_1(A), \tag{8.14}$$

where $\rho(A)$ denotes the spectral radius of a square complex matrix A.

Proof. It follows from (6.12) in Theorem 6.4 in Sec. 6.1.3 and (8.9) in Theorem 8.7 that $\rho(A) \leqq \|A\|_2 = \sigma_1(A)$. □

This theorem can be generalized as follows.[2]

Theorem 8.11. *Let A be an $n \times n$ complex matrix of rank r, with singular values $\sigma_1 \geqq \sigma_2 \geqq \cdots \geqq \sigma_r$ and nonzero eigenvalues $\lambda_1, \lambda_2, \ldots, \lambda_r$ where $|\lambda_1| \geqq |\lambda_2| \geqq \cdots \geqq |\lambda_r|$. Then we have*

$$\sigma_1\sigma_2\cdots\sigma_k \geqq |\lambda_1|\,|\lambda_2|\cdots|\lambda_k| \quad (k = 1, 2, \ldots, r) \tag{8.15}$$

and, for any $p > 0$, we have

$$\sigma_1{}^p + \sigma_2{}^p + \cdots + \sigma_k{}^p \geqq |\lambda_1|^p + |\lambda_2|^p + \cdots + |\lambda_k|^p \quad (k = 1, 2, \ldots, r). \tag{8.16}$$

The special case of (8.15) with $k = 1$ reduces to (8.14). The special case of (8.16) with $p = 2$ and $k = r$ is equivalent to the inequality $\|A\|_{\mathrm{F}}{}^2 \geqq \sum_{i=1}^{n} |\lambda_i|^2$ given in Theorem 6.28 in Sec. 6.5, since $\|A\|_{\mathrm{F}}{}^2 = \sum_{i=1}^{r} \sigma_i{}^2$ by (8.10) in Theorem 8.7 and $\lambda_i = 0$ for $i = r+1, \ldots, n$.

Example 8.1. Theorem 8.11 is illustrated for a matrix $A = \begin{bmatrix} a & 1 \\ 0 & a \end{bmatrix}$ containing a complex-valued parameter a. The eigenvalues are $\lambda_1 = \lambda_2 = a$ (with multiplicity 2) and the spectral radius $\rho(A)$ is equal to $|a|$. We have

$$A^*A = \begin{bmatrix} |a|^2 & \overline{a} \\ a & |a|^2 + 1 \end{bmatrix}, \qquad AA^* = \begin{bmatrix} |a|^2 + 1 & \overline{a} \\ a & |a|^2 \end{bmatrix},$$

which show that A is not a normal matrix. By calculating the eigenvalues of A^*A (or AA^*), we obtain the following singular values of A:

$$\sigma_1(A) = \left(|a|^2 + 1/2 + \left(|a|^2 + 1/4\right)^{1/2}\right)^{1/2},$$

$$\sigma_2(A) = \left(|a|^2 + 1/2 - \left(|a|^2 + 1/4\right)^{1/2}\right)^{1/2}.$$

---

[2]Theorem 8.11 is due to H. Weyl ("Inequalities between the two kinds of eigenvalues of a linear transformation," *Proceedings of the National Academy of Sciences of the United States of America*, **35** (1949), 408–411). The proof of this theorem can also be found in [15, 34, 35].

The equality (8.15) for $k = 1, 2$ shows

$$\sigma_1(A) > |a| = |\lambda_1|, \qquad \sigma_1(A)\sigma_2(A) = |a|^2 = |\lambda_1| \cdot |\lambda_2|.$$

For the second singular value, we have $\sigma_2(A) < |a| = |\lambda_2| = \rho(A)$. ■

Theorem 8.12. *The singular values of the Kronecker product $A \otimes B$ of complex matrices A and B are the products of a singular value of A and a singular value of B. To be more specific, let $\sigma_1, \sigma_2, \ldots, \sigma_r$ and $\tau_1, \tau_2, \ldots, \tau_s$ be the singular values of A and B, respectively. Then the singular values of $A \otimes B$ are given precisely by $\sigma_i \tau_j$ $(i = 1, 2, \ldots, r; j = 1, 2, \ldots, s)$.*

Proof. Let $A = U\Sigma V^*$ and $B = X\Gamma Y^*$ be the singular value decompositions, where U, V, X, and Y are unitary matrices and

$$\Sigma = \begin{bmatrix} \mathrm{diag}\,(\sigma_1, \sigma_2, \ldots, \sigma_r) & O \\ O & O \end{bmatrix}, \quad \Gamma = \begin{bmatrix} \mathrm{diag}\,(\tau_1, \tau_2, \ldots, \tau_s) & O \\ O & O \end{bmatrix}.$$

Then we have $A \otimes B = (U \otimes X)(\Sigma \otimes \Gamma)(V \otimes Y)^*$, in which $U \otimes X$ and $V \otimes Y$ are unitary and the diagonal entries of $\Sigma \otimes \Gamma$ are equal to $\sigma_i \tau_j$ $(i = 1, 2, \ldots, r; j = 1, 2, \ldots, s)$. □

8.3 Method of Least Squares

In this section we consider the problem of finding a vector $\boldsymbol{x}$ that minimizes

$$\|A\boldsymbol{x} - \boldsymbol{b}\|_2{}^2 = (A\boldsymbol{x} - \boldsymbol{b})^\top (A\boldsymbol{x} - \boldsymbol{b}) \tag{8.17}$$

for a given real matrix A and a real vector $\boldsymbol{b}$, where A is not necessarily a square matrix.[3] Such a problem is called a *least-square problem*, and arises, for example, in the *method of least squares* in statistical data analysis [36,38,40].

Let A be an $m \times n$ matrix, which implies that $\boldsymbol{b}$ is an m-dimensional vector and $\boldsymbol{x}$ is an n-dimensional vector. In applications we usually have $m > n$, that is, A has more rows than columns, but this is not assumed here. Indeed the argument below works also when $m \leqq n$.

A solution $\boldsymbol{x}$ to the least-square problem is characterized as a solution to a system of linear equations

$$A^\top A\boldsymbol{x} = A^\top \boldsymbol{b}. \tag{8.18}$$

[3]We assume here that A and $\boldsymbol{b}$ are real, but a similar problem can be formulated for complex data. The least-square problem for complex A and $\boldsymbol{b}$ asks for a complex vector $\boldsymbol{x}$ that minimizes $\|A\boldsymbol{x} - \boldsymbol{b}\|_2{}^2 = (A\boldsymbol{x} - \boldsymbol{b})^*(A\boldsymbol{x} - \boldsymbol{b})$. The argument of this section can be adapted to the complex case with minor modifications.

This equation is called the *normal equation.*

Theorem 8.13. *A vector $\boldsymbol{x}$ is a solution to the least-square problem if and only if it satisfies the normal equation* (8.18).

Proof. Fix a vector $\hat{\boldsymbol{x}} \in \mathbb{R}^n$. We derive a condition that is necessary and sufficient for the inequality

$$\|A\hat{\boldsymbol{x}} - \boldsymbol{b}\|_2{}^2 \leqq \|A\boldsymbol{x} - \boldsymbol{b}\|_2{}^2 \tag{8.19}$$

to hold for all $\boldsymbol{x} \in \mathbb{R}^n$. On expressing $\boldsymbol{x} = \boldsymbol{w} + \hat{\boldsymbol{x}}$ with a new variable $\boldsymbol{w}$, we can rewrite this right-hand side as

$$\|A\boldsymbol{x} - \boldsymbol{b}\|_2{}^2 = \|A\hat{\boldsymbol{x}} - \boldsymbol{b}\|_2{}^2 + 2\boldsymbol{w}^\top A^\top(A\hat{\boldsymbol{x}} - \boldsymbol{b}) + \boldsymbol{w}^\top A^\top A\boldsymbol{w}.$$

Hence the condition (8.19) is equivalent to requiring

$$0 \leqq 2\boldsymbol{w}^\top A^\top(A\hat{\boldsymbol{x}} - \boldsymbol{b}) + \boldsymbol{w}^\top A^\top A\boldsymbol{w} \tag{8.20}$$

for all $\boldsymbol{w} \in \mathbb{R}^n$. Here the quadratic term $\boldsymbol{w}^\top A^\top A\boldsymbol{w}$ is nonnegative, since $\boldsymbol{w}^\top A^\top A\boldsymbol{w} = (A\boldsymbol{w})^\top A\boldsymbol{w} \geqq 0$. Therefore, (8.20) holds for all $\boldsymbol{w}$ if and only if the linear term in $\boldsymbol{w}$ vanishes, which is equivalent to

$$A^\top(A\hat{\boldsymbol{x}} - \boldsymbol{b}) = \boldsymbol{0}. \tag{8.21}$$

This equation gives the normal equation (8.18) for $\hat{\boldsymbol{x}}$. □

The normal equation (8.18) always has a solution $\boldsymbol{x}$, which can be shown as follows. Obviously, we have $A^\top\boldsymbol{b} \in \mathrm{Im}(A^\top)$, whereas $\mathrm{Im}(A^\top) = \mathrm{Im}(A^\top A)$ by Theorem 9.71(2) in Sec. 9.8.1. Therefore, $A^\top\boldsymbol{b} \in \mathrm{Im}(A^\top A)$. Then the solvability of (8.18) follows from Theorem 5.2 in Sec. 5.1.

If the matrix A has linearly independent column vectors (that is, A is of full column rank, or $\mathrm{rank}\, A = n$), the matrix $A^\top A$ is nonsingular, and hence the normal equation has a unique solution

$$\boldsymbol{x} = (A^\top A)^{-1}A^\top\boldsymbol{b}. \tag{8.22}$$

In the general case where the column vectors of a matrix A may possibly be linearly dependent, the singular value decomposition offers an effective means to the solution of the normal equation. It follows from the singular value decomposition $A = U\Sigma V^\top$ in (8.1) that

$$A^\top A = V\Sigma^\top U^\top \cdot U\Sigma V^\top = V \cdot \Sigma^\top\Sigma V^\top, \qquad A^\top\boldsymbol{b} = V \cdot \Sigma^\top U^\top\boldsymbol{b}.$$

Since V is nonsingular, the normal equation $A^\top A\boldsymbol{x} = A^\top\boldsymbol{b}$ can be rewritten as

$$\Sigma^\top\Sigma\, V^\top\boldsymbol{x} = \Sigma^\top\, U^\top\, \boldsymbol{b}. \tag{8.23}$$

By defining $\boldsymbol{c} = U^\top \boldsymbol{b}$ and $\boldsymbol{y} = V^\top \boldsymbol{x}$, we can express this equation (8.23) as

$$\left[\begin{array}{ccc|c} {\sigma_1}^2 & & & \\ & \ddots & & O_{r,n-r} \\ & & {\sigma_r}^2 & \\ \hline & O_{n-r,r} & & O_{n-r,n-r} \end{array}\right] \left[\begin{array}{c} y_1 \\ \vdots \\ y_r \\ \hline y_{r+1} \\ \vdots \\ y_n \end{array}\right] = \left[\begin{array}{ccc|c} \sigma_1 & & & \\ & \ddots & & O_{r,m-r} \\ & & \sigma_r & \\ \hline & O_{n-r,r} & & O_{n-r,m-r} \end{array}\right] \left[\begin{array}{c} c_1 \\ \vdots \\ c_r \\ \hline c_{r+1} \\ \vdots \\ c_m \end{array}\right].$$

Therefore, the set of the solutions of (8.23), that is, the set of the solutions of the normal equation (8.18), is described as

$$\{\boldsymbol{x} = V\boldsymbol{y} \mid y_j = c_j/\sigma_j \ (1 \leqq j \leqq r);\ y_{r+1}, y_{r+2}, \ldots, y_n \text{ are arbitrary}\}. \tag{8.24}$$

When the solution $\boldsymbol{x}$ is not unique, it is customary to adopt the solution with minimum Euclidean norm $\|\boldsymbol{x}\|_2$, which is called the *minimum-norm solution.* Since $\|\boldsymbol{x}\|_2 = \|V\boldsymbol{x}\|_2 = \|\boldsymbol{y}\|_2$ by the orthogonality of matrix V, the minimum-norm solution $\boldsymbol{x}$ corresponds to a solution $\boldsymbol{y}$ with $y_j = 0$ $(r < j \leqq n)$ in (8.24). Therefore, the minimum-norm solution $\boldsymbol{x}$ is given by

$$\boldsymbol{x} = V\,\Sigma^+\,U^\top\,\boldsymbol{b}, \tag{8.25}$$

where

$$\Sigma^+ = \begin{bmatrix} \operatorname{diag}(1/\sigma_1, 1/\sigma_2, \ldots, 1/\sigma_r) & O_{r,m-r} \\ O_{n-r,r} & O_{n-r,m-r} \end{bmatrix}.$$

The matrix $A^+ = V\,\Sigma^+\,U^\top$ appearing in (8.25) is known as the *Moore–Penrose generalized inverse* of matrix A (see Chapter 6 of Volume II [9]).

Remark 8.1. In choosing a numerical method for solving the least-square problem, we should take into consideration both accuracy of the computed solution and amount of the computation. The method based on the singular value decomposition is most reliable with respect to numerical accuracy, especially when the column vectors of A are (close to be) linearly dependent. In its practical use we usually incorporate an additional technique to cope with numerical errors. While the formula (8.25) involves the rank r of matrix A, it is practically impossible to numerically determine r in the presence of rounding errors. In addition, division by too small a singular value σ_i is not desirable because it is likely to magnify rounding errors. Therefore, it is customary in practice to regard a singular values σ_i as being zero if it is too small compared with the largest singular value σ_1.

In so doing we can determine an "effective rank" and obtain a solution $\boldsymbol{x}$ that is more stable and reliable with respect to numerical accuracy. There is another reliable numerical method based on the QR decomposition [36–38, 43]. Finally, it is mentioned that the method of first computing the explicit forms of $A^\top A$ and $A^\top \boldsymbol{b}$ and then solving the normal equation (8.18) by the Gaussian method may be advantageous with respect to amount of computation, but it often suffers from numerical inaccuracy. ■

Chapter 9

Vector Spaces

In this chapter we present fundamental concepts and standard results concerning vector spaces such as subspace, linear independence, basis, linear mapping, inner product, and dual space. The rank of a matrix is considered again from the geometrical view of vector spaces and the concept of matroid is introduced as a combinatorial essence inherent in linear independence. This chapter, dealing with the general mathematical theory including infinite dimensional spaces, is (partly) more sophisticated than other chapters. The average reader, who is mainly interested in engineering applications of linear algebra, is expected to grasp a rough idea of the mathematical arguments without entering into technical details. Theorems are most often stated without proofs, and the reader is referred to other textbooks for mathematical details.

9.1 Vector Spaces

9.1.1 *Definition*

Definition 9.1. Let K be a field.[1] A nonempty set V is called a *vector space* (or *linear space*) over K if it is equipped with two operations, called vector addition and scalar multiplication, satisfying the following conditions. Here K is called the *field of scalars*, an element of K is a *scalar*, and an element of V is a *vector*.

[I] (*Vector addition*) For any two elements $\boldsymbol{x}$ and $\boldsymbol{y}$ of V, an element of

[1] We have assumed $K = \mathbb{R}$ or $K = \mathbb{C}$ till Chapter 8. In this chapter, K will denote a general field, excepting Secs. 9.6 and 9.8. A field means a set of numbers in which addition, subtraction, multiplication, and divisions are defined (see Remark 9.2).

V, denoted by $\boldsymbol{x} + \boldsymbol{y}$, is defined as their *sum*, and this operation has the following properties.

(1) $(\boldsymbol{x} + \boldsymbol{y}) + \boldsymbol{z} = \boldsymbol{x} + (\boldsymbol{y} + \boldsymbol{z})$ for all $\boldsymbol{x}, \boldsymbol{y}, \boldsymbol{z} \in V$.
(2) $\boldsymbol{x} + \boldsymbol{y} = \boldsymbol{y} + \boldsymbol{x}$ for all $\boldsymbol{x}, \boldsymbol{y} \in V$.
(3) There exists an element of V, denoted by $\mathbf{0}$, such that $\boldsymbol{x} + \mathbf{0} = \boldsymbol{x}$ holds for all $\boldsymbol{x} \in V$. (This element $\mathbf{0}$ is called the *zero vector*.)
(4) For each $\boldsymbol{x} \in V$, there exists an element of V, denoted by $-\boldsymbol{x}$, that satisfies $\boldsymbol{x} + (-\boldsymbol{x}) = \mathbf{0}$. (This element $-\boldsymbol{x}$ is called the *inverse vector* of $\boldsymbol{x}$.)

[II] (*Scalar multiplication*) For any element $\boldsymbol{x}$ of V and any scalar a in K, an element of V, denoted by $a\boldsymbol{x}$, is defined as the *scalar multiple* of $\boldsymbol{x}$, and this operation has the following properties.

(1) $(a + b)\boldsymbol{x} = a\boldsymbol{x} + b\boldsymbol{x}$ for all $a, b \in K$ and $\boldsymbol{x} \in V$.
(2) $a(\boldsymbol{x} + \boldsymbol{y}) = a\boldsymbol{x} + a\boldsymbol{y}$ for all $a \in K$ and $\boldsymbol{x}, \boldsymbol{y} \in V$.
(3) $(ab)\boldsymbol{x} = a(b\boldsymbol{x})$ for all $a, b \in K$ and $\boldsymbol{x} \in V$.
(4) $1\,\boldsymbol{x} = \boldsymbol{x}$ for all $\boldsymbol{x} \in V$, where 1 denotes the unit element of K.

If $K = \mathbb{R}$, V is called a *real vector space*. It is called a *complex vector space* if $K = \mathbb{C}$. ■

Remark 9.1. The conditions in Definition 9.1 guarantee that the operations valid for ordinary vectors (in the Euclidean geometry) are allowed in a general vector space V. In addition, we have the following.

- The zero vector is unique.
- For any vector $\boldsymbol{x}$, its inverse vector $-\boldsymbol{x}$ is unique.
- As a consequence of condition (I-1), the sum of vectors is not affected by the order of making their sum. Therefore, we may write $\boldsymbol{x}+\boldsymbol{y}+\boldsymbol{z}$ by omitting parentheses in $(\boldsymbol{x}+\boldsymbol{y})+\boldsymbol{z}$. Similarly, we may write $\boldsymbol{x}+\boldsymbol{y}+\boldsymbol{z}+\boldsymbol{u}$ for $(\boldsymbol{x} + \boldsymbol{y}) + (\boldsymbol{z} + \boldsymbol{u})$.
- $0\,\boldsymbol{x} = \mathbf{0}, \quad a\,\mathbf{0} = \mathbf{0}, \quad (-1)\,\boldsymbol{x} = -\boldsymbol{x}$. ■

Remark 9.2. A nonempty set K is called a *field* if it is equipped with two operations, called addition and multiplication, satisfying the conditions [I] to [IV] below. For any two elements a and b of K, the outcome of addition operation is an element of K, denoted by $a + b$, which is called the *sum* of a and b. The outcome of multiplication operation is an element of K, denoted by $a \cdot b$, which is called the *product* of a and b.

[I] (Conditions for addition)

(1) $(a+b)+c = a+(b+c)$ for all $a, b, c \in K$.
(2) $a+b = b+a$ for all $a, b \in K$.
(3) There exists an element of K, denoted by 0, such that $a + 0 = a$ holds for all $a \in K$. (This element 0 is called the *zero element.*)
(4) For each $a \in K$, there exists an element of K, denoted by $-a$, that satisfies $a + (-a) = 0$.

[II] (Conditions for multiplication)

(1) $(a \cdot b) \cdot c = a \cdot (b \cdot c)$ for all $a, b, c \in K$.
(2) There exists an element of K, denoted by 1, such that $1 \cdot a = a \cdot 1 = a$ holds for all $a \in K$. (This element 1 is called the *unit element* or *multiplicative identity.*)
(3) $a \cdot b = b \cdot a$ for all $a, b \in K$.
(4) For each $a \in K$ with $a \neq 0$, there exists an element of K, denoted by a^{-1}, that satisfies $a \cdot a^{-1} = a^{-1} \cdot a = 1$. (This element a^{-1} is called the *inverse element* of a.)

[III] (Conditions for combination of addition and multiplication)

(1) $a \cdot (b+c) = a \cdot b + a \cdot c$ for all $a, b, c \in K$.
(2) $(a+b) \cdot c = a \cdot c + b \cdot c$ for all $a, b, c \in K$.

[IV] The zero element and the unit element are distinct, that is, $0 \neq 1$.

If K satisfies all conditions other than the *commutative law* of product in (II-3) and the existence of the inverse element in (II-4), then K is called a *ring.* That is, a ring is an algebraic system satisfying [I], [III], [IV], (II-1), and (II-2). A ring is called a *commutative ring* if it satisfies the commutative law (II-3); otherwise it is called a *noncommutative ring.* In a ring we may have $a \cdot b = 0$ for $a \neq 0$ and $b \neq 0$; such elements a, b are called a *zero divisor.* Some books adopt the convention that the condition (II-2) is not imposed in the definition of a ring, and a ring satisfying (II-2) is called a *ring with unity* or *ring with identity.* ■

9.1.2 *Examples*

We show examples of vector spaces.

Example 9.1. The set $V = \{\mathbf{0}\}$ consisting of the zero vector $\mathbf{0}$ alone is a vector space. The sum is given by $\mathbf{0} + \mathbf{0} = \mathbf{0}$, and the scalar multiple is given by $a\,\mathbf{0} = \mathbf{0}$ for $a \in K$. ■

Example 9.2. The set of all geometric vectors in a plane (or space) is a vector space. The sum and scalar multiple are given by the (usual) sum of vectors and a multiple of a vector with a real number. ■

Example 9.3. The set K^n of all column vectors with n components is a vector space, which is sometimes called a space of numerical vectors. The sum and scalar multiple are given by the (usual) vector sum and scalar multiple. ■

Example 9.4. The set $\mathcal{M}_{m,n}(K)$ of all $m \times n$ matrices over K is a vector space. The sum and scalar multiple are given by the sum of matrices and scalar multiple of a matrix. ■

Example 9.5. The set S of all sequences $\{x_n\}_{n=0}^{\infty}$ consisting of elements of K is a vector space:

$$\text{sum:} \qquad \{x_n\}_{n=0}^{\infty} + \{y_n\}_{n=0}^{\infty} = \{x_n + y_n\}_{n=0}^{\infty},$$
$$\text{scalar multiple:} \qquad a\{x_n\}_{n=0}^{\infty} = \{ax_n\}_{n=0}^{\infty} \qquad (a \in K).$$

Moreover, if $K = \mathbb{R}$, the set ℓ^{∞} of all bounded real sequences $\{x_n\}_{n=0}^{\infty}$ is a vector space. For $p \geqq 1$, a real sequence $\{x_n\}_{n=0}^{\infty}$ is called *p-summable* if $\sum_{n=0}^{\infty} |x_n|^p$ is finite, and the set ℓ^p of all p-summable real sequences is a vector space. Indeed, if $\sum |x_n|^p < \infty$ and $\sum |y_n|^p < \infty$, then we have $\sum |x_n + y_n|^p < \infty$ from *Minkowski's inequality*:

$$\left(\sum_{n=0}^{\infty} |x_n + y_n|^p\right)^{1/p} \leqq \left(\sum_{n=0}^{\infty} |x_n|^p\right)^{1/p} + \left(\sum_{n=0}^{\infty} |y_n|^p\right)^{1/p},$$

showing that $\{x_n + y_n\}_{n=0}^{\infty}$ belongs to ℓ^p. We can similarly show that the scalar multiple $a\{x_n\}_{n=0}^{\infty}$ belongs to ℓ^p. ■

Example 9.6. The set of all mappings from a set D to $\mathbb{R}$ is a vector space over $\mathbb{R}$:

$$\text{sum:} \qquad (f+g)(x) = f(x) + g(x),$$
$$\text{scalar multiple:} \qquad (af)(x) = af(x) \qquad (a \in \mathbb{R}).$$

If D is an interval I of real numbers, the set $\mathrm{C}^0(I, \mathbb{R})$ of all continuous functions on I and the set $\mathrm{C}^m(I, \mathbb{R})$ of all m times continuously differentiable functions (functions that are differentiable m times with continuous mth order derivatives) on I are a vector space over $\mathbb{R}$ with the sum and scalar multiple operations above. ■

Example 9.7. The set $\mathcal{P}(K)$ of all polynomials in one variable with coefficients from K is a vector space over K. The sum and scalar multiple are given by the (usual) sum of polynomials and constant multiple of a polynomial. The set $\mathcal{P}_n(K)$ of all polynomials whose degrees are less than or equal to n is also a vector space over K. ■

9.2 Subspaces

9.2.1 *Definition*

Definition 9.2. Let V be a vector space over K. A (nonempty) subset W of V is called a *subspace* (or *linear subspace, vector subspace*) of V if W is a vector space over K under the operations of V. ■

Theorem 9.1. *A (nonempty) subset W of a vector space V over K is a subspace of V if and only if the following two conditions hold:*

(1) *If $\boldsymbol{x}, \boldsymbol{y} \in W$, then $\boldsymbol{x} + \boldsymbol{y} \in W$.*
(2) *If $\boldsymbol{x} \in W$, $a \in K$, then $a\boldsymbol{x} \in W$.*

Remark 9.3. When we want to show that a subset W of V is a vector space, it is not necessary to check all the conditions listed in Definition 9.1. According to Theorem 9.1, it suffices to verify that W is closed under the vector addition and scalar multiplication, that is, that the outcomes of these operations for elements of W also belong to W. The conditions (1) and (2) together can be expressed compactly as:

(3) If $\boldsymbol{x}, \boldsymbol{y} \in W$, $a, b \in K$, then $a\boldsymbol{x} + b\boldsymbol{y} \in W$. ■

Remark 9.4. The singleton set $\{\boldsymbol{0}\}$ and the whole set V are subspaces of V. A subspace other than those is called a *proper subspace.* ■

Definition 9.3. For vectors $\boldsymbol{x}_1, \boldsymbol{x}_2, \ldots, \boldsymbol{x}_k$ in a vector space V over K, a vector of the form

$$c_1\boldsymbol{x}_1 + c_2\boldsymbol{x}_2 + \cdots + c_k\boldsymbol{x}_k \qquad (c_1, c_2, \ldots, c_k \in K)$$

is called a *linear combination* of $\boldsymbol{x}_1, \boldsymbol{x}_2, \ldots, \boldsymbol{x}_k$. When $k = 0$, we assume $c_1\boldsymbol{x}_1 + c_2\boldsymbol{x}_2 + \cdots + c_k\boldsymbol{x}_k = \boldsymbol{0}$ (by convention). ■

Definition 9.4. Let S be a subset of a vector space V over K, where S may possibly be an infinite set. The set of all linear combinations of a finite number of vectors in S is a subspace of V. This is the smallest subspace containing S, called the *subspace generated* (or *spanned*) by S and denoted by $\mathrm{Span}(S)$. The subspace generated by a finite set $S = \{\boldsymbol{x}_1, \boldsymbol{x}_2, \ldots, \boldsymbol{x}_k\}$ is given by

$$\{\boldsymbol{x} \in V \mid \boldsymbol{x} = c_1\boldsymbol{x}_1 + c_2\boldsymbol{x}_2 + \cdots + c_k\boldsymbol{x}_k \quad (c_1, c_2, \ldots, c_k \in K)\}.$$

If S is the empty set, we have $\mathrm{Span}(S) = \{\mathbf{0}\}$. ■

9.2.2 *Examples*

We show examples of subspaces.

Example 9.8. In the three-dimensional space $\mathbb{R}^3$, a plane or a line containing the origin is a subspace of $\mathbb{R}^3$. ■

Example 9.9. For an $m \times n$ matrix A over K, the set of all $\boldsymbol{x} \in K^n$ satisfying $A\boldsymbol{x} = \mathbf{0}$ is a subspace of K^n. ■

Example 9.10. In the vector space $\mathcal{M}_{n,n}(K)$ of square matrices of order n, the set of all *symmetric matrices* (*i.e.*, $A = (a_{ij})$ satisfying $a_{ji} = a_{ij}$ for all (i, j)) is a subspace. The set of all *anti-symmetric matrices* (*i.e.*, $A = (a_{ij})$ satisfying $a_{ji} = -a_{ij}$ for all (i, j) and $a_{ii} = 0$ for all i) is also a subspace. ■

Example 9.11. Let S denote the vector space of all sequences with terms taken from K. For any given elements $a_0, a_1, \ldots, a_k$ of K, where $a_0 \neq 0$, the set of all sequences $\{x_n\}_{n=0}^{\infty}$ that satisfy the difference equation

$$a_0x_{n+k} + a_1x_{n+k-1} + \cdots + a_kx_n = 0 \qquad (n = 0, 1, 2, \ldots)$$

is a subspace of S. ■

Example 9.12. Let $\mathrm{C}^k(I, \mathbb{R})$ denote the vector space of all k times continuously differentiable real-valued functions on a real interval I of length > 0. For any given k continuous functions $a_1(x), a_2(x), \ldots, a_k(x)$, the set of all k times continuously differentiable functions $y(x)$ that satisfy the kth order homogeneous ordinary differential equation

$$\frac{\mathrm{d}^k y}{\mathrm{d}x^k} + a_1(x)\frac{\mathrm{d}^{k-1} y}{\mathrm{d}x^{k-1}} + \cdots + a_k(x)y = 0$$

is a subspace of $\mathrm{C}^k(I, \mathbb{R})$. ■

9.2.3 *Intersection and Sum of Subspaces*

Definition 9.5. Let V be a vector space over K and $\{W_\alpha\}$ a family of its subspaces indexed by α. Their intersection $\bigcap_\alpha W_\alpha$ (the set of elements belonging to every W_α) is a subspace of V, which is referred to as the *intersection* of the subspaces $\{W_\alpha\}$. ■

Remark 9.5. For two subspaces W_1 and W_2, their union $W_1 \cup W_2$ is not a subspace in general. Furthermore, the following is true: If $W_1 \cup W_2$ is a subspace, then $W_1 \subseteqq W_2$ or $W_1 \supseteqq W_2$ holds. The subspace generated by $W_1 \cup W_2$, denoted by $\mathrm{Span}(W_1 \cup W_2)$, coincides with their sum $W_1 + W_2$ introduced in the next definition. ■

Definition 9.6. Let V be a vector space over K, and $W_1, W_2, \ldots, W_k$ be a family of subspaces of V. The set W defined by

$$W = \{\boldsymbol{x} \in V \mid \boldsymbol{x} = \boldsymbol{x}_1 + \boldsymbol{x}_2 + \cdots + \boldsymbol{x}_k,\ \boldsymbol{x}_i \in W_i \quad (i = 1, 2, \ldots, k)\}$$

is a subspace of V. This is called the *sum* of $W_1, W_2, \ldots, W_k$, and denoted as

$$W_1 + W_2 + \cdots + W_k.$$

■

Remark 9.6. In general, the following identities are not true:

$$W_1 \cap (W_2 + W_3) = (W_1 \cap W_2) + (W_1 \cap W_3),$$
$$W_1 + (W_2 \cap W_3) = (W_1 + W_2) \cap (W_1 + W_3).$$

In $\mathbb{R}^2$, for example, both equalities fail for

$$W_1 = \{(x_1, x_2) \mid x_1 = x_2\},$$
$$W_2 = \{(x_1, x_2) \mid x_2 = 0\},$$
$$W_3 = \{(x_1, x_2) \mid x_1 = 0\}.$$

It is noted in this connection that the following inclusion relations are always true:

$$W_1 \cap (W_2 + W_3) \supseteqq (W_1 \cap W_2) + (W_1 \cap W_3),$$
$$W_1 + (W_2 \cap W_3) \subseteqq (W_1 + W_2) \cap (W_1 + W_3).$$

■

Definition 9.7. Let V be a vector space over K and $W = W_1 + W_2 + \cdots + W_k$ a sum of subspaces $W_1, W_2, \ldots, W_k$. If every vector of W can be represented uniquely as a sum of vectors each belonging to W_i, then W is called the *direct sum* of $W_1, W_2, \ldots, W_k$, and denoted as

$$W = W_1 \oplus W_2 \oplus \cdots \oplus W_k.$$

■

Theorem 9.2. *Let V be a vector space over K. A sum $W = W_1 + W_2 + \cdots + W_k$ of subspaces $W_1, W_2, \ldots, W_k$ is a direct sum if and only if the following holds:*

(a)
$$\boldsymbol{x}_1 + \boldsymbol{x}_2 + \cdots + \boldsymbol{x}_k = \boldsymbol{0},\ \boldsymbol{x}_i \in W_i\ (i = 1, 2, \ldots, k)$$
$$\Longrightarrow \boldsymbol{x}_i = \boldsymbol{0}\quad (i = 1, 2, \ldots, k).$$

Furthermore, this condition (a) *is equivalent to*

(b)
$$W_1 \cap W_2 = \{\boldsymbol{0}\},\quad (W_1 + W_2) \cap W_3 = \{\boldsymbol{0}\},\quad \ldots,$$
$$(W_1 + \cdots + W_{k-1}) \cap W_k = \{\boldsymbol{0}\}.$$

Example 9.13. The vector space $\mathcal{M}_{n,n}(K)$ of square matrices is a direct sum of the subspace of symmetric matrices and that of anti-symmetric matrices (see Example 9.10 in Sec. 9.2.2). ■

9.2.4 *Complement and Quotient*

Definition 9.8. When a vector space V over K is a direct sum of two subspaces W and W', that is, when $V = W \oplus W'$, W' is said to be a *complement* (or *complementary subspace*) of W. ■

Theorem 9.3. *For any subspace W, there exists a complement W' of W (which is not unique).*

Definition 9.9. Let W be a subspace of a vector space V over K. Define a binary relation $\sim$ on V by

$$\boldsymbol{x} \sim \boldsymbol{y} \iff \boldsymbol{x} - \boldsymbol{y} \in W,$$

where $\boldsymbol{x}, \boldsymbol{y} \in V$. This is an equivalence relation (see Remark 9.7 for the definition), and the quotient set $V/\!\sim$ is called the *quotient space* of V by W, denoted by V/W. Denote the equivalence class of $\boldsymbol{x} \in V$ by $[\boldsymbol{x}]$. Then we can define the operations of sum and scalar multiple on V/W by

$$[\boldsymbol{x}] + [\boldsymbol{y}] = [\boldsymbol{x} + \boldsymbol{y}] \qquad (\boldsymbol{x}, \boldsymbol{y} \in V),$$
$$a[\boldsymbol{x}] = [a\boldsymbol{x}] \qquad (\boldsymbol{x} \in V,\ a \in K).$$

With these operations, V/W is a vector space over K. ■

Remark 9.7. A binary relation $\sim$ on a set V in general (not necessarily a vector space) is called an *equivalence relation* if the following conditions are satisfied:

- *Reflexive law:* For any $v \in V$, we have $v \sim v$.
- *Symmetric law:* For any $u, v \in V$, if $u \sim v$, then $v \sim u$.
- *Transitive law:* For any $u, v, w \in V$, if $u \sim v$ and $v \sim w$, then $u \sim w$.

For each $v \in V$, the *equivalence class* $[v]$ is defined by

$$[v] = \{u \in V \mid u \sim v\},$$

which has the following properties:

- For any $v \in V$, v belongs to $[v]$, that is, $v \in [v]$.
- For any $u, v \in V$, $[u] \cap [v] \neq \emptyset$ implies $[u] = [v]$.

Therefore, the set V is partitioned into equivalence classes. The set of the equivalence classes is called a *quotient set*, and denoted by $V/\sim$. Each element of $V/\sim$ is an equivalence class (a subset of V), and hence $V/\sim$ is a set of subsets of V. ■

9.3 Linear Independence and Dependence

Definition 9.10. Let V be a vector space over K. A set of vectors $\boldsymbol{x}_1, \boldsymbol{x}_2, \ldots, \boldsymbol{x}_k$ in V is called *linearly independent* if the following implication holds:

$$c_1\boldsymbol{x}_1 + c_2\boldsymbol{x}_2 + \cdots + c_k\boldsymbol{x}_k = \mathbf{0} \quad (c_1, c_2, \ldots, c_k \in K)$$
$$\Longrightarrow \quad c_1 = c_2 = \cdots = c_k = 0.$$

They are called *linearly dependent* if they are not linearly independent. A subset X of V is said to be linearly independent if any finite number of vectors in X are linearly independent; otherwise X is called linearly dependent. The empty set is a linearly independent set (by convention). ■

Theorem 9.4. *The following hold in a vector space V over K.*

(1) *A set $X = \{\boldsymbol{x}\}$ consisting of a single vector $\boldsymbol{x}$ is linearly independent if $\boldsymbol{x} \neq \mathbf{0}$, and linearly dependent if $\boldsymbol{x} = \mathbf{0}$.*
(2) *Assume that X is a subset of V containing more than one vector. Then X is linearly dependent if and only if there exists a vector in X that can be represented as a linear combination of a finite number of other vectors in X. Equivalently, X is linearly independent if and only if none of the vectors of X can be represented as a linear combination of a finite number of other vectors in X.*

Theorem 9.5. *Linear independence in a vector space enjoys the following three properties* (a), (b), *and* (c).[2]

(a) *The empty set* $\emptyset$ *is linearly independent.*
(b) *If* X *is linearly independent, then any subset* Y *of* X *is linearly independent.*
(c) *If* $X = \{\boldsymbol{x}_1, \boldsymbol{x}_2, \ldots, \boldsymbol{x}_k\}$ *and* $Y = \{\boldsymbol{y}_1, \boldsymbol{y}_2, \ldots, \boldsymbol{y}_l\}$ *are linearly independent, and* $|X| < |Y|$ *(i.e.,* $k < l$*), then there exists a vector* $\boldsymbol{y}_j$ *in* $Y \setminus X$ *for which* $X \cup \{\boldsymbol{y}_j\}$ *is linearly independent.*

9.4 Basis

9.4.1 *Definition*

Definition 9.11. A subset B of a vector space V over K is called a *basis* of V if B is linearly independent and generates V. That is, B is a basis of V if (and only if) any finite set of vectors in B is linearly independent and every vector $\boldsymbol{x}$ of V can be represented as

$$\boldsymbol{x} = x_1\boldsymbol{b}_1 + x_2\boldsymbol{b}_2 + \cdots + x_k\boldsymbol{b}_k \tag{9.1}$$

with a finite number of $\boldsymbol{b}_1, \boldsymbol{b}_2, \ldots, \boldsymbol{b}_k \in B$ and $x_1, x_2, \ldots, x_k \in K$. ■

In the representation (9.1), the vectors $\boldsymbol{b}_1, \boldsymbol{b}_2, \ldots, \boldsymbol{b}_k$ and the coefficients $x_1, x_2, \ldots, x_k$ are uniquely determined by $\boldsymbol{x}$, as long as $x_1, x_2, \ldots, x_k \neq 0$.[3]

Remark 9.8. According to our definition, $V = \{\boldsymbol{0}\}$ has a basis, which is the empty set. It is noted, however, some books adopt a different convention that $V = \{\boldsymbol{0}\}$ has no basis. ■

Theorem 9.6. *Every vector space* V *over* K *has a basis.*

Theorem 9.7. *If a subset* X *of a vector space* V *over* K *is linearly independent, there exists a basis containing* X.

Theorem 9.8. *The cardinality*[4] *of a basis of a vector space* V *over* K *is determined uniquely by* V.

---

[2] These three properties lead to the concept of matroid; see Sec. 9.9.
[3] We mean the uniqueness up to the obvious permutations (reorderings) of $\boldsymbol{b}_1, \boldsymbol{b}_2, \ldots, \boldsymbol{b}_k$.
[4] *Cardinality* is a concept for measuring the size of a set. The cardinality of a finite set is equal to the number of the elements of the set.

Definition 9.12. The cardinality of a basis of a vector space V over K is called the *dimension* of V, and denoted by $\dim V$. If the cardinality of a basis is finite, V is called a *finite-dimensional vector space*. Otherwise, it is called an *infinite-dimensional vector space*. For $V = \{\mathbf{0}\}$ we have $\dim V = 0$. ■

Remark 9.9. The proofs of Theorems 9.6 and 9.7, for the most general case including infinite-dimensional spaces, consist of the following two steps:

(1) First, we note that B is a basis if and only if B is a maximal independent set. (This equivalence should be obvious from Theorem 9.4(2) and the definitions of the terms.) Then we reduce the existence proof of a basis to the existence proof of a maximal independent set. Here a subset B of V is called a *maximal independent set* if B is linearly independent and, for every $\boldsymbol{c} \notin B$, the set $B \cup \{\boldsymbol{c}\}$ is linearly dependent.
(2) Next, we show the existence of a maximal independent set in V on the basis of a proposition (axiom) in set theory, called *Zorn's lemma*. The lemma states that an inductive ordered set (a partially ordered set in which every chain has an upper bound in the set) has at least one maximal element. Such a sophisticated mathematical argument is necessary because, in general, a maximal independent set can be an infinite set. We do not have to rely on Zorn's lemma if a maximal independent set is known *a priori* to be a finite set (*e.g.*, if V is generated by a finite number of vectors). ■

Remark 9.10. We encounter the term "basis" also in the context of Hilbert spaces and Banach spaces (Secs. 9.6.8 and 9.6.9). However, this is a concept different from the one treated in this section (different in infinite-dimensional cases). A *basis* of an infinite-dimensional Banach space V means a countably infinite subset $B = \{\boldsymbol{b}_1, \boldsymbol{b}_2, \boldsymbol{b}_3, \ldots\}$ of V that satisfies the following two conditions:

(1) Every $\boldsymbol{x} \in V$ can be represented as $\boldsymbol{x} = \sum_n x_n \boldsymbol{b}_n$.

(2) The representation above is unique for each $\boldsymbol{x}$.

In a Banach space, the concept of convergence is defined, and hence the infinite sum in (1) makes sense. In a general vector space, in contrast, there is no way to define an infinite sum, and accordingly, Definition 9.11 for a basis of a general vector space refers only to a finite sum. Another difference is that there is no restriction on the cardinality of a basis of a general vector space, whereas a basis of a Banach space consists of at most countably many vectors. Since a Banach space is a vector space, we could also consider its basis as a vector space. However, a basis of a Banach space usually means the one satisfying (1) and (2) above. ■

9.4.2 *Examples*

Example 9.14. In the space of numerical vectors K^n (Example 9.3 in Sec. 9.1.2), a basis is given by

$$\begin{aligned}
e_1 &= (1, 0, 0, \ldots, 0, 0)^\top, \\
e_2 &= (0, 1, 0, \ldots, 0, 0)^\top, \\
&\cdots\cdots \\
e_i &= (0, \ldots, 0, \overset{\overset{i}{\vee}}{1}, 0, \ldots, 0)^\top \quad \text{(the } i\text{th component is 1; others are 0)}, \\
&\cdots\cdots \\
e_n &= (0, 0, 0, \ldots, 0, 1)^\top,
\end{aligned}$$

and hence $\dim K^n = n$. The set of vectors $\{e_1, e_2, \ldots, e_n\}$ is called the *standard basis* of K^n. If a set of n vectors $\{a_1, a_2, \ldots, a_n\}$ is linearly independent, it is a basis of K^n. ■

Example 9.15. In the space $\mathcal{M}_{m,n}(K)$ of all $m \times n$ matrices (Example 9.4 in Sec. 9.1.2), a basis is given by $\{E_{ij} \mid 1 \leqq i \leqq m, 1 \leqq j \leqq n\}$, where E_{ij} denotes the matrix in which the (i, j) entry is 1 and the other entries are all equal to 0. We have $\dim \mathcal{M}_{m,n}(K) = mn$. ■

Example 9.16. In the space $\mathcal{M}_{n,n}(K)$ of all square matrices of order n, the set of all symmetric matrices is a subspace (Example 9.10 in Sec. 9.2.2). A basis of this subspace is given by $\{E_{ij} + E_{ji} \mid 1 \leqq i \leqq n, 1 \leqq j \leqq i\}$ (see Example 9.15 for the definition of E_{ij}), and hence its dimension is equal to $n(n+1)/2$. In the space of all anti-symmetric matrices of order n (Example 9.10), a basis is given by $\{E_{ij} - E_{ji} \mid 2 \leqq i \leqq n, 1 \leqq j \leqq i-1\}$ and hence its dimension is equal to $n(n-1)/2$ if $n \geqq 2$. In the case of $n = 1$, this space consists only of the zero matrix (0), with dimension 0. Hence, the dimension of this subspace is given by $n(n-1)/2$ for all $n \geqq 1$. ■

Example 9.17. For an $m \times n$ matrix A over K, the set of all $x \in K^n$ satisfying $Ax = \mathbf{0}$ is a subspace of K^n (Example 9.9 in Sec. 9.2.2). The dimension of this subspace is equal to $n - \operatorname{rank} A$ by Theorem 9.69(2) in Sec. 9.8.1, with a basis given by $\{\boldsymbol{t}_{r+1}, \boldsymbol{t}_{r+2}, \ldots, \boldsymbol{t}_n\}$ in (5.8) in Sec. 5.2, where $r = \operatorname{rank} A$. ■

Example 9.18. In the space $\mathcal{P}(K)$ of all polynomials in one variable with coefficients from K (Example 9.7 in Sec. 9.1.2), a basis is naturally given by $\{1, x, x^2, \ldots, x^n, \ldots\}$. This shows that $\mathcal{P}(K)$ is infinite-dimensional, with the dimension being equal to the *cardinality of countable infinity*.[5] For the space $\mathcal{P}_n(K)$ of all polynomials with degrees bounded by n, a basis is given by $\{1, x, x^2, \ldots, x^n\}$, and hence the dimension of $\mathcal{P}_n(K)$ is equal to $n + 1$. ■

Example 9.19. Consider the vector space of all sequences $\{x_n\}_{n=0}^{\infty}$ consisting of elements of K (Example 9.5 in Sec. 9.1.2). For each $i = 0, 1, 2, \ldots$, let $\boldsymbol{e}_i$ denote the sequence in which the ith term is 1 and all other terms are 0. The set of such vectors $\{\boldsymbol{e}_0, \boldsymbol{e}_1, \boldsymbol{e}_2, \ldots\}$ is linearly independent, and hence this vector space is an infinite-dimensional space. However, $\{\boldsymbol{e}_0, \boldsymbol{e}_1, \boldsymbol{e}_2, \ldots\}$ is not a basis, because the sequence $(1, 1, 1, \ldots)$ with 1 for all terms cannot be represented as a linear combination of the vectors in $\{\boldsymbol{e}_0, \boldsymbol{e}_1, \boldsymbol{e}_2, \ldots\}$. Note that a linear combination is a finite sum, by definition.

For given elements $a_0, a_1, \ldots, a_k$ of K, where $a_0 \neq 0$, the set of all sequences $\{x_n\}_{n=0}^{\infty}$ that satisfy the difference equation

$$a_0 x_{n+k} + a_1 x_{n+k-1} + \cdots + a_k x_n = 0 \qquad (n = 0, 1, 2, \ldots)$$

is a vector space (Example 9.11 in Sec. 9.2.2). A member of this vector space is determined from the first k terms of the sequence. For each i with $0 \leqq i \leqq k - 1$, let $\hat{\boldsymbol{e}}_i$ denote the sequence satisfying the difference equation with $x_i = 1$, $x_j = 0$ $(j = 0, 1, \ldots, k-1; j \neq i)$. Then $\{\hat{\boldsymbol{e}}_0, \hat{\boldsymbol{e}}_1, \ldots, \hat{\boldsymbol{e}}_{k-1}\}$ is a basis, and the dimension of this space is equal to k. ■

Example 9.20. Consider the vector space of all mappings from a set D to $\mathbb{R}$ (Example 9.6 in Sec. 9.1.2). For each $p \in D$, let f_p denote the mapping defined by

$$f_p(p) = 1, \qquad f_p(x) = 0 \quad (x \neq p).$$

If D is a finite set given as $D = \{p_1, p_2, \ldots, p_n\}$, the set $\{f_{p_1}, f_{p_2}, \ldots, f_{p_n}\}$ is a basis, and the dimension of this space is equal to n. If D is an infinite set, $\{f_p \mid p \in D\}$ is linearly independent, and hence this space is an infinite-dimensional space, whose dimension is equal to or larger than

[5]This is the cardinality of the natural numbers.

the cardinality of D. Note that, in case of infinite D, the set of vectors $\{f_p \mid p \in D\}$ is not a basis. ■

Example 9.21. Consider the vector space of all continuous functions over a real interval I. The set $\{1, x, x^2, \ldots, x^n, \ldots\}$ is linearly independent, and hence this space is an infinite-dimensional space. However, $\{1, x, x^2, \ldots, x^n, \ldots\}$ is not a basis. Similarly for the vector space of all m times continuously differentiable functions over a real interval I. ■

Example 9.22. Suppose that we are given k continuous functions $a_1(x), a_2(x), \ldots, a_k(x)$ on a real interval I, and consider the vector space of all k times continuously differentiable functions $y(x)$ that satisfy the kth order homogeneous ordinary differential equation

$$\frac{\mathrm{d}^k y}{\mathrm{d}x^k} + a_1(x)\frac{\mathrm{d}^{k-1} y}{\mathrm{d}x^{k-1}} + \cdots + a_k(x)y = 0$$

(Example 9.12 in Sec. 9.2.2). Fix a point x_0 in the interval I and let $y_1(x), y_2(x), \ldots, y_k(x)$ be the solutions of this differential equation satisfying the conditions

$$\frac{\mathrm{d}^{j-1} y_i}{\mathrm{d}x^{j-1}}(x_0) = \delta_{ij} \qquad (i, j = 1, 2, \ldots, k),$$

where δ_{ij} is the Kronecker delta (cf., Remark 1.3 in Sec. 1.1.1). Then $\{y_1(x), y_2(x), \ldots, y_k(x)\}$ is a basis of this vector space, and the dimension of this vector space is equal to k. ■

Example 9.23. When we discuss the *Fourier expansion*

$$f(x) \sim \frac{a_0}{2} + \sum_{n=1}^{\infty}\left(a_n \cos\left(\frac{n\pi x}{l}\right) + b_n \sin\left(\frac{n\pi x}{l}\right)\right)$$

for a function $f(x)$ $(-l \leqq x \leqq l)$, we sometimes refer to

$$\frac{1}{\sqrt{2l}}, \ \frac{1}{\sqrt{l}}\cos\left(\frac{\pi x}{l}\right), \ \frac{1}{\sqrt{l}}\sin\left(\frac{\pi x}{l}\right), \ \frac{1}{\sqrt{l}}\cos\left(\frac{2\pi x}{l}\right), \ \frac{1}{\sqrt{l}}\sin\left(\frac{2\pi x}{l}\right), \ \ldots$$

as an orthonormal basis. However, this is not a basis in the sense of Definition 9.11, but a basis in the sense of Remark 9.10. See also Example 9.56 in Sec. 9.6.8. ■

Example 9.24. The field of real numbers $\mathbb{R}$ can be regarded as a vector space over the field of rational numbers $\mathbb{Q}$. The basis of this vector space is called a *Hamel basis* [57, 62, 64]. The cardinality of a Hamel basis is equal to the *cardinality of the continuum* (the same cardinality as $\mathbb{R}$) and hence the dimension of this space is equal to the cardinality of the continuum. ■

9.4.3 *Formulas Concerning Dimensions*

Throughout Sec. 9.4.3, vector spaces are assumed to be finite-dimensional.

Theorem 9.9. *The following hold for subspaces W_1 and W_2 of a vector space V over K.*

(1) $W_1 \subseteqq W_2 \implies \dim W_1 \leqq \dim W_2$.
(2) $W_1 \subseteqq W_2,\ \dim W_1 = \dim W_2 \implies W_1 = W_2$.

Theorem 9.10. *The following identity holds for subspaces W_1 and W_2 of a vector space V over K:*

$$\dim W_1 + \dim W_2 = \dim(W_1 + W_2) + \dim(W_1 \cap W_2).$$

Remark 9.11. It follows from Theorem 9.10 and Theorem 9.9(1) that[6]

$$\dim(W_1 \cap W_2) \geqq \dim W_1 + \dim W_2 - \dim V,$$
$$\dim(W_1 \cap W_2 \cap W_3) \geqq \dim W_1 + \dim W_2 + \dim W_3 - 2 \cdot \dim V.$$

■

Remark 9.12. It should be noted that the identity in Theorem 9.10 cannot be extended to three subspaces, that is, for three subspaces W_1, W_2, W_3, the identity

$$\begin{aligned}\dim W_1 + \dim W_2 + \dim W_3 = {} & \dim(W_1 + W_2 + W_3) + \dim(W_1 \cap W_2) \\ & + \dim(W_1 \cap W_3) + \dim(W_2 \cap W_3) \\ & - \dim(W_1 \cap W_2 \cap W_3)\end{aligned}$$

is not true. For example, consider the subspaces W_1, W_2, W_3 in Remark 9.6 in Sec. 9.2.3, for which the left-hand side is equal to 3 and the right-hand side is 2. It is also noted that the following inequality

$$\begin{aligned}\dim W_1 + \dim W_2 + \dim W_3 \geqq {} & \dim(W_1 + W_2 + W_3) + \dim(W_1 \cap W_2) \\ & + \dim(W_1 \cap W_3) + \dim(W_2 \cap W_3) \\ & - \dim(W_1 \cap W_2 \cap W_3)\end{aligned}$$

can be proved from Theorem 9.10 and Remark 9.6. ■

Theorem 9.11. *Let $W_1, W_2, \ldots, W_k$ be subspaces of a vector space V over K. Their sum $W = W_1 + W_2 + \cdots + W_k$ is a direct sum if and only if*

$$\dim W = \dim W_1 + \dim W_2 + \cdots + \dim W_k.$$

[6] These inequalities are used in the proofs of Theorems 6.12, 6.15, and 8.9.

Remark 9.13. We have already seen necessary and sufficient conditions for a sum of subspaces to be a direct sum in Theorem 9.2 in Sec. 9.2.3, where V may be infinite-dimensional. ■

Theorem 9.12. *For a subspace W of a vector space V over K, the dimension of the quotient space V/W is given as*

$$\dim(V/W) = \dim V - \dim W.$$

9.4.4 *Change of Coordinates and Basis*

Throughout Sec. 9.4.4, vector spaces are assumed to be finite-dimensional. So far we have regarded a basis as a set of vectors, but sometimes it is convenient to consider the ordering of the basis vectors. To emphasize the ordering of basis vectors, we use a matrix-like notation $[\boldsymbol{b}_1, \boldsymbol{b}_2, \ldots, \boldsymbol{b}_k]$, which enables us to carry out computations on basis vectors through matrix operations.

Definition 9.13. Let $B = [\boldsymbol{b}_1, \boldsymbol{b}_2, \ldots, \boldsymbol{b}_k]$ be a basis of a vector space V over K. Every vector $\boldsymbol{x} \in V$ can be represented uniquely as

$$\boldsymbol{x} = \sum_{i=1}^{k} x_i \boldsymbol{b}_i = [\boldsymbol{b}_1, \boldsymbol{b}_2, \ldots, \boldsymbol{b}_k] \begin{bmatrix} x_1 \\ x_2 \\ \vdots \\ x_k \end{bmatrix} \qquad (x_1, x_2, \ldots, x_k \in K).$$

We call $(x_1, x_2, \ldots, x_k)^\top \in K^k$ the *coordinates* of vector $\boldsymbol{x}$ with respect to basis B. ■

Definition 9.14. Let $B = [\boldsymbol{b}_1, \boldsymbol{b}_2, \ldots, \boldsymbol{b}_k]$ be a basis of a vector space V over K. If $B' = [\boldsymbol{b}'_1, \boldsymbol{b}'_2, \ldots, \boldsymbol{b}'_k]$ is another basis, then each $\boldsymbol{b}'_j$ can be represented as a linear combination of $\boldsymbol{b}_1, \boldsymbol{b}_2, \ldots, \boldsymbol{b}_k$ as

$$\boldsymbol{b}'_j = \sum_{i=1}^{k} a_{ij} \boldsymbol{b}_i \qquad (j = 1, 2, \ldots, k),$$

where $a_{ij} \in K$ $(i, j = 1, 2, \ldots, k)$. Equivalently, we can express this in the form of a matrix product as

$$[\boldsymbol{b}'_1, \boldsymbol{b}'_2, \ldots, \boldsymbol{b}'_k] = [\boldsymbol{b}_1, \boldsymbol{b}_2, \ldots, \boldsymbol{b}_k] \begin{bmatrix} a_{11} & a_{12} & \cdots & a_{1k} \\ a_{21} & a_{22} & \cdots & a_{2k} \\ \vdots & \vdots & & \vdots \\ a_{k1} & a_{k2} & \cdots & a_{kk} \end{bmatrix}.$$

The matrix $A = (a_{ij})$ is called the *basis change matrix* for the change of basis $B \to B'$. ■

Theorem 9.13.

(1) *A basis change matrix is nonsingular.*
(2) *Let A and A' be the basis change matrices for $B \to B'$ and $B' \to B''$, respectively. Then the basis change matrix for $B \to B''$ is given by AA'. That is,*

$$[\boldsymbol{b}'_1, \boldsymbol{b}'_2, \ldots, \boldsymbol{b}'_k] = [\boldsymbol{b}_1, \boldsymbol{b}_2, \ldots, \boldsymbol{b}_k]A, \quad [\boldsymbol{b}''_1, \boldsymbol{b}''_2, \ldots, \boldsymbol{b}''_k] = [\boldsymbol{b}'_1, \boldsymbol{b}'_2, \ldots, \boldsymbol{b}'_k]A'$$
$$\Longrightarrow \quad [\boldsymbol{b}''_1, \boldsymbol{b}''_2, \ldots, \boldsymbol{b}''_k] = [\boldsymbol{b}_1, \boldsymbol{b}_2, \ldots, \boldsymbol{b}_k]AA'.$$

(3) *If A is the basis change matrix for $B \to B'$, then the basis change matrix for $B' \to B$ is given by A^{-1}. That is,*

$$[\boldsymbol{b}'_1, \boldsymbol{b}'_2, \ldots, \boldsymbol{b}'_k] = [\boldsymbol{b}_1, \boldsymbol{b}_2, \ldots, \boldsymbol{b}_k]A$$
$$\Longrightarrow \quad [\boldsymbol{b}_1, \boldsymbol{b}_2, \ldots, \boldsymbol{b}_k] = [\boldsymbol{b}'_1, \boldsymbol{b}'_2, \ldots, \boldsymbol{b}'_k]A^{-1}.$$

Theorem 9.14. *Let $B = [\boldsymbol{b}_1, \boldsymbol{b}_2, \ldots, \boldsymbol{b}_k]$ and $B' = [\boldsymbol{b}'_1, \boldsymbol{b}'_2, \ldots, \boldsymbol{b}'_k]$ be bases of a vector space V over K, and let A denote the basis change matrix for $B \to B'$, that is,*

$$[\boldsymbol{b}'_1, \boldsymbol{b}'_2, \ldots, \boldsymbol{b}'_k] = [\boldsymbol{b}_1, \boldsymbol{b}_2, \ldots, \boldsymbol{b}_k]A.$$

For any vector $\boldsymbol{x} \in V$, let $(x_1, x_2, \ldots, x_k)^\top$ and $(x'_1, x'_2, \ldots, x'_k)^\top$ be the coordinates of $\boldsymbol{x}$ with respect to bases B and B', respectively. These coordinates are transformed as

$$\begin{bmatrix} x'_1 \\ x'_2 \\ \vdots \\ x'_k \end{bmatrix} = A^{-1} \begin{bmatrix} x_1 \\ x_2 \\ \vdots \\ x_k \end{bmatrix}.$$

9.5 Linear Mapping

9.5.1 *Definition*

Definition 9.15. Let V and V' be vector spaces over K. A mapping $T : V \to V'$ is called a *linear mapping* from V to V' if the following conditions are satisfied:

(1) $T(\boldsymbol{x} + \boldsymbol{y}) = T(\boldsymbol{x}) + T(\boldsymbol{y}) \qquad (\boldsymbol{x}, \boldsymbol{y} \in V)$,
(2) $T(a\boldsymbol{x}) = aT(\boldsymbol{x}) \qquad (\boldsymbol{x} \in V,\ a \in K)$.

In particular, a linear mapping from V to V itself is called a *linear transformation* in V. ■

The mapping that maps every $\boldsymbol{x} \in V$ to $\mathbf{0} \in V'$ is a linear mapping, which is called the *null mapping* (or *null transformation* in the case of $V = V'$) and denoted by O. The mapping that maps every $\boldsymbol{x} \in V$ to itself is a linear transformation, which is called the *identity transformation* and denoted by I.

Theorem 9.15. *Let V and V' be vector spaces over K, and T a linear mapping from V to V'.*

(1) *Let W be a subspace of V. The image of W by T:*

$$T(W) = \{T(\boldsymbol{x}) \mid \boldsymbol{x} \in W\}$$

is a subspace of V'.

(2) *Let W' be a subspace of V'. The inverse image of W' by T:*

$$T^{-1}(W') = \{\boldsymbol{x} \in V \mid T(\boldsymbol{x}) \in W'\}$$

is a subspace of V.

It follows from this theorem that $T(V)$ and $T^{-1}(\{\mathbf{0}\})$ are subspaces of V, where $\mathbf{0}$ denotes the zero vector of V'.

Definition 9.16. $T(V)$ is called the *image* of T and denoted by $\operatorname{Im} T$. We also call it the *image space* to emphasize that it is a subspace. ■

Definition 9.17. $T^{-1}(\{\mathbf{0}\})$ is called the *kernel* of T and denoted by $\operatorname{Ker} T$. We also call it the *kernel space* to emphasize that it is a subspace. ■

Theorem 9.16. *If V is finite-dimensional in Theorem 9.15(1), we have*[7]

$$\dim W = \dim(W \cap (\operatorname{Ker} T)) + \dim(T(W)). \tag{9.2}$$

In particular,

$$\dim V = \dim(\operatorname{Ker} T) + \dim(\operatorname{Im} T). \tag{9.3}$$

9.5.2 *Examples*

Example 9.25. Let A be an $m \times n$ matrix over K. The mapping T that multiplies $\boldsymbol{x} \in K^n$ with A:

$$\begin{array}{rccc} T: & K^n & \to & K^m \\ & \rotatebox{90}{$\in$} & & \rotatebox{90}{$\in$} \\ & \boldsymbol{x} & \mapsto & A\boldsymbol{x} \end{array}$$

[7]The proof of (9.2) is given in Remark 9.27 in Sec. 9.8.2.

is a linear mapping from K^n to K^m. The image of T coincides with $\operatorname{Im} A$ in Definition 5.1 in Sec. 5.1, whereas the kernel of T coincides with $\operatorname{Ker} A$ in Definition 5.2 in Sec. 5.1. That is, we have $\operatorname{Im} T = \operatorname{Im} A$ and $\operatorname{Ker} T = \operatorname{Ker} A$. ■

Example 9.26. In the space $\mathcal{P}(K)$ of all polynomials in one variable with coefficients from K, the mapping T that translates the variable by $b \in K$:

$$\begin{array}{rccc} T: & \mathcal{P}(K) & \to & \mathcal{P}(K) \\ & \Downarrow & & \Downarrow \\ & P(x) & \mapsto & P(x+b) \end{array}$$

is a linear transformation. This is also a linear transformation in the space $\mathcal{P}_n(K)$ of all polynomials with degrees bounded by n. ■

Example 9.27. In the vector space S of all sequences $\{x_n\}_{n=0}^{\infty}$ consisting of elements of K, the mapping T that shifts the terms forward by one:

$$\begin{array}{rccc} T: & S & \to & S \\ & \Downarrow & & \Downarrow \\ & \{x_n\}_{n=0}^{\infty} & \mapsto & \{x_{n+1}\}_{n=0}^{\infty} \end{array}$$

is a linear transformation. This is also a linear transformation in the space of all sequences satisfying the difference equation

$$a_0 x_{n+k} + a_1 x_{n+k-1} + \cdots + a_k x_n = 0 \qquad (n = 0, 1, 2, \ldots).$$ ■

Example 9.28. Let I be a real interval. The operation of differentiation

$$\begin{array}{rccc} T: & \mathrm{C}^{m+1}(I, \mathbb{R}) & \to & \mathrm{C}^m(I, \mathbb{R}) \\ & \Downarrow & & \Downarrow \\ & f(x) & \mapsto & \dfrac{\mathrm{d}f}{\mathrm{d}x}(x) \end{array}$$

is a linear mapping from $\mathrm{C}^{m+1}(I, \mathbb{R})$ to $\mathrm{C}^m(I, \mathbb{R})$.[8] ■

Example 9.29. Let $a_1(x), a_2(x), \ldots, a_k(x)$ be continuous functions defined on a real interval I. The mapping T defined by

$$\begin{array}{rccc} T: & \mathrm{C}^k(I, \mathbb{R}) & \to & \mathrm{C}^0(I, \mathbb{R}) \\ & \Downarrow & & \Downarrow \\ & f(x) & \mapsto & \dfrac{\mathrm{d}^k f}{\mathrm{d}x^k}(x) + a_1(x)\dfrac{\mathrm{d}^{k-1} f}{\mathrm{d}x^{k-1}}(x) + \cdots + a_k(x) f(x) \end{array}$$

is a linear mapping from $\mathrm{C}^k(I, \mathbb{R})$ to $\mathrm{C}^0(I, \mathbb{R})$. The kernel of this mapping, $\operatorname{Ker} T$, is given by the set of solutions that satisfy the kth order homogeneous ordinary differential equation in Example 9.12 in Sec. 9.2.2. ■

[8]For the notation $\mathrm{C}^m(I, \mathbb{R})$, see Example 9.6 in Sec. 9.1.2.

Example 9.30. Let $k(x,y)$ be a real-valued continuous function on an interval $[a,b] \times [a,b]$ in $\mathbb{R}^2$. Then the mapping T defined by

$$\begin{array}{cccc} T: & \mathrm{C}^0([a,b],\mathbb{R}) & \to & \mathrm{C}^0([a,b],\mathbb{R}) \\ & \Psi & & \Psi \\ & f(x) & \mapsto & \displaystyle\int_a^b k(x,y)f(y)\mathrm{d}y \end{array}$$

is a linear transformation. ■

Example 9.31. Let $k(x,y)$ be a real-valued continuous function on a closed domain $\{(x,y) \in \mathbb{R}^2 \mid a \leqq y \leqq x \leqq b\}$ in $\mathbb{R}^2$. Then the mapping T defined by

$$\begin{array}{cccc} T: & \mathrm{C}^0([a,b],\mathbb{R}) & \to & \mathrm{C}^0([a,b],\mathbb{R}) \\ & \Psi & & \Psi \\ & f(x) & \mapsto & \displaystyle\int_a^x k(x,y)f(y)\mathrm{d}y \end{array}$$

is a linear transformation.[9] ■

9.5.3 *Matrix Representations*

Throughout Sec. 9.5.3, vector spaces are assumed to be finite-dimensional.

Definition 9.18. Let V and V' be vector spaces over K, and $B = [\boldsymbol{b}_1, \boldsymbol{b}_2, \ldots, \boldsymbol{b}_k]$ and $B' = [\boldsymbol{b}'_1, \boldsymbol{b}'_2, \ldots, \boldsymbol{b}'_l]$ be bases of V and V', respectively. Furthermore, let T be a linear mapping from V to V'. Then each $T(\boldsymbol{b}_j)$ can be represented as a linear combination of $\boldsymbol{b}'_1, \boldsymbol{b}'_2, \ldots, \boldsymbol{b}'_l$, *i.e.*,

$$T(\boldsymbol{b}_j) = \sum_{i=1}^{l} a_{ij}\boldsymbol{b}'_i \qquad (j = 1, 2, \ldots, k),$$

where $a_{ij} \in K$ $(i = 1, 2, \ldots, l; j = 1, 2, \ldots, k)$. Equivalently, we can express this in the form of a matrix product as

$$[T(\boldsymbol{b}_1), T(\boldsymbol{b}_2), \ldots, T(\boldsymbol{b}_k)] = [\boldsymbol{b}'_1, \boldsymbol{b}'_2, \ldots, \boldsymbol{b}'_l] \begin{bmatrix} a_{11} & a_{12} & \cdots & a_{1k} \\ a_{21} & a_{22} & \cdots & a_{2k} \\ \vdots & \vdots & & \vdots \\ a_{l1} & a_{l2} & \cdots & a_{lk} \end{bmatrix}.$$

This $l \times k$ matrix $A = (a_{ij})$ is called the *representation matrix* of linear mapping T with respect to bases B and B'. Such representation of a linear

[9] Note that we have a definite integral in Example 9.30 and an indefinite integral in Example 9.31.

mapping in terms of a matrix is called a *matrix representation* of a linear mapping. If T is a linear transformation (with $V = V'$), it is natural to choose $B = B'$ and in this case the matrix A is called the representation matrix of a linear transformation T with respect to a basis B. ■

Theorem 9.17. *Let V and V' be vector spaces over K, and $B = [\boldsymbol{b}_1, \boldsymbol{b}_2, \ldots, \boldsymbol{b}_k]$ and $B' = [\boldsymbol{b}'_1, \boldsymbol{b}'_2, \ldots, \boldsymbol{b}'_l]$ be bases of V and V', respectively. Furthermore, let T be a linear mapping from V to V', and A be the representation matrix of T with respect to bases B and B'. For any vector $\boldsymbol{x} \in V$ with coordinates $(x_1, x_2, \ldots, x_k)^\top$ with respect to B, the coordinates $(y_1, y_2, \ldots, y_l)^\top$ of $\boldsymbol{y} = T(\boldsymbol{x})$ with respect to B' is given as*

$$\begin{bmatrix} y_1 \\ y_2 \\ \vdots \\ y_l \end{bmatrix} = A \begin{bmatrix} x_1 \\ x_2 \\ \vdots \\ x_k \end{bmatrix}.$$

Example 9.32. Let A be an $m \times n$ matrix over K, and T be the linear mapping that maps $\boldsymbol{x} \in K^n$ to $A\boldsymbol{x} \in K^m$ (Example 9.25 in Sec. 9.5.2). For the standard bases B and B' of K^n and K^m, respectively (Example 9.14 in Sec. 9.4.2), the representation matrix of T is the matrix A itself. ■

Example 9.33. Consider the linear transformation $T : P(x) \mapsto P(x + b)$ in the space $\mathcal{P}_n(K)$ of all polynomials with degrees bounded by n (Example 9.26 in Sec. 9.5.2). For the basis $B = [1, x, x^2, \ldots, x^n]$ of $\mathcal{P}_n(K)$, the representation matrix $A = (a_{ij})$ of T is an $(n+1) \times (n+1)$ matrix defined by

$$a_{ij} = \begin{cases} 0 & (i > j), \\ \dbinom{j-1}{i-1} b^{j-i} = \dfrac{(j-1)!}{(i-1)!\,(j-i)!} b^{j-i} & (i \leqq j). \end{cases}$$

■

Example 9.34. In the vector space of all sequences satisfying the difference equation

$$a_0 x_{n+k} + a_1 x_{n+k-1} + \cdots + a_k x_n = 0 \qquad (n = 0, 1, 2, \ldots),$$

let $T : \{x_n\}_{n=0}^{\infty} \mapsto \{x_{n+1}\}_{n=0}^{\infty}$ be the linear mapping to shift the terms (Example 9.27 in Sec. 9.5.2). For the basis $B = [\hat{\boldsymbol{e}}_0, \hat{\boldsymbol{e}}_1, \ldots, \hat{\boldsymbol{e}}_{k-1}]$ (Example 9.19 in Sec. 9.4.2), the representation matrix of the linear transformation

T is given by[10]

$$\begin{bmatrix} 0 & 1 & 0 & \cdots & 0 \\ 0 & 0 & 1 & \ddots & \vdots \\ \vdots & \vdots & \ddots & \ddots & 0 \\ 0 & 0 & \cdots & 0 & 1 \\ -\dfrac{a_k}{a_0} & -\dfrac{a_{k-1}}{a_0} & \cdots & -\dfrac{a_2}{a_0} & -\dfrac{a_1}{a_0} \end{bmatrix}.$$

This is derived as follows. It follows from the difference equation that

$$x_{n+k} = -\frac{a_1}{a_0}x_{n+k-1} - \frac{a_2}{a_0}x_{n+k-2} - \cdots - \frac{a_{k-1}}{a_0}x_{n+1} - \frac{a_k}{a_0}x_n,$$

which shows

$$\hat{\mathbf{e}}_0 = \{\overbrace{1, 0, 0, \ldots, 0}^{k}, -\frac{a_k}{a_0}, \ldots\ldots\},$$

$$\hat{\mathbf{e}}_1 = \{\overbrace{0, 1, 0, \ldots, 0}^{k}, -\frac{a_{k-1}}{a_0}, \ldots\ldots\},$$

$$\vdots$$

$$\hat{\mathbf{e}}_{k-1} = \{\overbrace{0, 0, 0, \ldots, 1}^{k}, -\frac{a_1}{a_0}, \ldots\ldots\}.$$

Therefore, $T(\hat{\mathbf{e}}_0), T(\hat{\mathbf{e}}_1), \ldots, T(\hat{\mathbf{e}}_{k-1})$ are given by

$$T(\hat{\mathbf{e}}_0) = \{\overbrace{0, 0, \ldots, 0}^{k-1}, -\frac{a_k}{a_0}, \ldots\ldots\} = -\frac{a_k}{a_0}\hat{\mathbf{e}}_{k-1},$$

$$T(\hat{\mathbf{e}}_1) = \{\overbrace{1, 0, \ldots, 0}^{k-1}, -\frac{a_{k-1}}{a_0}, \ldots\ldots\} = \hat{\mathbf{e}}_0 - \frac{a_{k-1}}{a_0}\hat{\mathbf{e}}_{k-1},$$

$$\vdots$$

$$T(\hat{\mathbf{e}}_{k-1}) = \{\overbrace{0, 0, \ldots, 1}^{k-1}, -\frac{a_1}{a_0}, \ldots\ldots\} = \hat{\mathbf{e}}_{k-2} - \frac{a_1}{a_0}\hat{\mathbf{e}}_{k-1},$$

and the representation matrix above is obtained. ■

The representation matrix of a linear mapping depends on the chosen bases. The following theorem shows how the representation matrix changes when the basis is changed.

---

[10]This matrix coincides with the transpose of the companion matrix (2.14) associated with the polynomial $P(x) = x^k + (a_1/a_0)x^{k-1} + \cdots + (a_{k-1}/a_0)x + (a_k/a_0)$.

Theorem 9.18. *Let V and V' be vector spaces over K, and T a linear mapping from V to V'. Let B and $\hat{B}$ be two bases of V, and similarly, B' and $\hat{B}'$ be two bases of V'. Furthermore, let P denote the basis change matrix for $B \to \hat{B}$, and Q the basis change matrix for $B' \to \hat{B}'$. Then the representation matrix A of T with respect to B and B' is transformed to the representation matrix $\hat{A}$ of T with respect to $\hat{B}$ and $\hat{B}'$ as*

$$\hat{A} = Q^{-1}AP. \tag{9.4}$$

If T is a linear transformation (with $V = V'$), $B = B'$, and $\hat{B} = \hat{B}'$ (where $P = Q$), then the representation matrix A of T with respect to B is transformed to the representation matrix $\hat{A}$ of T with respect to $\hat{B}$ as

$$\hat{A} = P^{-1}AP. \tag{9.5}$$

Remark 9.14. Equation (9.4) shows an operation that multiplies a matrix with a nonsingular matrix P from the right and the inverse Q^{-1} of another nonsingular matrix Q from the left. Such operation (transformation) is called an *equivalence transformation.* Two matrices are said to be *equivalent* if they are connected by an equivalence transformation with some nonsingular matrices P and Q. In contrast, (9.5) shows an operation that multiplies a matrix with a nonsingular matrix P from the right and P^{-1} from the left. Such operation (transformation) is called a *similarity transformation.* Two matrices are said to be *similar* if they are connected by a similarity transformation with some nonsingular matrix P. ■

9.5.4 *Spaces of Linear Mappings*

So far we have discussed properties of a linear mapping T from V to V' in general. In the following we are concerned with the properties of the set of all linear mappings from V to V'.

Theorem 9.19. *Let V and V' be vector spaces over K, and $\{\boldsymbol{b}_\lambda\}_{\lambda \in \Lambda}$ a basis of V, where the index set Λ may possibly be an infinite set. For any family $\{\boldsymbol{y}_\lambda\}_{\lambda \in \Lambda}$ of vectors in V', there uniquely exists a linear mapping T from V to V' that satisfies*

$$T(\boldsymbol{b}_\lambda) = \boldsymbol{y}_\lambda \qquad (\lambda \in \Lambda).$$

If V and V' are finite-dimensional, we have the following statement.

Theorem 9.20. *Let V and V' be vector spaces over K, and $B = [\boldsymbol{b}_1, \boldsymbol{b}_2, \ldots, \boldsymbol{b}_k]$ and $B' = [\boldsymbol{b}'_1, \boldsymbol{b}'_2, \ldots, \boldsymbol{b}'_l]$ be bases of V and V', respectively.*

Then the matrix representation with respect to B and B' establishes a one-to-one correspondence between the set of linear mappings from V to V' and the set of $l \times k$ matrices. That is, any linear mapping from V to V' determines an $l \times k$ representation matrix, and conversely, for any $l \times k$ matrix, there uniquely exists a linear mapping whose representation matrix coincides with the given matrix.

Next we consider the structure of the set of all linear mappings by introducing operations on linear mappings. For vector spaces V and V' over K, we denote by $\mathcal{L}(V, V')$ the set of all linear mappings from V to V'.

Theorem 9.21. *Let V and V' be vector spaces over K. For $T_1, T_2, T \in \mathcal{L}(V, V')$ and $a \in K$, define the sum $T_1 + T_2$ and the scalar multiple aT by*

$$(T_1 + T_2)(\boldsymbol{x}) = T_1(\boldsymbol{x}) + T_2(\boldsymbol{x}) \qquad (\boldsymbol{x} \in V),$$
$$(a\,T)(\boldsymbol{x}) = a\,T(\boldsymbol{x}) \qquad (\boldsymbol{x} \in V).$$

Then $T_1 + T_2$ and $a\,T$ are linear mappings from V to V'. Hence, $\mathcal{L}(V, V')$ forms a vector space over K with respect to these operations.

Theorem 9.22. *Let V, V', and V'' be vector spaces over K. For $S \in \mathcal{L}(V, V')$ and $T \in \mathcal{L}(V', V'')$, the composition $T \circ S$ defined by*

$$(T \circ S)(\boldsymbol{x}) = T(S(\boldsymbol{x})) \qquad (\boldsymbol{x} \in V)$$

is a linear mapping from V to V'', that is, $T \circ S \in \mathcal{L}(V, V'')$.

Theorem 9.23. *Let V be a vector space over K. The set $\mathcal{L}(V, V)$ of all linear transformations in V forms a ring*[11] *with respect to the addition $T_1 + T_2$ in Theorem 9.21 and the multiplication $T_1 T_2$ defined by*

$$T_1 T_2 = T_1 \circ T_2$$

for $T_1, T_2 \in \mathcal{L}(V, V)$. The zero element of this ring is the null transformation $O : \boldsymbol{x} \mapsto \boldsymbol{0}$ and the unit element is the identity transformation $I : \boldsymbol{x} \mapsto \boldsymbol{x}$.

9.5.5 *Invariant Subspaces*

Definition 9.19. Let V be a vector space over K and T a linear transformation in V. A subspace W of V is called an *invariant subspace* of T if $T(W) \subseteq W$. ■

[11] A ring is a set (algebraic system) equipped with operations of addition, subtraction, and multiplication. See Remark 9.2 in Sec. 9.1.1 for the precise definition.

Theorem 9.24. *Let V be a vector space over K and T a linear transformation in V.*

(1) *Both V and $\{\mathbf{0}\}$ are invariant subspaces of T.*
(2) *If W_1 and W_2 are invariant subspaces of T, the intersection $W_1 \cap W_2$ and the sum $W_1 + W_2$ are invariant subspaces of T.*

Theorem 9.25. *Let V be a finite-dimensional vector space over K, T a linear transformation in V, and W ($\subseteqq V$) an invariant subspace of T. For any complementary subspace W' of W, let $B_1 = [\boldsymbol{b}_1, \boldsymbol{b}_2, \ldots, \boldsymbol{b}_m]$ be a basis of W and $B_2 = [\boldsymbol{b}_{m+1}, \boldsymbol{b}_{m+2}, \ldots, \boldsymbol{b}_n]$ a basis of W', where $\dim V = n$, $\dim W = m$, and $\dim W' = n - m$. Then the union $B = [\boldsymbol{b}_1, \boldsymbol{b}_2, \ldots, \boldsymbol{b}_n]$ of B_1 and B_2 is a basis of V and the representation matrix of T with respect to B is a block-triangular matrix*

$$\begin{bmatrix} A_{11} & A_{12} \\ O & A_{22} \end{bmatrix},$$

where A_{11} is an $m \times m$ matrix, A_{12} is an $m \times (n-m)$ matrix, and A_{22} is an $(n-m) \times (n-m)$ matrix. If, in addition, W' is also an invariant subspace of T, then $A_{12} = O$ and the representation matrix of T is a block-diagonal matrix

$$\begin{bmatrix} A_{11} & O \\ O & A_{22} \end{bmatrix}.$$

9.5.6 *Isomorphism*

Definition 9.20. Let V and V' be vector spaces over K. They are said to be *isomorphic*, denoted as

$$V \cong V',$$

if there exists a bijective linear mapping $T : V \to V'$. Such linear mapping T is called an *isomorphism*. ■

Theorem 9.26. *The relation of being isomorphic is an equivalence relation among vector spaces over K.*

Theorem 9.27. *Two vector spaces V and V' over K are isomorphic if and only if* $\dim V = \dim V'$.

Example 9.35. An n-dimensional vector space V over K is isomorphic to K^n, that is, $V \cong K^n$. ■

Example 9.36 (*Homomorphism theorem*). Let V and V' be vector spaces over K. For a linear mapping $T : V \to V'$ we have[12]

$$V/(\operatorname{Ker} T) \cong \operatorname{Im} T.$$

■

Example 9.37. Let V be a vector space over K. For subspaces W_1 and W_2 of V we have

$$W_1/(W_1 \cap W_2) \cong (W_1 + W_2)/W_2.$$

■

Example 9.38. Let V and V' be vector spaces over K with $\dim V = n$ and $\dim V' = m$. Then we have

$$\mathcal{L}(V, V') \cong \mathcal{M}_{m,n}(K),$$

where an isomorphism is given by the one-to-one correspondence between linear mappings and matrices stated in Theorem 9.20 in Sec. 9.5.4. ■

Remark 9.15. Example 9.38 above implies that $\mathcal{L}(V, V)$ and $\mathcal{M}_{n,n}(K)$ are isomorphic as vector spaces, whereas Theorem 9.23 in Sec. 9.5.4 states that $\mathcal{L}(V, V)$ is a ring. It is easy to see that $\mathcal{M}_{n,n}(K)$ is also a ring with respect to the operations of matrix addition and multiplication, and furthermore, $\mathcal{L}(V, V)$ and $\mathcal{M}_{n,n}(K)$ are isomorphic as rings. Here we recall that two rings R and R' in general are said to be *isomorphic* if there exists a bijective mapping φ from R to R' satisfying

$$\varphi(a + b) = \varphi(a) +' \varphi(b), \qquad \varphi(a \cdot b) = \varphi(a) \cdot' \varphi(b),$$

where $+$ and $\cdot$ denote the addition and multiplication in R, and $+'$ and $\cdot'$ are those operations in R'. ■

9.5.7 *Projection*

Definition 9.21. A linear transformation T in a vector space V over K is called a *projection* (or *projection operator*) if

$$T^2 = T.$$

■

Projections are closely related to direct sum decompositions of a vector space.

---

[12] $V/(\operatorname{Ker} T)$ denotes the quotient space of V by $\operatorname{Ker} T$ (Definition 9.9 in Sec. 9.2.4).

Definition 9.22. When a vector space V over K is represented as a direct sum $V = W_1 \oplus W_2$ of two subspaces W_1 and W_2, each vector $\boldsymbol{x} \in V$ is represented uniquely as

$$\boldsymbol{x} = \boldsymbol{x}_1 + \boldsymbol{x}_2, \qquad \boldsymbol{x}_1 \in W_1, \quad \boldsymbol{x}_2 \in W_2.$$

The mapping T that maps $\boldsymbol{x}$ to $\boldsymbol{x}_1$ is a linear transformation in V, called the *projection along* W_2 *on* W_1.[13] ■

Theorem 9.28. *Let V be a vector space over K. For a direct sum decomposition*

$$V = W_1 \oplus W_2,$$

the projection T along W_2 on W_1 satisfies $T^2 = T$, and hence is a projection in the sense of Definition 9.21. Conversely, if a linear transformation T in V satisfies $T^2 = T$ (that is, a projection), then $\mathrm{Im}(I - T) = \mathrm{Ker}\, T$ holds, V is decomposed into a direct sum as

$$V = (\mathrm{Im}\, T) \oplus (\mathrm{Im}(I - T)) = (\mathrm{Im}\, T) \oplus (\mathrm{Ker}\, T),$$

and T is the projection along $\mathrm{Im}(I - T)$ on $\mathrm{Im}\, T$.

Theorem 9.28 can be extended for more than two subspaces.

Theorem 9.29. *Let V be a vector space over K and $k \geqq 2$.*

(1) *Suppose that V is a direct sum of subspaces $W_1, W_2, \ldots, W_k$, that is,*

$$V = W_1 \oplus W_2 \oplus \cdots \oplus W_k.$$

For each $i = 1, 2, \ldots, k$, let T_i denote the projection along $W_1 \oplus \cdots \oplus W_{i-1} \oplus W_{i+1} \oplus \cdots \oplus W_k$ on W_i. Then the following hold for $T_1, T_2, \ldots, T_k$:

$$I = T_1 + T_2 + \cdots + T_k, \tag{9.6}$$

$$T_i{}^2 = T_i \qquad (1 \leqq i \leqq k), \tag{9.7}$$

$$T_i T_j = O \qquad (i \neq j; 1 \leqq i, j \leqq k). \tag{9.8}$$

(2) *If (9.6), (9.7), and (9.8) hold for linear transformations $T_1, T_2, \ldots, T_k$ in V, we have a direct sum decomposition*

$$V = (\mathrm{Im}\, T_1) \oplus (\mathrm{Im}\, T_2) \oplus \cdots \oplus (\mathrm{Im}\, T_k),$$

and each T_i is the projection along $(\mathrm{Im}\, T_1) \oplus \cdots \oplus (\mathrm{Im}\, T_{i-1}) \oplus (\mathrm{Im}\, T_{i+1}) \oplus \cdots \oplus (\mathrm{Im}\, T_k)$ on $\mathrm{Im}\, T_i$.

[13] In Definition 9.22, $\boldsymbol{x}_1$ is also called the *projection* of $\boldsymbol{x}$ along W_2 on W_1.

Remark 9.16. If $k = 2$ in Theorem 9.29, the conditions (9.6), (9.7), and (9.8) together are equivalent to

$$I = T_1 + T_2, \qquad {T_1}^2 = T_1.$$

Therefore, Theorem 9.28 is precisely the special case of Theorem 9.29 for $k = 2$. ■

Remark 9.17. The condition (9.7) is not completely independent of (9.8), but (9.7) can be derived from (9.8) under (9.6). Therefore, three conditions (9.6), (9.7), and (9.8) may be replaced by two conditions (9.6) and (9.8). Furthermore, in the case of a finite-dimensional V, (9.7) and (9.8) are equivalent under (9.6), and also the dimension condition

$$\dim(\operatorname{Im} T_1) + \dim(\operatorname{Im} T_2) + \cdots + \dim(\operatorname{Im} T_k) = \dim V$$

is equivalent to (9.8). ■

Definition 9.23. A square matrix P over K is called a *projection matrix* if

$$P^2 = P.$$ ■

Theorem 9.30. *In a finite-dimensional vector space over K, the representation matrix of a projection T (with respect to any basis) is a projection matrix.*[14] *Conversely, if the representation matrix of a linear transformation (with respect to an arbitrarily chosen basis) is a projection matrix, then the linear transformation is a projection.*

Theorem 9.31. *A projection matrix P is a diagonalizable matrix*[15] *with eigenvalues* 1 *or* 0. *Conversely, such square matrix is a projection matrix.*

Theorem 9.32. *The rank of a projection matrix P is equal to its trace, that is,*

$$\operatorname{rank} P = \operatorname{Tr} P.$$

9.6 Inner Product

Throughout Sec. 9.6, we assume that $K = \mathbb{R}$ or $\mathbb{C}$.

---

[14] The concrete forms of representation matrices for $\mathbb{R}^n$ and $\mathbb{C}^n$ are given in Theorem 9.51 in Sec. 9.6.6.

[15] That is, there exists a nonsingular matrix S for which $S^{-1}PS$ is a diagonal matrix; see Sec. 6.7.1.

9.6.1 *Definition*

Definition 9.24. Let V be a vector space over $\mathbb{R}$. A function $(\cdot\,,\cdot) : V \times V \to \mathbb{R}$, which assigns a real number $(\boldsymbol{x}, \boldsymbol{y})$ to every pair of vectors $\boldsymbol{x}, \boldsymbol{y}$ in V, is called an *inner product* on V if it satisfies the following conditions (1) to (4).

(1) $(\boldsymbol{x}, \boldsymbol{x}) \geqq 0$ for all $\boldsymbol{x} \in V$, where the equality holds only for $\boldsymbol{x} = \mathbf{0}$.
(2) $(\boldsymbol{x}, \boldsymbol{y}) = (\boldsymbol{y}, \boldsymbol{x})$ for all $\boldsymbol{x}, \boldsymbol{y} \in V$.
(3) $(\boldsymbol{x} + \boldsymbol{y}, \boldsymbol{z}) = (\boldsymbol{x}, \boldsymbol{z}) + (\boldsymbol{y}, \boldsymbol{z})$ for all $\boldsymbol{x}, \boldsymbol{y}, \boldsymbol{z} \in V$.
(4) $(a\boldsymbol{x}, \boldsymbol{y}) = a(\boldsymbol{x}, \boldsymbol{y})$ for all $a \in \mathbb{R}$ and $\boldsymbol{x}, \boldsymbol{y} \in V$.

If V is a vector space over $\mathbb{C}$, a mapping $(\cdot\,,\cdot) : V \times V \to \mathbb{C}$, which assigns a complex number $(\boldsymbol{x}, \boldsymbol{y})$ to every pair of vectors $\boldsymbol{x}, \boldsymbol{y}$ in V, is called an *inner product* on V if it satisfies (1) and (3) above and (2′) and (4′) below.

(2′) $(\boldsymbol{x}, \boldsymbol{y}) = \overline{(\boldsymbol{y}, \boldsymbol{x})}$ for all $\boldsymbol{x}, \boldsymbol{y} \in V$.
(4′) $(a\boldsymbol{x}, \boldsymbol{y}) = a(\boldsymbol{x}, \boldsymbol{y})$ for all $a \in \mathbb{C}$ and $\boldsymbol{x}, \boldsymbol{y} \in V$.

A vector space equipped with an inner product is called a *metric vector space* (or *inner product space*). In particular, if V is a vector space over $\mathbb{R}$, it is called a *real metric vector space* (or *real inner product space*). If V is a vector space over $\mathbb{C}$, it is called a *complex metric vector space* (or *complex inner product space*). ■

Theorem 9.33. *Let V be a metric vector space. A subspace of V is a metric vector space with respect to the inner product of V.*

Definition 9.25. Let V be a metric vector space. For any vector $\boldsymbol{x} \in V$, $\sqrt{(\boldsymbol{x}, \boldsymbol{x})}$ is called the *length* (or *norm*) of $\boldsymbol{x}$, and is denoted by $\|\boldsymbol{x}\|$. A vector of length 1 is called a *unit-length vector* (or *unit vector*). ■

Theorem 9.34. *Let V be a metric vector space. The length $\|\boldsymbol{x}\|$ has the following properties.*

(1) $\|\boldsymbol{x}\| \geqq 0$ *for all $\boldsymbol{x} \in V$, where the equality holds only for $\boldsymbol{x} = \mathbf{0}$.*
(2) $\|a\boldsymbol{x}\| = |a| \cdot \|\boldsymbol{x}\|$ *for all $a \in K$, $\boldsymbol{x} \in V$.*
(3) $|(\boldsymbol{x}, \boldsymbol{y})| \leqq \|\boldsymbol{x}\| \cdot \|\boldsymbol{y}\|$ *for all $\boldsymbol{x}, \boldsymbol{y} \in V$.*
(4) $\|\boldsymbol{x} + \boldsymbol{y}\| \leqq \|\boldsymbol{x}\| + \|\boldsymbol{y}\|$ *for all $\boldsymbol{x}, \boldsymbol{y} \in V$* (triangle inequality).

The property (3) above is a generalization of the Cauchy–Schwarz inequality in Remark 1.18 in Sec. 1.6.

Remark 9.18. In this section we restrict ourselves to $K = \mathbb{R}$ or $\mathbb{C}$. The main reason for this restriction is that, in a general field, we cannot consider nonnegativity $(\boldsymbol{x}, \boldsymbol{x}) \geqq 0$ in the definition of an inner product and square root $\|\boldsymbol{x}\| = \sqrt{(\boldsymbol{x}, \boldsymbol{x})}$ in the definition of a norm. ■

9.6.2 *Examples*

Example 9.39. In the real vector space $\mathbb{R}^n$, $(\boldsymbol{x}, \boldsymbol{y})$ defined by

$$(\boldsymbol{x}, \boldsymbol{y}) = \boldsymbol{x}^\top \boldsymbol{y} = x_1 y_1 + x_2 y_2 + \cdots + x_n y_n$$

for $\boldsymbol{x} = (x_1, x_2, \ldots, x_n)^\top$, $\boldsymbol{y} = (y_1, y_2, \ldots, y_n)^\top \in \mathbb{R}^n$ satisfies the conditions for an inner product, and renders $\mathbb{R}^n$ a real metric vector space. This inner product is called the *standard inner product* of $\mathbb{R}^n$. The length of a vector $\boldsymbol{x}$ is given by $\|\boldsymbol{x}\| = \sqrt{{x_1}^2 + {x_2}^2 + \cdots + {x_n}^2}$. ■

Example 9.40. In the complex vector space $\mathbb{C}^n$, $(\boldsymbol{z}, \boldsymbol{w})$ defined by

$$(\boldsymbol{z}, \boldsymbol{w}) = \boldsymbol{z}^\top \overline{\boldsymbol{w}} = z_1 \overline{w_1} + z_2 \overline{w_2} + \cdots + z_n \overline{w_n}$$

for $\boldsymbol{z} = (z_1, z_2, \ldots, z_n)^\top$, $\boldsymbol{w} = (w_1, w_2, \ldots, w_n)^\top \in \mathbb{C}^n$ satisfies the conditions for an inner product, and renders $\mathbb{C}^n$ a complex metric vector space. This inner product is called the *standard inner product* of $\mathbb{C}^n$. The length of a vector $\boldsymbol{z}$ is given by $\|\boldsymbol{z}\| = \sqrt{|z_1|^2 + |z_2|^2 + \cdots + |z_n|^2}$. ■

Remark 9.19. In physics, for example, an inner product in $\mathbb{C}^n$ is often defined as

$$(\boldsymbol{z}, \boldsymbol{w}) = \overline{\boldsymbol{z}}^\top \boldsymbol{w} = \overline{z_1} w_1 + \overline{z_2} w_2 + \cdots + \overline{z_n} w_n.$$

In this case, the condition (4′) in Definition 9.24 is replaced by

(4″) $(a\boldsymbol{x}, \boldsymbol{y}) = \overline{a}(\boldsymbol{x}, \boldsymbol{y})$ for all $a \in \mathbb{C}$ and $\boldsymbol{x}, \boldsymbol{y} \in V$. ■

Example 9.41. In the real vector space $\mathcal{M}_{m,n}(\mathbb{R})$ of all $m \times n$ real matrices (Example 9.4 in Sec. 9.1.2), (X, Y) defined by

$$(X, Y) = \mathrm{Tr}\,(X^\top Y) = \sum_{i=1}^{m} \sum_{j=1}^{n} x_{ij} y_{ij}$$

for $X = (x_{ij}), Y = (y_{ij}) \in \mathcal{M}_{m,n}(\mathbb{R})$ satisfies the conditions for an inner product, and renders $\mathcal{M}_{m,n}(\mathbb{R})$ a real metric vector space. The length of a matrix X is given by $\|X\| = (\mathrm{Tr}\,(X^\top X))^{1/2}$, which is nothing but the Frobenius norm of X. ■

Example 9.42. In the space ℓ^2 of all 2-summable real sequences (Example 9.5 in Sec. 9.1.2), $(\boldsymbol{x}, \boldsymbol{y})$ defined by

$$(\boldsymbol{x}, \boldsymbol{y}) = \sum_{n=0}^{\infty} x_n y_n$$

for $\boldsymbol{x} = \{x_n\}_{n=0}^{\infty}$, $\boldsymbol{y} = \{y_n\}_{n=0}^{\infty} \in \ell^2$ satisfies the conditions for an inner product, and renders ℓ^2 a real metric vector space. The length of a sequence $\boldsymbol{x}$ is given by $\|\boldsymbol{x}\| = \big(\sum_{n=0}^{\infty} x_n{}^2\big)^{1/2}$. Note that the infinite series defining $(\boldsymbol{x}, \boldsymbol{y})$ is absolutely convergent (and hence convergent), because, for any N, we have

$$\sum_{n=0}^{N} |x_n y_n| \leqq \Big(\sum_{n=0}^{N} x_n{}^2\Big)^{1/2} \Big(\sum_{n=0}^{N} y_n{}^2\Big)^{1/2} \leqq \|\boldsymbol{x}\| \cdot \|\boldsymbol{y}\|$$

from the Cauchy–Schwarz inequality. ■

Example 9.43. In the real vector space $\mathrm{C}^0([a, b], \mathbb{R})$ of all continuous functions on a finite interval $[a, b]$ with $a < b$, (f, g) defined by

$$(f, g) = \int_a^b f(x)g(x)\mathrm{d}x$$

for $f, g \in \mathrm{C}^0([a, b], \mathbb{R})$ satisfies the conditions for an inner product, and renders $\mathrm{C}^0([a, b], \mathbb{R})$ a real metric vector space. The length of a continuous function f is given by $\|f\| = \big(\int_a^b (f(x))^2 \mathrm{d}x\big)^{1/2}$. ■

Example 9.44. In the real vector space $\mathcal{P}(\mathbb{R})$ of all polynomials, (P, Q) defined by

$$(P, Q) = \int_a^b P(x)Q(x)\mathrm{d}x$$

for $P, Q \in \mathcal{P}(\mathbb{R})$, where a and b are finite with $a < b$, satisfies the conditions for an inner product. With this inner product, $\mathcal{P}(\mathbb{R})$ is a real metric vector space. The length of a polynomial P is given by $\|P\| = \big(\int_a^b (P(x))^2 \mathrm{d}x\big)^{1/2}$. ■

9.6.3 *Orthonormal Basis*

Definition 9.26. Let V be a metric vector space with an inner product $(\cdot\,, \cdot)$. Two vectors $\boldsymbol{x}, \boldsymbol{y} \in V$ are said to be *orthogonal* to each other, denoted by $\boldsymbol{x} \perp \boldsymbol{y}$, if $(\boldsymbol{x}, \boldsymbol{y}) = 0$. Two subspaces W_1 and W_2 of V are said to be orthogonal to each other, denoted by $W_1 \perp W_2$, if $\boldsymbol{x} \perp \boldsymbol{y}$ for all $\boldsymbol{x} \in W_1$ and $\boldsymbol{y} \in W_2$. ■

Definition 9.27. Let V be a metric vector space and S a subset of V.

- S is called an *orthogonal system* if every two (distinct) vectors in S are orthogonal to each other.
- S is called an *orthonormal system* if it is an orthogonal system and every vector in S has length 1.
- S is called an *orthonormal basis* if it is an orthonormal system and is a basis of V. ■

Theorem 9.35. *An orthogonal system is linearly independent, provided that it does not contain the zero vector.*

Theorem 9.36. *Let V be a metric vector space. Given linearly independent vectors $\boldsymbol{x}_1, \boldsymbol{x}_2, \ldots, \boldsymbol{x}_k$, we can construct an orthonormal system $\{\mathbf{e}_1, \mathbf{e}_2, \ldots, \mathbf{e}_k\}$ according to the recurrence relation*

$$\tilde{\mathbf{e}}_j = \boldsymbol{x}_j - \sum_{i=1}^{j-1} (\boldsymbol{x}_j, \mathbf{e}_i)\mathbf{e}_i, \quad \mathbf{e}_j = \tilde{\mathbf{e}}_j / \|\tilde{\mathbf{e}}_j\| \qquad (j = 1, 2, \ldots, k).$$

The procedure given in Theorem 9.36 is known as the *Gram–Schmidt orthogonalization.*

Theorem 9.37. *Let V be a metric vector space whose dimension is finite or countably infinite.*

(1) *An orthonormal basis exists in V.*
(2) *If $\{\mathbf{e}_1, \mathbf{e}_2, \ldots, \mathbf{e}_k\}$ is an orthonormal system, there exists an orthonormal basis of V that contains $\{\mathbf{e}_1, \mathbf{e}_2, \ldots, \mathbf{e}_k\}$.*

Theorem 9.38. *Let V be a finite-dimensional metric vector space and $\{\mathbf{e}_1, \mathbf{e}_2, \ldots, \mathbf{e}_n\}$ an orthonormal basis of V. For any vector $\boldsymbol{x} \in V$, let $(x_1, x_2, \ldots, x_n)^\top$ be its coordinates with respect to the basis $\{\mathbf{e}_1, \mathbf{e}_2, \ldots, \mathbf{e}_n\}$. Then we have the following.*

(1) $x_i = (\boldsymbol{x}, \mathbf{e}_i) \quad (i = 1, 2, \ldots, n)$.
(2) $\|\boldsymbol{x}\|^2 = \sum_{i=1}^{n} |x_i|^2$.

Theorem 9.39. *In a finite-dimensional real metric vector space, the basis change matrix between two orthonormal bases is an orthogonal matrix. In the case of a complex metric vector space, the matrix is a unitary matrix.*

9.6.4 *Orthogonal Complement*

Throughout Sec. 9.6.4, vector spaces are assumed to be finite-dimensional.

Definition 9.28. Let W be a subspace of a metric vector space V. The set of the vectors in V that are orthogonal to every vector in W:

$$W^{\perp} = \{\boldsymbol{y} \in V \mid (\boldsymbol{y}, \boldsymbol{x}) = 0 \ (\boldsymbol{x} \in W)\} \tag{9.9}$$

is a subspace of V, called the *orthogonal complement* of W in V. ■

Theorem 9.40. *The following hold for a subspace W of a metric vector space V.*

(1) $V = W \oplus W^{\perp}$.
(2) $\dim V = \dim W + \dim W^{\perp}$.

Theorem 9.41. *The following hold for subspaces W, W_1, and W_2 of a metric vector space V.*

(1) $(W^{\perp})^{\perp} = W$.
(2) $(W_1 + W_2)^{\perp} = {W_1}^{\perp} \cap {W_2}^{\perp}$.
(3) $(W_1 \cap W_2)^{\perp} = {W_1}^{\perp} + {W_2}^{\perp}$.

Definition 9.29. For a metric vector space V, an *orthogonal direct sum decomposition* of V means a representation of V as a direct sum of pairwise orthogonal subspaces $W_1, W_2, \ldots, W_k$:

$$V = W_1 \oplus W_2 \oplus \cdots \oplus W_k, \qquad W_i \perp W_j \quad (i \neq j).$$

■

9.6.5 *Adjoint Mapping*

Throughout Sec. 9.6.5, vector spaces are assumed to be finite-dimensional.

Definition 9.30. Let V and V' be metric vector spaces. For any linear mapping T from V to V', there uniquely exists[16] a linear mapping T^* from V' to V such that, for any $\boldsymbol{x} \in V$ and $\boldsymbol{y} \in V'$, the equality

$$(T(\boldsymbol{x}), \boldsymbol{y})_{V'} = (\boldsymbol{x}, T^*(\boldsymbol{y}))_V \tag{9.10}$$

holds, where $(\cdot\,,\cdot)_{V'}$ and $(\cdot\,,\cdot)_V$ are inner products in V' and V, respectively. This T^* is called the *adjoint mapping* of T. If $V = V'$, it is called the *adjoint transformation.* ■

[16]The existence of such T^* is not guaranteed for infinite-dimensional spaces. This is why we restrict ourselves to finite-dimensional spaces in this section.

Theorem 9.42. *Let V, V', and V'' be metric vector spaces, and T a linear mapping from V to V'.*

(1) $(T^*)^* = T$.
(2) $(T+S)^* = T^* + S^*$, *where* $S : V \to V'$.
(3) $(cT)^* = \bar{c}T^*$. *In a real metric vector space,* $(cT)^* = cT^*$.
(4) $(S \circ T)^* = T^* \circ S^*$, *where* $S : V' \to V''$.

Theorem 9.43. *Let V and V' be metric vector spaces, and T a linear mapping from V to V'.*

(1) $\mathrm{Im}(T^*) = (\mathrm{Ker}\, T)^{\perp}$.
(2) $\mathrm{Ker}(T^*) = (\mathrm{Im}\, T)^{\perp}$.
(3) $\dim(\mathrm{Im}(T^*)) = \dim(\mathrm{Im}\, T)$.

Theorem 9.44. *Let V and V' be real metric vector spaces with respective orthonormal bases B and B', and T a linear mapping from V to V'. If A is the representation matrix of T with respect to B and B', the representation matrix of T^* with respect to B' and B is given by $A^\top$. If V and V' are complex metric vector spaces, the representation matrix of T^* is given by $A^* = (\overline{A})^\top$.*

Definition 9.31. A linear transformation T in a real metric vector space V is called a *symmetric transformation* in V if $(T(\boldsymbol{x}), \boldsymbol{y}) = (\boldsymbol{x}, T(\boldsymbol{y}))$ for all $\boldsymbol{x}, \boldsymbol{y} \in V$, that is, if $T^* = T$. In the case of a complex metric vector space, it is called a *Hermitian transformation.* ■

Theorem 9.45. *In a real metric vector space, the representation matrix of a symmetric transformation T with respect to an orthonormal basis is a symmetric matrix. In the case of a complex metric vector space, the representation matrix of a Hermitian transformation is a Hermitian matrix.*

Definition 9.32. A linear transformation T in a real metric vector space V is called an *orthogonal transformation* in V if $(T(\boldsymbol{x}), T(\boldsymbol{y})) = (\boldsymbol{x}, \boldsymbol{y})$ for all $\boldsymbol{x}, \boldsymbol{y} \in V$, that is, if $T^*T = I$. In the case of a complex metric vector space, it is called a *unitary transformation.* ■

Theorem 9.46. *In a real metric vector space, the representation matrix of an orthogonal transformation T with respect to an orthonormal basis is an orthogonal matrix. In the case of a complex metric vector space, the representation matrix of a unitary transformation is a unitary matrix.*

9.6.6 *Orthogonal Projection*

Throughout Sec. 9.6.6, vector spaces are assumed to be finite-dimensional.

Definition 9.33. A linear transformation T in a metric vector space V is called an *orthogonal projection* if

$$T^2 = T, \qquad T^* = T.$$

■

Orthogonal projections are closely related to orthogonal direct sum decompositions of a vector space.

Definition 9.34. Let V be a metric vector space. For any subspace W of V, we have an orthogonal direct sum decomposition $V = W \oplus W^{\perp}$ (Theorem 9.40(1) in Sec. 9.6.4). The projection along $W^{\perp}$ on W is called the *orthogonal projection on* W.[17] ■

Remark 9.20. Let V be a metric vector space and T the orthogonal projection on a subspace W of V. For any vector $\boldsymbol{x}$ in V, $T(\boldsymbol{x})$ is the unique vector in W that minimizes the distance $\|\boldsymbol{x} - \boldsymbol{y}\|$ between $\boldsymbol{x}$ and a vector $\boldsymbol{y}$ in W. That is,

$$\|\boldsymbol{x} - T(\boldsymbol{x})\| = \min_{\boldsymbol{y} \in W} \|\boldsymbol{x} - \boldsymbol{y}\|$$

and $T(\boldsymbol{x})$ is the unique vector $\boldsymbol{y} \in W$ that attains the minimum on the right-hand side. ■

Theorem 9.47. *Let V be a metric vector space. For a subspace W of V, the orthogonal projection T on W satisfies $T^2 = T$ and $T^* = T$, and hence is an orthogonal projection in the sense of Definition 9.33. Conversely, if a linear transformation T in V satisfies $T^2 = T$ and $T^* = T$ (that is, an orthogonal projection), then T is the orthogonal projection on $\operatorname{Im} T$, $(\operatorname{Im} T)^{\perp} = \operatorname{Im}(I - T) = \operatorname{Ker} T$ holds, and V is decomposed into an orthogonal direct sum as*

$$V = (\operatorname{Im} T) \oplus (\operatorname{Im}(I - T)) = (\operatorname{Im} T) \oplus (\operatorname{Ker} T).$$

Theorem 9.48. *Let V be a metric vector space and $k \geqq 2$.*

(1) *Suppose that V is an orthogonal direct sum of subspaces $W_1, W_2, \ldots, W_k$ as $V = W_1 \oplus W_2 \oplus \cdots \oplus W_k$. For each $i = 1, 2, \ldots, k$, let*

[17] We do not have to say "along $W^{\perp}$" since $W^{\perp}$ is determined uniquely by W.

T_i denote the orthogonal projection on W_i. Then the following hold for $T_1, T_2, \ldots, T_k$:

$$I = T_1 + T_2 + \cdots + T_k, \tag{9.11}$$

$$T_i^2 = T_i, \qquad T_i^* = T_i \qquad (1 \leqq i \leqq k), \tag{9.12}$$

$$T_i T_j = O \qquad (i \neq j; 1 \leqq i, j \leqq k). \tag{9.13}$$

(2) *If (9.11), (9.12), and (9.13) hold for linear transformations $T_1, T_2, \ldots, T_k$ in V, then $\operatorname{Im} T_1, \operatorname{Im} T_2, \ldots, \operatorname{Im} T_k$ are pairwise orthogonal, V is decomposed into an orthogonal direct sum*

$$V = (\operatorname{Im} T_1) \oplus (\operatorname{Im} T_2) \oplus \cdots \oplus (\operatorname{Im} T_k),$$

and each T_i is the orthogonal projection on $\operatorname{Im} T_i$.

Definition 9.35. A real square matrix P is called an *orthogonal projection matrix* if

$$P^2 = P, \qquad P^\top = P.$$

Similarly, a complex square matrix P is called an *orthogonal projection matrix* if

$$P^2 = P, \qquad P^* = P.$$

■

Theorem 9.49. *In a metric vector space, the representation matrix of an orthogonal projection T with respect to any orthonormal basis is an orthogonal projection matrix.*

Theorem 9.50. *For a subspace W in the real vector space $\mathbb{R}^n$, a representation matrix of the orthogonal projection T on W is given by*

$$P = X(X^\top X)^{-1} X^\top,$$

where X is an $n \times (\dim W)$ matrix consisting of the (column) vectors of any basis of W (which is not necessarily an orthogonal basis). In the complex vector space $\mathbb{C}^n$, we have

$$P = X(X^* X)^{-1} X^*.$$

For a general projection (not necessarily orthogonal), the representation matrix is given as follows.

Theorem 9.51. *For a direct sum decomposition $W_1 \oplus W_2$ of the real vector space $\mathbb{R}^n$, a representation matrix of the projection T along W_2 on W_1 is given by*

$$P = X(Y^\top X)^{-1} Y^\top,$$

where X is an $n \times (\dim W_1)$ matrix consisting of the (column) vectors of any basis of W_1 and Y is an $n \times (\dim W_1)$ matrix consisting of the (column) vectors of any basis of $W_2^\perp$. In the complex vector space $\mathbb{C}^n$, we have

$$P = X(Y^* X)^{-1} Y^*.$$

9.6.7 *Normed Space*

In Sec. 9.6.1 we have defined the norm (length) of a vector from an inner product and presented its fundamental properties in Theorem 9.34. In this section we shall introduce the concept of a norm (length) in a general vector space that is not equipped with an inner product. The definition below is based on the properties stated in Theorem 9.34(1), (2), and (4).

Definition 9.36. Let V be a vector space over K. A function $\|\cdot\| : V \to \mathbb{R}$, which assigns a real number $\|\boldsymbol{x}\|$ to every vector $\boldsymbol{x}$ in V, is called a *norm* on V if it satisfies the following conditions (1), (2), and (3).

(1) $\|\boldsymbol{x}\| \geqq 0$ for all $\boldsymbol{x} \in V$, where the equality holds only for $\boldsymbol{x} = \mathbf{0}$.
(2) $\|a\boldsymbol{x}\| = |a| \cdot \|\boldsymbol{x}\|$ for all $a \in K$, $\boldsymbol{x} \in V$.
(3) $\|\boldsymbol{x} + \boldsymbol{y}\| \leqq \|\boldsymbol{x}\| + \|\boldsymbol{y}\|$ for all $\boldsymbol{x}, \boldsymbol{y} \in V$ (*triangle inequality*).

A vector space equipped with a norm is called a *normed space.* ■

As an inner product naturally induces a norm, every metric vector space can be regarded as a normed space. However, the converse is not true, that is, not every norm can be induced from an inner product. An additional condition for this is given in the following theorem.

Theorem 9.52. *Let* $\|\cdot\|$ *be a norm on* V. *There exists an inner product* $(\cdot,\cdot)$ *on* V *for which* $\|\boldsymbol{x}\| = \sqrt{(\boldsymbol{x}, \boldsymbol{x})}$ *for all* $\boldsymbol{x} \in V$ *if and only if*

$$\|\boldsymbol{x} + \boldsymbol{y}\|^2 + \|\boldsymbol{x} - \boldsymbol{y}\|^2 = 2(\|\boldsymbol{x}\|^2 + \|\boldsymbol{y}\|^2)$$

holds[18] *for all* $\boldsymbol{x}, \boldsymbol{y} \in V$.

In Examples 9.45 to 9.49 below, we assume that $p \geqq 1$ or $p = \infty$.

Example 9.45. In the real vector space $\mathbb{R}^n$, the function $\|\cdot\|_p$ defined by

$$\|\boldsymbol{x}\|_p = \begin{cases} (\sum_{j=1}^{n} |x_j|^p)^{1/p} & (1 \leqq p < \infty), \\ \max_{1 \leqq j \leqq n} |x_j| & (p = \infty) \end{cases}$$

for $\boldsymbol{x} = (x_1, x_2, \ldots, x_n)^\top \in \mathbb{R}^n$ satisfies the conditions for a norm, and renders $\mathbb{R}^n$ a normed space. This norm (depending on p) is called the *p-norm*, which is the same as the one introduced in Definition 1.43 in Sec. 1.6. ■

[18] This identity corresponds to the *parallelogram law* in the elementary Euclidean geometry.

Example 9.46. In the real vector space $\mathcal{M}_{m,n}(\mathbb{R})$ (Example 9.41 in Sec. 9.6.2), the p-norm $\|X\|_p$ for $X \in \mathcal{M}_{m,n}(\mathbb{R})$ in Definition 1.44 in Sec. 1.6 satisfies the conditions for a norm (Theorem 1.11), and renders $\mathcal{M}_{m,n}(\mathbb{R})$ a normed space. The Frobenius norm $\|X\|_{\mathrm{F}}$ also satisfies the conditions for a norm (Theorem 1.14), and therefore $\mathcal{M}_{m,n}(\mathbb{R})$ is a normed space with respect to the Frobenius norm $\|X\|_{\mathrm{F}}$. The Frobenius norm is induced from the inner product $(X,Y) = \mathrm{Tr}\,(X^\top Y)$ defined in Example 9.41 as $\|X\| = \sqrt{(X,X)}$. ■

Example 9.47. In the real vector space ℓ^p, the function $\|\cdot\|_p$ defined by

$$\|\boldsymbol{x}\|_p = \begin{cases} \left(\sum_{n=0}^{\infty} |x_n|^p\right)^{1/p} & (1 \leqq p < \infty), \\ \sup_{n \geqq 0} |x_n| & (p = \infty) \end{cases}$$

for $\boldsymbol{x} = \{x_n\}_{n=0}^{\infty} \in \ell^p$ satisfies the conditions for a norm, and renders ℓ^p a normed space.[19] This norm is called the p-norm, similarly to Example 9.45. ■

Example 9.48. In the real vector space $\mathrm{C}^0([a,b],\mathbb{R})$ of all continuous functions on a finite interval $[a,b]$ with $a < b$, the function $\|\cdot\|_p$ defined by

$$\|f\|_p = \begin{cases} \left(\int_a^b |f(x)|^p \mathrm{d}x\right)^{1/p} & (1 \leqq p < \infty), \\ \max_{x \in [a,b]} |f(x)| & (p = \infty) \end{cases}$$

for $f \in \mathrm{C}^0([a,b],\mathbb{R})$ satisfies the conditions for a norm, and renders $\mathrm{C}^0([a,b],\mathbb{R})$ a normed space. This norm is called the p-norm. ■

Example 9.49. In the real vector space $\mathcal{P}(\mathbb{R})$ of all polynomials, the function $\|\cdot\|_p$ defined by

$$\|P\|_p = \begin{cases} \left(\int_a^b |P(x)|^p \mathrm{d}x\right)^{1/p} & (1 \leqq p < \infty), \\ \max_{x \in [a,b]} |P(x)| & (p = \infty) \end{cases}$$

for $P \in \mathcal{P}(\mathbb{R})$, where a and b are finite with $a < b$, satisfies the conditions for a norm. This norm is called the p-norm. With this norm, $\mathcal{P}(\mathbb{R})$ is a normed space. ■

[19] ℓ^p with $1 \leqq p < \infty$ denotes the set of all p-summable real sequences, whereas, for $p = \infty$, ℓ^p is the set of all bounded real sequences. See Example 9.5 in Sec. 9.1.2.

Remark 9.21. Among the p-norms $\|\boldsymbol{x}\|_p$ on $\mathbb{R}^n$ ($n \geqq 2$) shown in Example 9.45, the 2-norm is induced from an inner product, whereas, for $p \neq 2$, there is no inner product $(\cdot\,,\cdot)$ satisfying $\|\boldsymbol{x}\|_p = \sqrt{(\boldsymbol{x},\boldsymbol{x})}$, which we can prove from Theorem 9.52. Similarly for the p-norms stated in Examples 9.47 to 9.49. The p-norm on $\mathcal{M}_{m,n}(\mathbb{R})$ ($m, n \geqq 2$) in Example 9.46 cannot be induced from an inner product for any p. ■

Definition 9.37. Suppose that two norms $\|\cdot\|$ and $\|\cdot\|'$ are defined on a vector space V. We say that $\|\cdot\|'$ is *equivalent* to $\|\cdot\|$ if there exist positive numbers c_1 and c_2 such that

$$c_1 \|\boldsymbol{x}\| \leqq \|\boldsymbol{x}\|' \leqq c_2 \|\boldsymbol{x}\|$$

holds for all $\boldsymbol{x} \in V$. We denote this equivalence as $\|\cdot\|' \sim \|\cdot\|$. ■

Theorem 9.53. *The relation of equivalence $\sim$ among norms is an equivalence relation (that is, the binary relation $\sim$ satisfies reflexive law, symmetric law, and transitive law).*

Remark 9.22. Suppose that we are concerned with convergence of a sequence of vectors $\boldsymbol{x}_0, \boldsymbol{x}_1, \boldsymbol{x}_2, \ldots$ in V. The convergence/divergence property, that is, whether the sequence converges or not, is not affected by our choice of a norm among equivalent norms.[20] In some cases, however, the choice of a norm matters. For example, if we are interested in an error estimate for an approximation scheme, the magnitude of errors should be measured in terms of an appropriate norm. ■

Remark 9.23. In $\mathbb{R}^n$ of Example 9.45, any two norms are known to be equivalent, and therefore, the choice of a norm does not matter in discussing convergence (cf., Remark 9.22). In contrast, in $\mathrm{C}^0([a,b],\mathbb{R})$ of Example 9.48 as well as in $\mathcal{P}(\mathbb{R})$ of Example 9.49, p-norms with different values of p are not equivalent, which implies that the convergence/divergence property of a sequence depends on the choice of a norm.[21] This indicates that an appropriate norm should be chosen in applications. ■

---

[20] A more sophisticated mathematical expression of this fact would be: Equivalent norms on V determine the same topology.

[21] For example, let $f_n(x) = n^\alpha x^n$ ($n = 0, 1, 2, \ldots$), where α is a parameter with $0 < \alpha < 1$. Each f_n is a continuous function (polynomial) on the interval $[a, b] = [0, 1]$, and the limit of $\|f_n\|_p$ varies with p as $\lim\limits_{n\to\infty} \|f_n\|_p = \begin{cases} 0 & (1 \leqq p < 1/\alpha), \\ p^{-1/p} & (p = 1/\alpha), \\ +\infty & (p > 1/\alpha). \end{cases}$

9.6.8 *Hilbert Space*

In dealing with linear mappings on metric vector spaces, we often need to consider various kinds of limiting operations. The concept of Hilbert space, treated in this section, is a convenient and standard framework of metric vector spaces in which limiting operations can be carried out.

Definition 9.38. A sequence $\{\boldsymbol{x}_n\}$ in a normed space is called a *Cauchy sequence* if $\lim_{m,n\to\infty} \|\boldsymbol{x}_n - \boldsymbol{x}_m\| = 0$. A normed space V is said to be *complete* if every Cauchy sequence $\{\boldsymbol{x}_n\}$ in V has a limit in V (that is, there exists $\boldsymbol{y} \in V$ such that $\lim_{n\to\infty} \|\boldsymbol{x}_n - \boldsymbol{y}\| = 0$). ■

Definition 9.39. A metric vector space V is called a *Hilbert space* if it is a complete normed space with respect to the norm induced from its inner product. ■

A complete metric vector space is convenient for mathematical arguments. The following theorem states that every metric vector space can be embedded in a natural manner in a complete metric vector space (Hilbert space). This theorem often plays a critical role in developing mathematically rigorous theory in applications [72].

Theorem 9.54. *For any metric vector space V, there exist a Hilbert space $\tilde{V}$ and a linear mapping J from V to $\tilde{V}$ that have the following properties:*

(1) $J : V \to \tilde{V}$ *is injective.*
(2) $(J(\boldsymbol{x}), J(\boldsymbol{y}))_{\tilde{V}} = (\boldsymbol{x}, \boldsymbol{y})_V$ *holds for all* $\boldsymbol{x}, \boldsymbol{y} \in V$, *where* $(\cdot\,,\cdot)_{\tilde{V}}$ *and* $(\cdot\,,\cdot)_V$ *denote the inner products in* $\tilde{V}$ *and* V, *respectively.*
(3) *The image* $J(V)$ *is dense*[22] *in* $\tilde{V}$.

The space $\tilde{V}$ in the above theorem is called a *completion* of V.

Example 9.50. The metric vector space $\mathbb{R}^n$ (Example 9.39 in Sec. 9.6.2) is complete, and hence a Hilbert space. Similarly, $\mathbb{C}^n$ is a Hilbert space. ■

Example 9.51. The metric vector space $\mathcal{M}_{m,n}(\mathbb{R})$ (Example 9.41 in Sec. 9.6.2) is complete, and hence a Hilbert space. ■

Example 9.52. The metric vector space ℓ^2 (Example 9.42 in Sec. 9.6.2) is complete, and hence a Hilbert space. ■

---

[22]By definition, $J(V)$ is *dense* in $\tilde{V}$ if (and only if), for any $\boldsymbol{y} \in \tilde{V}$ and any $\varepsilon > 0$, there exists $\boldsymbol{x} \in J(V)$ such that $\|\boldsymbol{y} - \boldsymbol{x}\|_{\tilde{V}} < \varepsilon$.

Example 9.53. The metric vector space $\mathrm{C}^0([a,b],\mathbb{R})$ of all continuous functions on a finite interval $[a,b]$ with $a < b$ (Example 9.43 in Sec. 9.6.2) is not complete. The completion of $\mathrm{C}^0([a,b],\mathbb{R})$ is the metric vector space $\mathrm{L}^2((a,b))$ consisting of all *square integrable* functions, where the precise definition of $\mathrm{L}^2((a,b))$ is as follows. Consider the set of all measurable functions f on open interval (a,b) satisfying

$$\int_a^b |f(x)|^2 \mathrm{d}x < +\infty \quad \text{(finite)},$$

where two functions are identified if they take the same values almost everywhere. This set forms a vector space with respect to the ordinary addition and scalar multiplication for functions, and moreover, it is a metric vector space with the inner product

$$(f,g) = \int_a^b f(x)g(x)\mathrm{d}x.$$

This metric vector space is denoted by $\mathrm{L}^2((a,b))$. ■

Example 9.54. The metric vector space $\mathcal{P}(\mathbb{R})$ of all polynomials (Example 9.44 in Sec. 9.6.2) is not complete. The completion of $\mathcal{P}(\mathbb{R})$ coincides with $\mathrm{L}^2((a,b))$. ■

In our argument on metric vector spaces we often had to restrict ourselves to finite-dimensional spaces to avoid technical difficulties inherent in infinite-dimensional spaces. However, the framework of Hilbert spaces allows us, under some additional topological conditions, to introduce concepts similar to orthonormal basis, orthogonal complement, adjoint mapping, and orthogonal projection in the finite-dimensional case.

In the following we discuss orthonormal basis of a Hilbert space.

Definition 9.40. Let V be an infinite-dimensional Hilbert space. A countably infinite subset $\{\boldsymbol{u}_1, \boldsymbol{u}_2, \boldsymbol{u}_3, \ldots\}$ of V is called an *orthonormal basis* (or *complete orthonormal system*) of V if it satisfies the following two conditions:

(1) $\{\boldsymbol{u}_1, \boldsymbol{u}_2, \boldsymbol{u}_3, \ldots\}$ is an orthonormal system.

(2) Every $\boldsymbol{x} \in V$ can be represented as $\boldsymbol{x} = \sum_{n=1}^{\infty} x_n \boldsymbol{u}_n$, where $x_n \in K$ for all n. ■

Theorem 9.55. *Let V be an infinite-dimensional Hilbert space. If V has an orthonormal basis, then V is separable.*[23] *Conversely, if V is separable, it admits an orthonormal basis.*

Theorem 9.56. *Let V be a separable infinite-dimensional Hilbert space, and $\{\boldsymbol{u}_1, \boldsymbol{u}_2, \boldsymbol{u}_3, \ldots\}$ an orthonormal basis of V.*

(1) *Every $\boldsymbol{x} \in V$ can be represented as $\boldsymbol{x} = \displaystyle\sum_{n=1}^{\infty} (\boldsymbol{x}, \boldsymbol{u}_n)\boldsymbol{u}_n$.*

(2) $\|\boldsymbol{x}\|^2 = \displaystyle\sum_{n=1}^{\infty} |(\boldsymbol{x}, \boldsymbol{u}_n)|^2$.

Example 9.55. In the Hilbert space ℓ^2, define $\boldsymbol{e}_j = (x_{j0}, x_{j1}, x_{j2}, \ldots)$ by

$$x_{jn} = \begin{cases} 1 & (n = j), \\ 0 & (n \neq j) \end{cases}$$

for $j = 0, 1, 2, \ldots$, and let $\boldsymbol{u}_j = \boldsymbol{e}_{j-1}$ $(j = 1, 2, 3, \ldots)$. Then $\{\boldsymbol{u}_1, \boldsymbol{u}_2, \boldsymbol{u}_3, \ldots\}$ is an orthonormal basis of ℓ^2. ■

Example 9.56. In the Hilbert space $\mathrm{L}^2((-1, 1))$,

$$\left\{\frac{1}{\sqrt{2}}\right\} \cup \left\{\cos(n\pi x),\ \sin(n\pi x) \;\middle|\; n = 1, 2, \ldots\right\} \tag{9.14}$$

is an orthonormal basis. Denote this basis as $\{\boldsymbol{u}_1, \boldsymbol{u}_2, \boldsymbol{u}_3, \ldots\}$. Then the representation $\boldsymbol{x} = \displaystyle\sum_{n=1}^{\infty} (\boldsymbol{x}, \boldsymbol{u}_n)\boldsymbol{u}_n$ in Theorem 9.56(1) is nothing but the *Fourier expansion* of a function $\boldsymbol{x} \in \mathrm{L}^2((-1, 1))$. We can also take an orthonormal basis consisting of polynomials. Define $P_n(x)$ by

$$P_n(x) = \frac{1}{2^n n!} \frac{\mathrm{d}^n}{\mathrm{d}x^n} (x^2 - 1)^n, \tag{9.15}$$

which is the *Legendre polynomial* of degree n. Then

$$\left\{\sqrt{\frac{2n+1}{2}} P_n(x) \;\middle|\; n = 0, 1, 2, \ldots\right\}$$

is an orthonormal basis of $\mathrm{L}^2((-1, 1))$. ■

---

[23] We say that V is *separable* if there exists a countably infinite subset that is dense in V.

Example 9.57. The *Haar system* on the interval $(0, 1)$ is a family of functions $\{\psi_{j,k}(x) \mid j = 0, 1, 2, \ldots;\ k = 0, 1, \ldots, 2^j - 1\}$ defined by[24]

$$\psi_{j,k}(x) = \begin{cases} 2^{j/2} & \text{if } k2^{-j} < x \leqq (k + \frac{1}{2})2^{-j}, \\ -2^{j/2} & \text{if } (k + \frac{1}{2})2^{-j} < x \leqq (k+1)2^{-j}, \\ 0 & \text{otherwise.} \end{cases}$$

By adding a constant function $\psi_*(x) = 1$ $(0 < x < 1)$ to the Haar system, we obtain a set of functions $\{\psi_*(x)\} \cup \{\psi_{j,k}(x) \mid j = 0, 1, 2, \ldots;\ k = 0, 1, \ldots, 2^j - 1\}$, which is an orthonormal basis of $\mathrm{L}^2((0, 1))$.[25] ■

9.6.9 *Banach Space*

In this section we introduce the concept of Banach space as a normed space where limiting operations can be defined.

Definition 9.41. A complete normed space is called a *Banach space.* ■

In parallel with the embedding of a metric vector space to a Hilbert space (Theorem 9.54), the following theorem states that every normed space can be embedded in a natural manner in a complete normed space (Banach space).

Theorem 9.57. *For any normed space V, there exist a Banach space $\tilde{V}$ and a linear mapping J from V to $\tilde{V}$ that have the following properties:*

(1) $J : V \to \tilde{V}$ *is injective.*
(2) $\|J(\boldsymbol{x})\|_{\tilde{V}} = \|\boldsymbol{x}\|_V$ *holds for all $\boldsymbol{x} \in V$, where $\|\cdot\|_{\tilde{V}}$ and $\|\cdot\|_V$ denote the norms in $\tilde{V}$ and V, respectively.*
(3) *The image $J(V)$ is dense in $\tilde{V}$.*

The space $\tilde{V}$ in the above theorem is called a *completion* of V.

Example 9.58. $\mathbb{R}^n$ is complete with respect to the p-norm, and hence a Banach space, where $1 \leqq p \leqq \infty$. ■

Example 9.59. $\mathcal{M}_{m,n}(\mathbb{R})$ (Example 9.46 in Sec. 9.6.7) is complete with respect to the p-norm of matrices, and hence a Banach space, where $1 \leqq p \leqq \infty$. $\mathcal{M}_{m,n}(\mathbb{R})$ is also complete with respect to the Frobenius norm. ■

---

[24] When $k = 2^j - 1$, we have $(k+1)2^{-j} = 1$, and therefore, the range of x in the second equation should be modified to $(k + \frac{1}{2})2^{-j} < x < (k+1)2^{-j}$ (*i.e.*, $1 - 2^{-j-1} < x < 1$).

[25] We have considered $\mathrm{L}^2((0, 1))$ in place of $\mathrm{L}^2((-1, 1))$, because the functions of the orthonormal basis of $\mathrm{L}^2((0, 1))$ are easier to describe.

Example 9.60. ℓ^p is complete with respect to the p-norm, and hence a Banach space, where $1 \leqq p \leqq \infty$. ■

Example 9.61. $C^0([a,b],\mathbb{R})$ is complete with respect to the ∞-norm, and hence a Banach space. However, it is not complete with respect to the p-norm if $1 \leqq p < \infty$. The completion of $C^0([a,b],\mathbb{R})$ is $L^p((a,b))$, which denotes the set of all *p-integrable* functions. To be more precise, $L^p((a,b))$ is defined as follows. Consider the set of all measurable functions f on open interval (a,b) satisfying

$$\int_a^b |f(x)|^p \mathrm{d}x < +\infty \quad \text{(finite)},$$

where two functions are identified if they take the same values almost everywhere. This set forms a vector space with respect to the ordinary addition and scalar multiplication for functions, and moreover, it is a normed space with the norm

$$\|f\|_p = \left(\int_a^b |f(x)|^p \mathrm{d}x \right)^{1/p}.$$

This normed space is denoted by $L^p((a,b))$. ■

Example 9.62. $\mathcal{P}(\mathbb{R})$ is not complete with respect to the p-norm, where $1 \leqq p \leqq \infty$. The completion with respect to the ∞-norm is given by $C^0([a,b],\mathbb{R})$ (with the ∞-norm). If $1 \leqq p < \infty$, the completion of $\mathcal{P}(\mathbb{R})$ with respect to the p-norm is given by $L^p((a,b))$. ■

A Hilbert space, though infinite-dimensional, allows us to extend some results for a finite-dimensional space with the aid of some additional topological conditions (cf., Sec. 9.6.8). To be specific, concepts such as orthonormal basis, orthogonal complement, and orthogonal projection can be introduced for a Hilbert space. In this respect a Banach space differs considerably from a Hilbert space.

In the following we discuss basis of a Banach space.

Definition 9.42. Let V be an infinite-dimensional Banach space. A countably infinite subset $\{\boldsymbol{u}_1, \boldsymbol{u}_2, \boldsymbol{u}_3, \ldots\}$ of V is called a *basis* (or *Schauder basis*) of V if it satisfies the following two conditions:

(1) Every $\boldsymbol{x} \in V$ can be represented as $\boldsymbol{x} = \displaystyle\sum_{n=1}^{\infty} x_n \boldsymbol{u}_n$, where $x_n \in K$ for all n.

(2) The representation in (1) is unique.[26] ■

Theorem 9.58. *Let V be an infinite-dimensional Banach space. If V has a basis, then V is separable. However, the converse is not true, that is, there is a separable Banach space that admits no basis.*

Example 9.63. The Banach space ℓ^∞ is not separable, and hence has no basis. For each p with $1 \leqq p < \infty$, let $\boldsymbol{u}_j = \boldsymbol{e}_{j-1}$ $(j = 1, 2, \ldots)$ using $\boldsymbol{e}_j$ $(j = 0, 1, 2, \ldots)$ defined in Example 9.55 in Sec. 9.6.8. Then $\{\boldsymbol{u}_1, \boldsymbol{u}_2, \boldsymbol{u}_3, \ldots\}$ is a basis of ℓ^p. ■

Example 9.64. In the Banach space $\mathrm{C}^0([0,1], \mathbb{R})$ (equipped with ∞-norm), the union of x, $1 - x$, and the functions $\varphi_{j,k}(x)$ $(j = 0, 1, 2, \ldots;\ k = 0, 1, \ldots, 2^j - 1)$ is a basis, where

$$\varphi_{j,k}(x) = \begin{cases} 2^{j+1}(x - k2^{-j}) & \text{if } k2^{-j} < x \leqq (k + \frac{1}{2})2^{-j}, \\ 2^{j+1}((k+1)2^{-j} - x) & \text{if } (k + \frac{1}{2})2^{-j} < x < (k+1)2^{-j}, \\ 0 & \text{otherwise.} \end{cases}$$

■

Example 9.65. The set of trigonometric functions in (9.14) is a basis of the Banach space $\mathrm{L}^p((-1,1))$ if $1 < p < \infty$ (and not for $p = 1$). The set of Legendre polynomials in (9.15) is a basis of $\mathrm{L}^p((-1,1))$ if $4/3 < p < 4$ (and not if $1 \leqq p \leqq 4/3$ or $p \geqq 4$). The Haar system augmented by a constant function (Example 9.57 in Sec. 9.6.8) is a basis of $\mathrm{L}^p((0,1))$ when $1 \leqq p < \infty$. ■

As is seen from Example 9.65, bases of Banach spaces are highly complicated. The reader is referred to [67] for details about the cases where bases do not exist.

9.7 Dual Space

9.7.1 *Definition*

Definition 9.43. Let V be a vector space over K. The vector space $\mathcal{L}(V, K)$ of all linear mappings from V to K (cf., Sec. 9.5.4) is called the

[26] Uniqueness of the representation is not imposed in the definition of orthonormal basis of a Hilbert space (Definition 9.40 in Sec. 9.6.8). This is because the uniqueness is implied by the orthonormality.

dual space of V, and denoted[27] by V^{d}. An element of V^{d} is called a *linear form* on V. ∎

For $l \in V^{\mathrm{d}}$ and $\boldsymbol{x} \in V$, the image of vector $\boldsymbol{x}$ by mapping l is denoted by $l(\boldsymbol{x})$, which is an element of K. In discussing dual spaces we often write l as $\boldsymbol{l}$ (in boldface) to reflect our view that l is a vector belonging to vector space V^{d}. Then the image of $\boldsymbol{x}$ by l is considered a kind of product of $\boldsymbol{x}$ and $\boldsymbol{l}$, and denoted as $\langle \boldsymbol{l}, \boldsymbol{x} \rangle$, that is,

$$\langle \boldsymbol{l}, \boldsymbol{x} \rangle = l(\boldsymbol{x}).$$

This is called the *scalar product* of $\boldsymbol{x}$ and $\boldsymbol{l}$.

Example 9.66. Given a row vector $\boldsymbol{l} = [l_1, l_2, \ldots, l_n] \in \mathcal{M}_{1,n}(K)$, define a mapping $l : K^n \to K$ by

$$l(\boldsymbol{x}) = \boldsymbol{l}\boldsymbol{x} = [l_1, l_2, \ldots, l_n] \begin{bmatrix} x_1 \\ x_2 \\ \vdots \\ x_n \end{bmatrix} \qquad (\boldsymbol{x} \in K^n). \tag{9.16}$$

Then l is a linear form on K^n. ∎

When K is regarded as a vector space over K, a linear form $l : V \to K$ on a vector space V is a linear mapping from a vector space V to another vector space K. Therefore, if V is finite-dimensional, we can think of a matrix representation of l in the sense of Sec. 9.5.3. When V is n-dimensional, the representation matrix of l is an n-dimensional row vector as in (9.16). Accordingly, the dual space V^{d} of an n-dimensional vector space V is often identified with the vector space $\mathcal{M}_{1,n}(K)$ of all n-dimensional row vectors.

9.7.2 *Examples*

Example 9.67. Let V be the real vector space $\mathrm{C}^0([a, b], \mathbb{R})$ of all continuous functions on an interval $[a, b]$ with $a < b$. For any point x_1 in $[a, b]$, a real-valued mapping l on V defined by

$$l(f) = f(x_1) \qquad (f \in V) \tag{9.17}$$

[27]The dual space of V is sometimes denoted by V' or V^*, but these notations are also used in other meanings. In this book, we use V^{d} for the dual space, with "d" standing for "dual."

is a linear form on V. For any distinct points $x_1, x_2, \ldots, x_m$ in $[a, b]$ and real numbers $c_1, c_2, \ldots, c_m$,

$$l(f) = c_1 f(x_1) + c_2 f(x_2) + \cdots + c_m f(x_m) \qquad (f \in V) \tag{9.18}$$

is also a linear form on V. The definite integral

$$l(f) = \int_a^b f(x)\mathrm{d}x \qquad (f \in V) \tag{9.19}$$

is also a linear form on V. ∎

Example 9.68. Let V be the real vector space $\mathrm{C}^m((a,b),\mathbb{R})$ of all m times continuously differentiable functions on an open interval (a, b) with $a < b$. For any fixed point x_1 in (a, b) and real numbers $c_1, c_2, \ldots, c_m$, a real-valued mapping l on V defined by

$$l(f) = c_m \frac{\mathrm{d}^m f}{\mathrm{d}x^m}(x_1) + c_{m-1}\frac{\mathrm{d}^{m-1} f}{\mathrm{d}x^{m-1}}(x_1) + \cdots + c_1 \frac{\mathrm{d}f}{\mathrm{d}x}(x_1) + c_0 f(x_1) \quad (f \in V) \tag{9.20}$$

is a linear form on V. ∎

The expressions (9.17), (9.18), (9.19), and (9.20) in the above example also give linear forms on the space $\mathcal{P}(\mathbb{R})$ of all polynomials or $\mathcal{P}_n(\mathbb{R})$ of all polynomials with degrees at most n.

9.7.3 *Dual Basis*

Throughout Sec. 9.7.3, vector spaces are assumed to be finite-dimensional.

Definition 9.44. Let $\{\boldsymbol{b}_1, \boldsymbol{b}_2, \ldots, \boldsymbol{b}_n\}$ be a basis of a vector space V over K. For each $i = 1, 2, \ldots, n$, let $\boldsymbol{b}_i^{\mathrm{d}}$ be the vector in V^{d} satisfying the relation[28]

$$\langle \boldsymbol{b}_i^{\mathrm{d}}, \boldsymbol{b}_j \rangle = \delta_{ij} = \begin{cases} 1 & (j = i), \\ 0 & (j \neq i). \end{cases} \tag{9.21}$$

Then $\{\boldsymbol{b}_1^{\mathrm{d}}, \boldsymbol{b}_2^{\mathrm{d}}, \ldots, \boldsymbol{b}_n^{\mathrm{d}}\}$ is a basis of V^{d}, called the *dual basis* of $\{\boldsymbol{b}_1, \boldsymbol{b}_2, \ldots, \boldsymbol{b}_n\}$. ∎

Theorem 9.59. *For every* $\boldsymbol{l} \in V^{\mathrm{d}}$,

$$\boldsymbol{l} = \langle \boldsymbol{l}, \boldsymbol{b}_1 \rangle \boldsymbol{b}_1^{\mathrm{d}} + \langle \boldsymbol{l}, \boldsymbol{b}_2 \rangle \boldsymbol{b}_2^{\mathrm{d}} + \cdots + \langle \boldsymbol{l}, \boldsymbol{b}_n \rangle \boldsymbol{b}_n^{\mathrm{d}}. \tag{9.22}$$

Theorem 9.60. $\dim V^{\mathrm{d}} = \dim V$.

[28] The relation (9.21) uniquely determines vector $\boldsymbol{b}_i^{\mathrm{d}}$. In particular, the ordering (or numbering) of the dual vectors $\boldsymbol{b}_1^{\mathrm{d}}, \boldsymbol{b}_2^{\mathrm{d}}, \ldots, \boldsymbol{b}_n^{\mathrm{d}}$ is determined by the given basis $\boldsymbol{b}_1, \boldsymbol{b}_2, \ldots, \boldsymbol{b}_n$.

Theorem 9.61. *Let $B = [\boldsymbol{b}_1, \boldsymbol{b}_2, \ldots, \boldsymbol{b}_n]$ and $B' = [\boldsymbol{b}'_1, \boldsymbol{b}'_2, \ldots, \boldsymbol{b}'_n]$ be two bases of V, and consider the corresponding dual bases $D = \begin{bmatrix} \boldsymbol{b}_1^{\mathrm{d}} \\ \boldsymbol{b}_2^{\mathrm{d}} \\ \vdots \\ \boldsymbol{b}_n^{\mathrm{d}} \end{bmatrix}$ and $D' = \begin{bmatrix} \boldsymbol{b}'^{\mathrm{d}}_1 \\ \boldsymbol{b}'^{\mathrm{d}}_2 \\ \vdots \\ \boldsymbol{b}'^{\mathrm{d}}_n \end{bmatrix}$, where $\boldsymbol{b}_i^{\mathrm{d}}$ and $\boldsymbol{b}'^{\mathrm{d}}_i$ are regarded as row vectors. If the basis change $B \to B'$ is expressed by a matrix A, that is, if*

$$[\boldsymbol{b}'_1, \boldsymbol{b}'_2, \ldots, \boldsymbol{b}'_n] = [\boldsymbol{b}_1, \boldsymbol{b}_2, \ldots, \boldsymbol{b}_n]A$$

as in Definition 9.14 in Sec. 9.4.4, the change of the dual basis $D \to D'$ is given by

$$\begin{bmatrix} \boldsymbol{b}'^{\mathrm{d}}_1 \\ \boldsymbol{b}'^{\mathrm{d}}_2 \\ \vdots \\ \boldsymbol{b}'^{\mathrm{d}}_n \end{bmatrix} = A^{-1} \begin{bmatrix} \boldsymbol{b}_1^{\mathrm{d}} \\ \boldsymbol{b}_2^{\mathrm{d}} \\ \vdots \\ \boldsymbol{b}_n^{\mathrm{d}} \end{bmatrix}.$$

Theorem 9.62. *Let $B = [\boldsymbol{b}_1, \boldsymbol{b}_2, \ldots, \boldsymbol{b}_n]$ be a basis of V and $D = \begin{bmatrix} \boldsymbol{b}_1^{\mathrm{d}} \\ \boldsymbol{b}_2^{\mathrm{d}} \\ \vdots \\ \boldsymbol{b}_n^{\mathrm{d}} \end{bmatrix}$ its dual basis. For any $\boldsymbol{x} \in V$ and $\boldsymbol{l} \in V^{\mathrm{d}}$, let $(x_1, x_2, \ldots, x_n)^\top$ be the coordinates of $\boldsymbol{x}$ with respect to B and $(l_1, l_2, \ldots, l_n)$ be the coordinates of $\boldsymbol{l}$ with respect to D.*[29] *Then $\langle \boldsymbol{l}, \boldsymbol{x} \rangle$ is given by*

$$\langle \boldsymbol{l}, \boldsymbol{x} \rangle = [l_1, l_2, \ldots, l_n] \begin{bmatrix} x_1 \\ x_2 \\ \vdots \\ x_n \end{bmatrix}.$$

The coordinates of $\boldsymbol{x}$ and $\boldsymbol{l}$ change with a change of basis, but the value of $\langle \boldsymbol{l}, \boldsymbol{x} \rangle$ remains invariant.

[29] The coordinates of $\boldsymbol{l}$ are represented by a row vector, compatibly with our convention to regard the dual basis D as a set of row vectors.

Example 9.69. Let V be the real vector space $\mathcal{P}_n(\mathbb{R})$ of all polynomials over $\mathbb{R}$ whose degrees are at most n. The dual space V^{d} is the set of all linear mappings that assign a real value to each polynomial. Take $\{1, x, x^2, \ldots, x^n\}$ as a natural basis of V, and denote by $\{C_0, C_1, \ldots, C_n\}$ the corresponding dual basis. The element C_i of the dual basis is determined by the condition

$$C_i(x^j) = \begin{cases} 1 & (j = i), \\ 0 & (j \neq i). \end{cases}$$

This means that, for a polynomial $P(x) = a_0 + a_1 x + \cdots + a_n x^n$, $C_i(P)$ is given by $C_i(P) = a_i$. That is, C_i is a linear form that extracts the coefficient of x^i from a polynomial $P(x)$. ■

Example 9.70. We continue with $V = \mathcal{P}_n(\mathbb{R})$, but consider another dual space V^{d} different from the one in the above example. For any real number ξ, let δ_ξ denote the linear form that returns the value of $P \in V$ at ξ, that is, $\delta_\xi(P) = P(\xi)$. For any $n+1$ distinct numbers $\xi_0, \xi_1, \ldots, \xi_n$, the set $D = \{\delta_{\xi_0}, \delta_{\xi_1}, \ldots, \delta_{\xi_n}\}$ is a basis of V^{d} (shown later). Define $\omega_0, \omega_1, \ldots, \omega_n \in V$ by

$$Q_j(x) = \prod_{i \neq j} (x - \xi_i), \qquad \omega_j(x) = \frac{Q_j(x)}{Q_j(\xi_j)} \qquad (j = 0, 1, \ldots, n).$$

Then $B = \{\omega_0(x), \omega_1(x), \ldots, \omega_n(x)\}$ is a basis of V, and the dual basis of B coincides with $D = \{\delta_{\xi_0}, \delta_{\xi_1}, \ldots, \delta_{\xi_n}\}$ (shown later).

Theorem 9.59 shows that every $l \in V^{\mathrm{d}}$ can be represented as

$$l = l(\omega_0)\delta_{\xi_0} + l(\omega_1)\delta_{\xi_1} + \cdots + l(\omega_n)\delta_{\xi_n}. \qquad (9.23)$$

In the following we consider two particular choices of l with concrete meanings.

- Let $l = \delta_\xi$. By $l(\omega_j) = \omega_j(\xi)$ $(j = 0, 1, \ldots, n)$, the expression (9.23) gives

  $$\delta_\xi = \omega_0(\xi)\delta_{\xi_0} + \omega_1(\xi)\delta_{\xi_1} + \cdots + \omega_n(\xi)\delta_{\xi_n},$$

  which means that the identity

  $$P(\xi) = \omega_0(\xi)P(\xi_0) + \omega_1(\xi)P(\xi_1) + \cdots + \omega_n(\xi)P(\xi_n)$$

  holds for any polynomial $P(x)$ of degree $\leqq n$. This is nothing but the Lagrange interpolation [76, 77].

- Let $l(P) = \int_a^b P(x)\mathrm{d}x$. The expression (9.23) shows that the identity
$$\int_a^b P(x)\mathrm{d}x = \left(\int_a^b \omega_0(x)\mathrm{d}x\right)P(\xi_0) + \left(\int_a^b \omega_1(x)\mathrm{d}x\right)P(\xi_1) + \cdots + \left(\int_a^b \omega_n(x)\mathrm{d}x\right)P(\xi_n)$$
holds for any polynomial $P(x)$ of degree $\leqq n$. This leads to numerical integration formulas based on interpolating functions [76, 77].

It remains to show that $B = \{\omega_0(x), \omega_1(x), \ldots, \omega_n(x)\}$ is a basis of V and its dual basis is given by $D = \{\delta_{\xi_0}, \delta_{\xi_1}, \ldots, \delta_{\xi_n}\}$. First note that $\omega_j(\xi_i) = \delta_{ij}$ holds for $i, j = 0, 1, \ldots, n$. To show the linear independence of B, suppose that
$$c_0\omega_0(x) + c_1\omega_1(x) + \cdots + c_n\omega_n(x) = 0$$
holds with some $c_0, c_1, \ldots, c_n \in \mathbb{R}$. By substituting $x = \xi_i$ into this expression we obtain $c_i = 0$ for each $i = 0, 1, \ldots, n$. Therefore, $\omega_0(x), \omega_1(x), \ldots, \omega_n(x)$ are linearly independent. Since $\dim V = n + 1$, this implies that B is a basis of V. It follows from $\delta_{\xi_i}(\omega_j) = \omega_j(\xi_i) = \delta_{ij}$ $(i, j = 0, 1, \ldots, n)$ that D is the dual basis of B. ■

9.7.4 *Bidual Space*

The dual space V^{d} of a vector space V is again a vector space, for which we can think of the dual space $(V^{\mathrm{d}})^{\mathrm{d}}$, which is called the *bidual space* of V. For each $\boldsymbol{x} \in V$, the mapping φ defined by
$$\begin{array}{rccc} \varphi: & V^{\mathrm{d}} & \to & K \\ & \Psi & & \Psi \\ & \boldsymbol{l} & \mapsto & \langle \boldsymbol{l}, \boldsymbol{x} \rangle \end{array}$$
is a linear form on V^{d}, that is, $\varphi \in (V^{\mathrm{d}})^{\mathrm{d}}$. We write $\varphi_{\boldsymbol{x}}$ for φ to indicate the dependence of φ on $\boldsymbol{x} \in V$. Since $\boldsymbol{x} \neq \boldsymbol{y}$ implies $\varphi_{\boldsymbol{x}} \neq \varphi_{\boldsymbol{y}}$, we may identify $\boldsymbol{x}$ with $\varphi_{\boldsymbol{x}}$ to regard V as a subset of $(V^{\mathrm{d}})^{\mathrm{d}}$, that is,
$$V \subseteqq (V^{\mathrm{d}})^{\mathrm{d}}. \tag{9.24}$$
In general it can happen that $V \neq (V^{\mathrm{d}})^{\mathrm{d}}$, but the equality holds here if V is finite-dimensional, since $\dim (V^{\mathrm{d}})^{\mathrm{d}} = \dim V$ by Theorem 9.60. We state this fact as a theorem.

Theorem 9.63. *If V is finite-dimensional, then $(V^{\mathrm{d}})^{\mathrm{d}} = V$.*

9.7.5 *Annihilator*

Definition 9.45. Let V be a vector space and V^{d} its dual space. For any subspace W of V, the set of all elements of V^{d} that are equal to 0 for vectors in W, denoted as[30]

$$W^\circ = \{\boldsymbol{l} \in V^{\mathrm{d}} \mid \langle \boldsymbol{l}, \boldsymbol{x} \rangle = 0 \ (\boldsymbol{x} \in W)\}, \tag{9.25}$$

is a subspace of V^{d}. This is called the *annihilator* of W in V^{d}. ■

Theorem 9.64. *The following hold for any subspace W of a vector space V.*

(1) $W^\circ \cong (V/W)^{\mathrm{d}}$.
(2) *If V is finite-dimensional, then* $\dim W^\circ = \dim V - \dim W$.

Theorem 9.65. *The following hold for any subspaces W, W_1, and W_2 of a vector space V.*

(1) *If V is finite-dimensional, then* $(W^\circ)^\circ = W$.
(2) $(W_1 + W_2)^\circ = {W_1}^\circ \cap {W_2}^\circ$.
(3) $(W_1 \cap W_2)^\circ = {W_1}^\circ + {W_2}^\circ$.
(4) *If $V = W_1 \oplus W_2$, then* $V^{\mathrm{d}} = {W_1}^\circ \oplus {W_2}^\circ$.

9.7.6 *Dual Mapping*

Definition 9.46. Let V and V' be vector spaces over K, and V^{d} and $(V')^{\mathrm{d}}$ be their dual spaces, respectively. For any linear mapping T from V to V', the mapping T^{d} from $(V')^{\mathrm{d}}$ to V^{d} defined by[31]

$$T^{\mathrm{d}}(l') = l' \circ T \qquad (l' \in (V')^{\mathrm{d}}) \tag{9.26}$$

is a linear mapping. This mapping $T^{\mathrm{d}} : (V')^{\mathrm{d}} \to V^{\mathrm{d}}$ is called the *dual mapping* of T. ■

Remark 9.24. The defining condition (9.26) for T^{d} is equivalent to saying that the equality

$$\langle T^{\mathrm{d}}(\boldsymbol{l}'), \boldsymbol{x} \rangle_V = \langle \boldsymbol{l}', T(\boldsymbol{x}) \rangle_{V'} \tag{9.27}$$

holds for all $\boldsymbol{x} \in V$ and $\boldsymbol{l}' \in (V')^{\mathrm{d}}$, where $\langle \cdot, \cdot \rangle_V$ and $\langle \cdot, \cdot \rangle_{V'}$ are the scalar products for V and V', respectively. ■

---

[30] The annihilator is sometimes denoted as $W^\perp$. In this book, however, we adopt a different notation W° for the annihilator as we use $W^\perp$ for the orthogonal complement.
[31] The dual mapping is sometimes denoted by T' or T^*, but these notations are also used in other meanings. In this book, we use T^{d} for the dual mapping, with "d" standing for "dual." It should be clear that $\circ$ in (9.26) means the composition of mappings.

Theorem 9.66. *Let V, V', and V'' be vector spaces, and T a linear mapping from V to V'.*

(1) *If V is finite-dimensional, then* $(T^{\mathrm{d}})^{\mathrm{d}} = T$.
(2) $(T+S)^{\mathrm{d}} = T^{\mathrm{d}} + S^{\mathrm{d}}$, *where* $S : V \to V'$.
(3) $(cT)^{\mathrm{d}} = cT^{\mathrm{d}}$.
(4) $(S \circ T)^{\mathrm{d}} = T^{\mathrm{d}} \circ S^{\mathrm{d}}$, *where* $S : V' \to V''$.

Theorem 9.67. *Let V and V' be vector spaces, and T a linear mapping from V to V'.*

(1) $\mathrm{Im}(T^{\mathrm{d}}) = (\mathrm{Ker}\, T)^{\circ}$.
(2) $\mathrm{Ker}(T^{\mathrm{d}}) = (\mathrm{Im}\, T)^{\circ}$.
(3) *If V is finite-dimensional, then* $\dim(\mathrm{Im}(T^{\mathrm{d}})) = \dim(\mathrm{Im}\, T)$.

Theorem 9.68. *Let V and V' be finite-dimensional vector spaces, B and B' their bases, and D and D' the dual bases, respectively. Let T be a linear mapping from V to V'. If A is the representation matrix of T with respect to B and B', the representation matrix of T^{d} with respect to D' and D is given by $A^{\top}$.*

Remark 9.25. Here is a supplement to Theorem 9.68. Express the bases as $B = [\boldsymbol{b}_1, \boldsymbol{b}_2, \ldots, \boldsymbol{b}_n]$ and $B' = [\boldsymbol{b}'_1, \boldsymbol{b}'_2, \ldots, \boldsymbol{b}'_m]$, and their dual bases as $D = [\boldsymbol{b}^{\mathrm{d}}_1, \boldsymbol{b}^{\mathrm{d}}_2, \ldots, \boldsymbol{b}^{\mathrm{d}}_n]$ and $D' = [\boldsymbol{b}'^{\mathrm{d}}_1, \boldsymbol{b}'^{\mathrm{d}}_2, \ldots, \boldsymbol{b}'^{\mathrm{d}}_m]$. With these notations the statement of Theorem 9.68 reads

$$[T(\boldsymbol{b}_1), T(\boldsymbol{b}_2), \ldots, T(\boldsymbol{b}_n)] = [\boldsymbol{b}'_1, \boldsymbol{b}'_2, \ldots, \boldsymbol{b}'_m]A,$$
$$[T^{\mathrm{d}}(\boldsymbol{b}'^{\mathrm{d}}_1), T^{\mathrm{d}}(\boldsymbol{b}'^{\mathrm{d}}_2), \ldots, T^{\mathrm{d}}(\boldsymbol{b}'^{\mathrm{d}}_m)] = [\boldsymbol{b}^{\mathrm{d}}_1, \boldsymbol{b}^{\mathrm{d}}_2, \ldots, \boldsymbol{b}^{\mathrm{d}}_n]A^{\top}.$$

If the dual basis is regarded as a set of row vectors, the second expression above is replaced by

$$\begin{bmatrix} T^{\mathrm{d}}(\boldsymbol{b}'^{\mathrm{d}}_1) \\ T^{\mathrm{d}}(\boldsymbol{b}'^{\mathrm{d}}_2) \\ \vdots \\ T^{\mathrm{d}}(\boldsymbol{b}'^{\mathrm{d}}_m) \end{bmatrix} = A \begin{bmatrix} \boldsymbol{b}^{\mathrm{d}}_1 \\ \boldsymbol{b}^{\mathrm{d}}_2 \\ \vdots \\ \boldsymbol{b}^{\mathrm{d}}_n \end{bmatrix}.$$

The coordinates $(l'_1, l'_2, \ldots, l'_m)$ of $\boldsymbol{l}' \in (V')^{\mathrm{d}}$ with respect to D' and the coordinates $(l_1, l_2, \ldots, l_n)$ of $\boldsymbol{l} = T^{\mathrm{d}}(\boldsymbol{l}')$ with respect to D are related as

$$[l_1, l_2, \ldots, l_n] = [l'_1, l'_2, \ldots, l'_m]A.$$

■

As we have seen above, the annihilator has similar properties as the orthogonal complement, and the dual mapping is similar to the adjoint mapping.[32] However, it is important to recognize the fundamental difference that the concepts of orthogonal complement and adjoint mapping can be defined only for a metric vector space, which is equipped with an inner product, whereas those of annihilator and dual mapping are defined for any vector space.

9.8 Rank of Matrices: Vector Space Views

For an $m \times n$ matrix A, the image ($\operatorname{Im} A$) is a subspace of K^m and the kernel ($\operatorname{Ker} A$) is a subspace of K^n. The dimensions of these subspaces are closely related to the concept of rank ($\operatorname{rank} A$) introduced in Chapter 4. The objective of this section is to shed a light on the rank of a matrix from the viewpoint of vector spaces. Throughout Sec. 9.8 we assume that $K = \mathbb{R}$ or $\mathbb{C}$.

Recall the definitions of the image and the kernel:

$$\operatorname{Im} A = \{\boldsymbol{y} \mid \boldsymbol{y} = A\boldsymbol{x} \ \text{ for some } \boldsymbol{x}\}, \tag{9.28}$$

$$\operatorname{Ker} A = \{\boldsymbol{x} \mid A\boldsymbol{x} = \boldsymbol{0}\} \tag{9.29}$$

(Definitions 5.1 and 5.2 in Sec. 5.1) and the identity

$$n = \dim(\operatorname{Ker} A) + \dim(\operatorname{Im} A), \tag{9.30}$$

which is a reformulation of (9.3) given in Theorem 9.16 in Sec. 9.5.1.

9.8.1 *Rank and Dimension*

Theorem 9.69. *The following hold for an $m \times n$ matrix A.*

(1) $\dim(\operatorname{Im} A) = \operatorname{rank} A$.
(2) $\dim(\operatorname{Ker} A) = n - \operatorname{rank} A$.

Proof. (1) The rank of matrix $A = [\boldsymbol{a}_1, \boldsymbol{a}_2, \ldots, \boldsymbol{a}_n]$ is equal to the maximum number of linearly independent column vectors (Definition 4.2 in Sec. 4.1.2). Let $r = \operatorname{rank} A$ and take linearly independent r column vectors $\boldsymbol{a}_{i_1}, \boldsymbol{a}_{i_2}, \ldots, \boldsymbol{a}_{i_r}$. We shall show that these vectors form a basis of $\operatorname{Im} A$. Let $I = \{i_1, i_2, \ldots, i_r\}$ and $J = \{1, 2, \ldots, n\} \setminus I$.

[32] Compare (9.25) with (9.9), and (9.27) with (9.10).

First we note that, for every $j \in J$, the vector $\boldsymbol{a}_j$ can be represented as a linear combination of $\boldsymbol{a}_{i_1}, \boldsymbol{a}_{i_2}, \ldots, \boldsymbol{a}_{i_r}$ as

$$\boldsymbol{a}_j = \sum_{i \in I} d_{ij} \boldsymbol{a}_i.$$

This can be shown as follows. Since $\boldsymbol{a}_j$, $\boldsymbol{a}_{i_1}, \boldsymbol{a}_{i_2}, \ldots, \boldsymbol{a}_{i_r}$ are linearly dependent, we have

$$c_j \boldsymbol{a}_j + \sum_{i \in I} c_{ij} \boldsymbol{a}_i = \boldsymbol{0}$$

for some coefficients satisfying $c_j \neq 0$ or $(c_{ij} \mid i \in I) \neq \boldsymbol{0}$. If $c_j = 0$, this identity contradicts the linear independence of $\boldsymbol{a}_i$ $(i \in I)$. Therefore, $c_j \neq 0$ and we can take $d_{ij} = -c_{ij}/c_j$ to obtain $\boldsymbol{a}_j = \sum_{i \in I} d_{ij} \boldsymbol{a}_i$.

For every $\boldsymbol{y} \in \operatorname{Im} A$, there exists some $\boldsymbol{x}$ such that

$$\boldsymbol{y} = A\boldsymbol{x} = \sum_{k=1}^{n} x_k \boldsymbol{a}_k = \sum_{i \in I} x_i \boldsymbol{a}_i + \sum_{j \in J} x_j \boldsymbol{a}_j = \sum_{i \in I} (x_i + \sum_{j \in J} d_{ij} x_j) \boldsymbol{a}_i,$$

which shows that $\boldsymbol{a}_{i_1}, \boldsymbol{a}_{i_2}, \ldots, \boldsymbol{a}_{i_r}$ generate $\operatorname{Im} A$. On the other hand, $\boldsymbol{a}_{i_1}, \boldsymbol{a}_{i_2}, \ldots, \boldsymbol{a}_{i_r}$ are linearly independent (by definition). Therefore, $\boldsymbol{a}_{i_1}, \boldsymbol{a}_{i_2}, \ldots, \boldsymbol{a}_{i_r}$ is a basis of $\operatorname{Im} A$.

(2) This identity is immediate from (1) and (9.30). □

Theorem 9.70.[33]

(1) $(\operatorname{Ker} A)^{\perp} = \operatorname{Im}(A^*)$.
(2) $(\operatorname{Im} A)^{\perp} = \operatorname{Ker}(A^*)$.
(3) $\dim(\operatorname{Im} A) = \dim(\operatorname{Im}(A^*))$, *and hence* $\operatorname{rank} A = \operatorname{rank}(A^*)$.

If A is a real matrix, (1), (2), *and* (3) *hold with A^* replaced by $A^\top$.*

Proof. (1) We have the following equivalences: $\boldsymbol{x} \in \operatorname{Ker} A \iff A\boldsymbol{x} = \boldsymbol{0} \iff (A^*)^*\boldsymbol{x} = \boldsymbol{0} \iff \boldsymbol{x}$ is orthogonal to the column vectors of A^* $\iff \boldsymbol{x} \in (\operatorname{Im}(A^*))^{\perp}$. This implies $(\operatorname{Ker} A)^{\perp} = ((\operatorname{Im}(A^*))^{\perp})^{\perp}$, whereas $((\operatorname{Im}(A^*))^{\perp})^{\perp}$ is equal to $\operatorname{Im}(A^*)$ by Theorem 9.41(1) in Sec. 9.6.4.

(2) By replacing A in (1) with A^*, we obtain $(\operatorname{Ker}(A^*))^{\perp} = \operatorname{Im}((A^*)^*) = \operatorname{Im} A$. This implies $(\operatorname{Im} A)^{\perp} = ((\operatorname{Ker}(A^*))^{\perp})^{\perp}$, whereas $((\operatorname{Ker}(A^*))^{\perp})^{\perp} = \operatorname{Ker}(A^*)$ by Theorem 9.41(1).

(3) It follows from (1) above, Theorem 9.40(2) in Sec. 9.6.4, and (9.30) that

$$\dim(\operatorname{Im}(A^*)) = \dim((\operatorname{Ker} A)^{\perp}) = n - \dim(\operatorname{Ker} A) = \dim(\operatorname{Im} A).$$

This is equivalent to $\operatorname{rank}(A^*) = \operatorname{rank} A$ by Theorem 9.69(1). □

---

[33] Theorem 9.70 can be regarded as a translation of Theorem 9.43 (Sec. 9.6.5) into matrix terms via Theorem 9.44.

Theorem 9.71.

(1) $\operatorname{Ker} A = \operatorname{Ker}(A^* A)$, *and hence* $\operatorname{rank} A = \operatorname{rank}(A^* A)$.
(2) $\operatorname{Im} A = \operatorname{Im}(AA^*)$, *and hence* $\operatorname{rank} A = \operatorname{rank}(AA^*)$.

Proof. (1) Obviously, we have $\operatorname{Ker} A \subseteqq \operatorname{Ker}(A^* A)$, and the reverse inclusion $\operatorname{Ker} A \supseteqq \operatorname{Ker}(A^* A)$ can be shown as follows: $\boldsymbol{x} \in \operatorname{Ker}(A^* A) \iff A^* A\boldsymbol{x} = \mathbf{0} \implies \boldsymbol{x}^* A^* A\boldsymbol{x} = 0 \iff A\boldsymbol{x} = \mathbf{0} \iff \boldsymbol{x} \in \operatorname{Ker} A$. The equality of ranks follows from this by Theorem 9.69(2).

(2) Obviously, we have $\operatorname{Im} A \supseteqq \operatorname{Im}(AA^*)$. On the other hand, it follows from (9.30), (1) above, and Theorem 9.70(3) that

$$\begin{aligned}\dim(\operatorname{Im}(AA^*)) &= m - \dim(\operatorname{Ker}(AA^*)) = m - \dim(\operatorname{Ker}(A^*)) \\ &= \dim(\operatorname{Im}(A^*)) = \dim(\operatorname{Im} A).\end{aligned}$$

Hence we obtain $\operatorname{Im} A = \operatorname{Im}(AA^*)$ by Theorem 9.9(2) in Sec. 9.4.3. The equality of ranks follows from this by Theorem 9.69(1). □

Remark 9.26. In this remark we give an alternative proof of Theorem 9.69 based on the rank normal form (Sec. 3.2). Let A be an $m \times n$ matrix of rank r. For any matrix $\tilde{A} = SAT$ obtained from A with nonsingular matrices S and T, we have

$$\operatorname{rank} \tilde{A} = \operatorname{rank} A, \quad \dim(\operatorname{Im} \tilde{A}) = \dim(\operatorname{Im} A), \quad \dim(\operatorname{Ker} \tilde{A}) = \dim(\operatorname{Ker} A).$$

Here we choose the rank normal form

$$\tilde{A} = \begin{bmatrix} I_r & O_{r,n-r} \\ O_{m-r,r} & O_{m-r,n-r} \end{bmatrix},$$

for which

$$\begin{aligned}\operatorname{Im} \tilde{A} &= \{(y_1, y_2, \ldots, y_r, \overbrace{0, 0, \ldots, 0}^{m-r})^\top \mid y_1, y_2, \ldots, y_r \in K\}, \\ \operatorname{Ker} \tilde{A} &= \{(\overbrace{0, 0, \ldots, 0}^{r}, x_{r+1}, x_{r+2}, \ldots, x_n)^\top \mid x_{r+1}, x_{r+2}, \ldots, x_n \in K\}.\end{aligned}$$

In either expression, the number of parameters corresponds to the dimension, so that we obtain $\dim(\operatorname{Im} \tilde{A}) = r$ and $\dim(\operatorname{Ker} \tilde{A}) = n - r$. Thus Theorem 9.69 is proved. ■

9.8.2 *Rank Inequalities*

On the basis of Theorem 9.69 for the relation between the dimension of the image space and the rank of a matrix, we shall give alternative proofs to some of the rank inequalities presented in Sec. 4.3. Theorems 9.72, 9.73, and 9.74 below deal with the same inequalities as those in Theorems 4.14, 4.15, and 4.20.

Theorem 9.72.

(1) $\operatorname{rank}(AB) \leqq \operatorname{rank} A$.
(2) $\operatorname{rank}(AB) \leqq \operatorname{rank} B$.

Proof. (1) It follows from $\operatorname{Im}(AB) \subseteqq \operatorname{Im} A$ that

$$\operatorname{rank}(AB) = \dim(\operatorname{Im}(AB)) \leqq \dim(\operatorname{Im} A) = \operatorname{rank} A,$$

where Theorem 9.69(1) and Theorem 9.9(1) in Sec. 9.4.3 are used.

(2) Let B be an $n \times l$ matrix. It follows from $\operatorname{Ker}(AB) \supseteqq \operatorname{Ker} B$ that

$$\operatorname{rank}(AB) = l - \dim(\operatorname{Ker}(AB)) \leqq l - \dim(\operatorname{Ker} B) = \operatorname{rank} B,$$

where Theorem 9.69(2) and Theorem 9.9(1) in Sec. 9.4.3 are used. □

Theorem 9.73. $\operatorname{rank}(A+B) \leqq \operatorname{rank} A + \operatorname{rank} B$.

Proof. It follows from $\operatorname{Im}(A+B) \subseteqq \operatorname{Im} A + \operatorname{Im} B$ that

$$\begin{aligned}
\operatorname{rank}(A+B) &= \dim(\operatorname{Im}(A+B)) \\
&\leqq \dim(\operatorname{Im} A + \operatorname{Im} B) \\
&= \dim(\operatorname{Im} A) + \dim(\operatorname{Im} B) - \dim(\operatorname{Im} A \cap \operatorname{Im} B) \\
&\leqq \dim(\operatorname{Im} A) + \dim(\operatorname{Im} B) = \operatorname{rank} A + \operatorname{rank} B,
\end{aligned}$$

where Theorem 9.69(1), Theorem 9.9(1) in Sec. 9.4.3, and Theorem 9.10 are used. □

Theorem 9.74. *Let A be an $l \times m$ matrix, B an $m \times n$ matrix, and C an $n \times k$ matrix.*

(1) Sylvester inequality:

$$\operatorname{rank} A + \operatorname{rank} B \leqq \operatorname{rank}(AB) + m.$$

(2) Frobenius inequality:

$$\operatorname{rank}(AB) + \operatorname{rank}(BC) \leqq \operatorname{rank}(ABC) + \operatorname{rank} B.$$

Proof. It suffices to prove (2), since (1) is a special case of (2) with $B = I$ (identity matrix). Let T be the linear mapping represented by matrix A, and $W = \operatorname{Im} B$. Then $T(W) = \operatorname{Im}(AB)$, and the formula (9.2) in Theorem 9.16 (Sec. 9.5.1) implies

$$\dim(\operatorname{Im} B) - \dim(\operatorname{Im}(AB)) = \dim((\operatorname{Im} B) \cap (\operatorname{Ker} A)).$$

The left-hand side here is equal to $\operatorname{rank} B - \operatorname{rank}(AB)$ by Theorem 9.69(1), and hence we obtain

$$\operatorname{rank} B - \operatorname{rank}(AB) = \dim((\operatorname{Im} B) \cap (\operatorname{Ker} A)). \tag{9.31}$$

By replacing B with BC in this expression we also obtain

$$\operatorname{rank}(BC) - \operatorname{rank}(ABC) = \dim((\operatorname{Im}(BC)) \cap (\operatorname{Ker} A)). \tag{9.32}$$

Since $\operatorname{Im}(BC) \subseteq \operatorname{Im} B$, we have the inequality: [RHS of (9.32)] $\leqq$ [RHS of (9.31)]. Therefore,

$$\operatorname{rank}(BC) - \operatorname{rank}(ABC) \leqq \operatorname{rank} B - \operatorname{rank}(AB).$$

□

Remark 9.27. We give a proof to the identity

$$\dim W = \dim(W \cap (\operatorname{Ker} T)) + \dim(T(W)) \tag{9.33}$$

used in the proof of Theorem 9.74. (This identity has been presented in Theorem 9.16 of Sec. 9.5.1 without proof.) Let $m = \dim W$, $k = \dim(W \cap (\operatorname{Ker} T))$, and $\{\boldsymbol{b}_1, \boldsymbol{b}_2, \ldots, \boldsymbol{b}_k\}$ be a basis of $W \cap (\operatorname{Ker} T)$. By Theorem 9.7 (Sec. 9.4.1) we can enlarge this set to a basis of W by adding $m - k$ vectors $\boldsymbol{b}_{k+1}, \boldsymbol{b}_{k+2}, \ldots, \boldsymbol{b}_m$.

We show $\dim(T(W)) = m - k$ by proving that $T(\boldsymbol{b}_{k+1}), T(\boldsymbol{b}_{k+2}), \ldots, T(\boldsymbol{b}_m)$ form a basis of $T(W)$. That is, we prove (1) and (2) below:

(1) $T(\boldsymbol{b}_{k+1}), T(\boldsymbol{b}_{k+2}), \ldots, T(\boldsymbol{b}_m)$ generate $T(W)$.
(2) $T(\boldsymbol{b}_{k+1}), T(\boldsymbol{b}_{k+2}), \ldots, T(\boldsymbol{b}_m)$ are linearly independent.

Proof of (1): For any $\boldsymbol{y} \in T(W)$, there exists some $\boldsymbol{x} \in W$ satisfying $\boldsymbol{y} = T(\boldsymbol{x})$. On expressing $\boldsymbol{x} = \sum_{j=1}^{m} x_j \boldsymbol{b}_j$, we obtain

$$\boldsymbol{y} = T(\boldsymbol{x}) = \sum_{j=1}^{m} x_j T(\boldsymbol{b}_j) = \sum_{j=k+1}^{m} x_j T(\boldsymbol{b}_j).$$

Proof of (2): To prove the linear independence, assume $\sum_{j=k+1}^{m} c_j T(\boldsymbol{b}_j) = \mathbf{0}$, where $c_{k+1}, c_{k+2}, \ldots, c_m \in K$. Let $\boldsymbol{w} = \sum_{j=k+1}^{m} c_j \boldsymbol{b}_j$. This vector belongs to $W \cap (\mathrm{Ker}\, T)$, that is, $\boldsymbol{w} \in W \cap (\mathrm{Ker}\, T)$, since $T(\boldsymbol{w}) = \sum_{j=k+1}^{m} c_j T(\boldsymbol{b}_j) = \mathbf{0}$. Therefore, $\boldsymbol{w}$ can be represented as $\boldsymbol{w} = \sum_{i=1}^{k} c'_i \boldsymbol{b}_i$ with $c'_1, c'_2, \ldots, c'_k \in K$. Then we have

$$\mathbf{0} = \boldsymbol{w} - \boldsymbol{w} = \sum_{i=1}^{k} c'_i \boldsymbol{b}_i + \sum_{j=k+1}^{m} (-c_j) \boldsymbol{b}_j,$$

from which it follows, by the linear independence of $\boldsymbol{b}_1, \boldsymbol{b}_2, \ldots, \boldsymbol{b}_m$, that $c'_1 = c'_2 = \cdots = c'_k = 0$ and $c_{k+1} = c_{k+2} = \cdots = c_m = 0$. This establishes (2). ∎

9.9 Matroid: Combinatorial Structure of Linear Independence

In Theorem 9.5 (Sec. 9.3) we have seen three properties of linear independence: (a) the empty set $\emptyset$ is linearly independent, (b) if X is linearly independent, then any subset Y of X is linearly independent, and (c) if $X = \{\boldsymbol{x}_1, \boldsymbol{x}_2, \ldots, \boldsymbol{x}_k\}$ and $Y = \{\boldsymbol{y}_1, \boldsymbol{y}_2, \ldots, \boldsymbol{y}_l\}$ are linearly independent, and $|X| < |Y|$, then there exists a vector $\boldsymbol{y}_j$ in $Y \setminus X$ for which $X \cup \{\boldsymbol{y}_j\}$ is linearly independent. These properties are not concerned with quantitative properties such as the length of a vector or the inner product of vectors, but indicate combinatorial properties inherent in the concept of linear independence. Such combinatorial properties are extracted into an abstract discrete structure called matroid. Despite its simple defining conditions (axioms), a matroid is equipped with a rich structure and considered to be a fundamental and useful concept in various fields of mathematics [68, 73].

The objective of this section is to illustrate the concept of matroid in an informal manner with reference to a particular example of a matrix. Suppose that we are given a matrix and denote its column set by S. For example, when given a 3×5 matrix

$$A = \begin{array}{c} \begin{array}{ccccc} 1 & 2 & 3 & 4 & 5 \end{array} \\ \begin{array}{|c|c|c|c|c|} \hline 1 & 0 & 0 & 1 & 0 \\ 0 & 1 & 0 & 1 & 1 \\ 0 & 0 & 1 & 0 & 1 \\ \hline \end{array} \end{array} = [\boldsymbol{a}_1, \boldsymbol{a}_2, \boldsymbol{a}_3, \boldsymbol{a}_4, \boldsymbol{a}_5],$$

we have $S = \{1, 2, 3, 4, 5\}$.

We classify subsets of S into two categories on the basis of the linear independence of column vectors. We call a subset X of S an *independent set* if the set of vectors $\{\boldsymbol{a}_j \mid j \in X\}$ belonging to X is linearly independent, and a *dependent set* if $\{\boldsymbol{a}_j \mid j \in X\}$ is linearly dependent. The family of all the independent sets (for a fixed matrix) is denoted by $\mathcal{I}$. For our matrix A above, we have

$$\begin{aligned} &\mathcal{I} = \\ &\{\emptyset, \{1\}, \{2\}, \{3\}, \{4\}, \{5\}, \\ &\{1,2\}, \{1,3\}, \{1,4\}, \{1,5\}, \{2,3\}, \{2,4\}, \{2,5\}, \{3,4\}, \{3,5\}, \{4,5\}, \\ &\{1,2,3\}, \{1,2,5\}, \{1,3,4\}, \{1,3,5\}, \{1,4,5\}, \{2,3,4\}, \{2,4,5\}, \{3,4,5\}\}. \end{aligned}$$

The family $\mathcal{I}$ of independent sets enjoys the following properties (cf., Theorem 9.5):

(I1) The empty set $\emptyset$ belongs to $\mathcal{I}$, that is, $\emptyset \in \mathcal{I}$.
(I2) If $X \in \mathcal{I}$ and $Y \subseteqq X$, then $Y \in \mathcal{I}$.
(I3) If $X, Y \in \mathcal{I}$ and $|X| < |Y|$, then there exists $y \in Y \setminus X$ for which $X \cup \{y\} \in \mathcal{I}$.

By (I2), a subset of an independent set is an independent set. Hence the family $\mathcal{I}$ of independent sets is determined by its maximal members (with respect to set inclusion). A maximal independent set is called a *base* and we denote the family of bases by $\mathcal{B}$. In our example, we have

$$\begin{aligned} \mathcal{B} = \{&\{1,2,3\}, \{1,2,5\}, \{1,3,4\}, \{1,3,5\}, \\ &\{1,4,5\}, \{2,3,4\}, \{2,4,5\}, \{3,4,5\}\}. \end{aligned}$$

The basis family $\mathcal{B}$ has the following property, called the *exchange property* (or *exchange axiom*):

(B) For any $B, B' \in \mathcal{B}$ and $i \in B \setminus B'$, there exists $j \in B' \setminus B$ such that $(B \setminus \{i\}) \cup \{j\} \in \mathcal{B}$.

The property (I2) also implies that a subset of S containing a dependent set is a dependent set. Hence the family of the dependent sets is determined by its minimal members (with respect to set inclusion). A minimal dependent set is called a *circuit* and we denote the family of circuits by $\mathcal{C}$. In our example, we have

$$\mathcal{C} = \{\{1,2,4\},\{2,3,5\},\{1,3,4,5\}\}.$$

The circuit family $\mathcal{C}$ has the following properties:

(C1) The empty set $\emptyset$ does not belong to $\mathcal{C}$, that is, $\emptyset \notin \mathcal{C}$.
(C2) If $C \in \mathcal{C}$, $C' \in \mathcal{C}$, and $C \subseteqq C'$, then $C = C'$.
(C3) For every distinct $C, C' \in \mathcal{C}$ and every $i \in C \cap C'$, there exists $C'' \in \mathcal{C}$ such that $C'' \subseteqq (C \cup C') \setminus \{i\}$.

The linear independence among the column vectors can also be expressed by a function $\rho : 2^S \to \mathbb{Z}$ defined by

$$\rho(X) = \mathrm{rank}\,\{\boldsymbol{a}_j \mid j \in X\} \qquad (X \subseteqq S).$$

This function, called the *rank function*, has the following properties (Theorem 4.18 in Sec. 4.3.2):

(R1) $0 \leqq \rho(X) \leqq |X|$.
(R2) $X \subseteqq Y \implies \rho(X) \leqq \rho(Y)$.
(R3) $\rho(X) + \rho(Y) \geqq \rho(X \cup Y) + \rho(X \cap Y)$.

The inequality in (R3) is called the *submodular inequality*, whereas the property of satisfying this inequality is called *submodularity* of a function.

In this way, a matrix induces set families $\mathcal{I}$, $\mathcal{B}$, and $\mathcal{C}$ and a set function ρ representing combinatorial structures in the column vectors with respect to linear independence, and furthermore, these induced objects $\mathcal{I}$, $\mathcal{B}$, $\mathcal{C}$, and ρ are equipped with the properties (I), (B), (C), and (R), respectively. As these properties are stated without reference to the original matrix, the properties (I), (B), and (C) are meaningful for set families on a finite set in general and (R) is meaningful for a set function in general. Thus we arrive at a generalization or abstraction beyond matrices.

Furthermore, it is known that the four objects:

$$\mathcal{I} \text{ satisfying (I)}, \quad \mathcal{B} \text{ satisfying (B)}, \quad \mathcal{C} \text{ satisfying (C)}, \quad \rho \text{ satisfying (R)}$$

are equivalent in the sense that any one of them determines the others uniquely. For example, when $\mathcal{B}$ is given, $\mathcal{I}$ is determined by

$$\mathcal{I} = \{I \mid \text{There exists } B \in \mathcal{B} \text{ containing } I\},$$

$\mathcal{C}$ is determined as the family of the minimal members of

$$\{X \mid \text{There exists no } B \in \mathcal{B} \text{ containing } X\},$$

and ρ is determined by

$$\rho(X) = \max\{|X \cap B| \mid B \in \mathcal{B}\}.$$

In view of the above-mentioned equivalence among the properties (I), (B), (C), and (R), we can recognize them as different representations of one and the same discrete structure. A structure equipped with these properties is called a *matroid.* Formally, a matroid is defined to be a pair of a finite set S and a set family $\mathcal{I}$ satisfying (I). Similarly, a matroid can be defined in terms of $\mathcal{B}$ or $\mathcal{C}$. Or alternatively, a matroid is a pair of a finite set S and a set function ρ satisfying (R). Here the set S is called the *ground set,* $\mathcal{I}$ the *family of independent sets,* $\mathcal{B}$ the *base family,* $\mathcal{C}$ the *circuit family,* and ρ the *rank function.*

A matroidal structure appears not only in linear algebra but in many areas of mathematics. For example, suppose that a (connected) graph $G = (V, E)$ is given. The family $\mathcal{T}$ of (the edge sets of) spanning trees satisfies the condition (B), which implies that graph G gives rise to a matroid with ground set E and base family $\mathcal{T}$. A circuit in this matroid corresponds to a simple cycle (circuit) of graph G. In discrete optimization it is well known that the matroidal structure is closely related to the existence of efficient algorithms.

Bibliography

[**General (in English)**]
The following are textbooks for linear algebra in general written in English.

[1] P. J. Davis: *Circulant Matrices*, Wiley, New York, 1979; 2nd ed., Chelsea, New York, 1994.

[2] F. R. Gantmacher: *The Theory of Matrices, Vol. I, Vol. II*, Chelsea, New York, 1959. Also: *Applications of the Theory of Matrices*, Interscience Publishers, New York, 1959; Dover, Mineola, New York, 2005.

[3] R. M. Gray: *Toeplitz and Circulant Matrices: A Review*, Foundations and Trends in Communications and Information Theory, Vol. 2, No. 3, pp 155–239, 2006.

[4] R. A. Horn and C. R. Johnson: *Matrix Analysis*, Cambridge University Press, Cambridge, 1985; 2nd ed., 2013.

[5] R. A. Horn and C. R. Johnson: *Topics in Matrix Analysis*, Cambridge University Press, Cambridge, 1991.

[6] P. Lancaster and M. Tismenetsky: *The Theory of Matrices: with Applications*, 2nd ed., Academic Press, San Diego, 1985.

[7] P. D. Lax: *Linear Algebra and Its Applications*, 2nd ed., John Wiley & Sons, Inc., Hoboken, NJ, 2007.

[8] H. Lütkepohl: *Handbook of Matrices*, John Wiley & Sons, Inc., Chichester, 1996.

[9] K. Murota and M. Sugihara: *Linear Algebra II: Advanced Topics for Applications*, The University of Tokyo Engineering Course/Basic Mathematics, World Scientific and Maruzen Publishing, 2022.

[10] W.W. Sawyer: *An Engineering Approach to Linear Algebra*, Cambridge University Press, Cambridge, 1972.

[11] G. W. Stewart and Ji-guang Sun: *Matrix Perturbation Theory*, Academic Press, San Diego, 1990.

[12] G. Strang: *Linear Algebra and Its Applications*, Academic Press, New York, 1976; 4th ed., Thomson Brooks/Cole, 2006.

[13] G. Strang: *Introduction to Linear Algebra*, 4th ed., Wellesley-Cambridge Press, Wellesley, MA, 2009.

[14] R. S. Varga: *Geršgorin and His Circles*, Springer, Berlin, 2004.

[15] F. Zhang: *Matrix Theory: Basic Results and Techniques*, Springer, New York, 1999.

[General (in Japanese)]
The following are textbooks for linear algebra in general written in Japanese.

[16] H. Arai: *Linear Algebra—Fundamentals and Applications* (in Japanese), Nippon Hyoron Sha, Tokyo, 2006.
[17] T. Fujiwara: *Linear Algebra* (in Japanese), Iwanami Shoten, Tokyo, 1996.
[18] T. S. Han and M. Iri: *Jordan Canonical Form* (in Japanese), University of Tokyo Press, Tokyo, 1982.
[19] K. Hasegawa: *Linear Algebra* (in Japanese), Revised edition, Nippon Hyoron Sha, Tokyo, 2015.
[20] Y. Ikebe, Y. Ikebe, S. Asai, and Y. Miyazaki: *Modern Linear Algebra Through Decomposition Theorems* (in Japanese), Kyoritsu Shuppan, Tokyo, 2009.
[21] M. Iri: *General Linear Algebra* (in Japanese), Asakura Publishing, Tokyo, 2009.
[22] M. Iri and T. S. Han: *Linear Algebra—Matrices and Their Normal Forms* (in Japanese), Kyouiku Shuppan, Tokyo, 1977.
[23] S. Kakei: *Engineering Linear Algebra* (in Japanese), New edition, Suurikougaku-sha, Tokyo, 2014.
[24] A. Kaneko: *Lecture on Linear Algebra* (in Japanese), Saiensu-sha, Tokyo, 2004.
[25] H. Kimura: *Linear Algebra—Foundation of Mathematical Sciences* (in Japanese), University of Tokyo Press, Tokyo, 2003.
[26] S. Kodama and N. Suda: *Matrix Theory for Systems Control* (in Japanese), 2nd ed., The Society of Instrument and Control Engineers, ed., Corona Publishing, Tokyo, 1981.
[27] T. Kusaba: *Linear Algebra* (in Japanese), Extended edition, Asakura Publishing, Tokyo, 1988.
[28] M. Saito: *Introduction to Linear Algebra* (in Japanese), University of Tokyo Press, Tokyo, 1966.
[29] M. Saito: *Linear Algebra* (in Japanese), Tokyo Tosho, Tokyo, 2014.
[30] T. Saito: *Linear Algebra* (in Japanese), University of Tokyo Press, Tokyo, 2007.
[31] I. Satake: *Linear Algebra* (in Japanese), Shokabo Publishing, Tokyo, 1974.
[32] M. Sugiura: *Jordan Canonical Form and Theory of Elementary Divisors, I, II* (in Japanese), Iwanami Shoten, Tokyo, 1976.
[33] T. Tanino: *System Linear Algebra—Applications to Engineering Systems* (in Japanese), Asakura Publishing, Tokyo, 2013.
[34] T. Yamamoto: *Fundamentals of Matrix Analysis—Advanced Linear Algebra* (in Japanese), Saiensu-sha, Tokyo, 2010.
[35] T. Yamamoto: *Notes on Matrix Analysis—Beautiful Theorems and Selected Exercises* (in Japanese), Saiensu-sha, Tokyo, 2013.

[Numerical Methods]
The following are books for numerical linear algebra.

[36] Å. Björck: *Numerical Methods for Least Squares Problems*, SIAM, Philadelphia, 1996.
[37] G. H. Golub and C. F. Van Loan: *Matrix Computations*, 4th ed., Johns Hopkins University Press, Baltimore, 2013.
[38] C. L. Lawson and R. J. Hanson: *Solving Least Square Problems*, Prentice Hall, Englewood Cliffs, NJ, 1974; SIAM, Philadelphia, 1995.
[39] G. Meurant: *Computer Solution of Large Linear Systems*, Elsevier, Amsterdam, 1999.
[40] T. Nakagawa and Y. Oyanagi: *Least Squares Fitting in Experimental Sciences: Program SALS* (in Japanese), University of Tokyo Press, Tokyo, 1982.
[41] G. W. Stewart: *Matrix Algorithms, Vol I: Basic Decompositions*, SIAM, Philadelphia, 1998.
[42] G. W. Stewart: *Matrix Algorithms, Vol. II: Eigensystems*, SIAM, Philadelphia, 2001.
[43] M. Sugihara and K. Murota: *Theoretical Numerical Linear Algebra* (in Japanese), Iwanami Shoten, Tokyo, 2009.
[44] J. H. Wilkinson: *The Algebraic Eigenvalue Problem*, Oxford University Press, Oxford, 1965; Clarendon Press, Oxford, 1988.

[Related Areas]
The following books are references for topics mentioned in the text.

[45] T. W. Anderson: *An Introduction to Multivariate Statistical Analysis*, 3rd ed., Wiley-Interscience, Hoboken , N.J., 2003. (Sec. 6.2.2)
[46] M. F. Anjos and J. B. Lasserre (eds.): *Handbook on Semidefinite, Conic and Polynomial Optimization*, Springer, New York, 2012. (Sec. 7.2.2)
[47] P. Billingsley: *Probability and Measure*, 3rd ed., Wiley, New York, 1995. (Sec. 6.2.2, Sec. 7.4.1)
[48] S. Boyd, L.E. Ghaoui, E. Feron and V. Balakrishnan: *Linear Matrix Inequalities in System and Control Theory*, SIAM, Philadelphia, 1994. (Sec. 7.2.2)
[49] L. Breiman: *Probability*, SIAM, Philadelphia, 1992. (Sec. 6.2.2, Sec. 7.4.1)
[50] C.-T. Chen: *Linear System Theory and Design*, 4th ed., Oxford University Press, New York, 2013. (Sec. 4.4.2, Sec. 5.7, Sec. 6.2.1)
[51] K. L. Chung: *A Course in Probability Theory*, 3rd ed., Academic Press, San Diego, 2001. (Sec. 6.2.2, Sec. 7.4.1)
[52] R. C. Dorf and R. H. Bishop: *Modern Control Systems*, 12th ed., Prentice Hall, Upper Saddle River, NJ, 2011. (Sec. 4.4.2, Sec. 5.7, Sec. 6.2.1)
[53] Y. Ebihara: *System Control by LMI—A Systematic Approach for Robust Control System Design* (in Japanese), Morikita Publishing, Tokyo, 2012. (Sec. 7.2.2)
[54] A. J. M. Ferreira and N. Fantuzzi: *MATLAB Codes for Finite Element Analysis: Solids and Structures* (2nd ed.), Springer Nature, Heidelberg, 2020. (Sec. 7.4.2)

[55] M. Fukushima: *Fundamentals of Nonlinear Optimization* (in Japanese), Asakura Publishing, Tokyo, 2001. (Sec. 7.2.2)

[56] Z. Gajić and M. T. J. Qureshi: *Lyapunov Matrix Equation in System Stability and Control*, Academic Press, San Diego, 1995; Dover, Mineola, New York, 2008. (Sec. 5.7)

[57] S. Igari: *Introduction to Real Analysis* (in Japanese), Iwanami Shoten, Tokyo, 1996. (Sec. 9.4.2)

[58] K. Itô: *Probability Theory* (in Japanese), Iwanami Shoten, Tokyo, 1991. (Sec. 7.4.1)

[59] Y. Kanno and T. Tsuchiya: *Optimization and Variational Methods* (in Japanese), UTokyo Engineering Course, Maruzen Publishing, Tokyo, 2014. (Sec. 1.3.2, Sec. 7.2.2, Sec. 7.4.3)

[60] T. Katayama: *Fundamentals of Feedback Control Theory* (in Japanese), Asakura Publishing, Tokyo, 2002. (Sec. 4.4.2, Sec. 5.7, Sec. 6.2.1)

[61] H. Kogo and T. Mita: *Introduction to System Control Theory* (in Japanese), Jikkyo Shuppan, Tokyo, 1979. (Sec. 4.4.2, Sec. 5.7, Sec. 6.2.1)

[62] P. Komjáth and V. Totik: *Problems and Theorems in Classical Set Theory*, Springer, New York, 2006. (Sec. 9.4.2)

[63] S. Konishi: *Introduction to Multivariate Analysis: Linear and Nonlinear Modeling*, CRC, Chapman & Hall, New York, 2014. (Sec. 6.2.2)

[64] M. Kuczma: *An Introduction to the Theory of Functional Equations and Inequalities: Cauchy's Equation and Jensen's Inequality*, 2nd ed., Birkhäuser, Basel, 2009. (Sec. 9.4.2)

[65] Y. W. Kwon and H. Bang: *The Finite Element Method Using MATLAB* (2nd ed.), CRC Press, Boca Raton, 2000. (Sec. 7.4.2)

[66] J. W. Lamperti: *Probability: A Survey of the Mathematical Theory*, 2nd ed., Wiley, New York, 1996. (Sec. 6.2.2, Sec. 7.4.1)

[67] J. Lindenstrauss and L. Tzafriri: *Classical Banach Spaces, I: Sequence Spaces, II: Function Spaces*, Springer, Berlin, 1977. (Sec. 9.6.9)

[68] K. Murota: *Matrices and Matroids for Systems Analysis*, Springer, Berlin, 2000. (Sec. 4.3.2, Sec. 9.9)

[69] T. Nakamura: *Mechanics of Building Structures II: Illustrative Description and Exercises, I, II* (in Japanese), 2nd ed., Maruzen Publishing, Tokyo, 1994. (Sec. 7.4.2)

[70] F. Nishino and A. Hasegawa: *Elastic Analysis of Structures* (in Japanese), Japan Society of Civil Engineers, ed., Gihodo Publishing, Tokyo, 1983. (Sec. 7.4.2)

[71] M. Ohsaki and T. Honma: *Mechanics of Building Structures: Learning from Exercise, Vol.2: Statically Indeterminate Structures* (in Japanese), Corona Publishing, Tokyo, 2013. (Sec. 7.4.2)

[72] H. Okamoto and S. Nakamura: *Functional Analysis* (in Japanese), Iwanami Shoten, Tokyo, 2006. (Sec. 9.6)

[73] J. G. Oxley: *Matroid Theory*, Oxford University Press, Oxford, 1992. (Sec. 4.3.2, Sec. 9.9)

[74] L.S. Pontryagin: *Ordinary Differential Equations*, Addison-Wesley, Reading, 1962. (Sec. 6.8.6)

[75] T. Saito: *Ordinary Differential Equations* (in Japanese), Asakura Publishing, Tokyo, 1967. (Sec. 6.8.6)

[76] J. Stoer and R. Bulirsch: *Introduction to Numerical Analysis*, 3rd ed., Springer, Berlin, 2002. (Sec. 6.6, Sec. 9.7.3)

[77] M. Sugihara and K. Murota: *Elements of Numerical Computation* (in Japanese), Iwanami Shoten, Tokyo, 1994. (Sec. 6.6, Sec. 9.7.3)

[78] Y. Takahashi: *Real Functions and Fourier Analysis* (in Japanese), Iwanami Shoten, Tokyo, 2006. (Sec. 7.4.1)

[79] D. Williams: *Weighing the Odds: A Course in Probability and Statistics*, Cambridge University Press, Cambridge, 2001. (Sec. 6.2.2, Sec. 7.4.1)

[80] H. Wolkowicz, R. Saigal, and L. Vandenberghe (eds.): *Handbook on Semidefinite Programming: Theory, Algorithms and Applications*, Kluwer Academic Publishers, Boston, 2000. (Sec. 7.2.2)

[81] O. C. Zienkiewicz, R. L. Taylor, and D. D. Fox: *The Finite Element Method for Solid and Structural Mechanics* (7th ed.), Elsevier Butterworth-Heinemann, Burlington, 2014. (Sec. 7.4.2)

Index

Printed in the United States
by Baker & Taylor Publisher Services